Society, Culture, and Urbanization

Society, Culture, and Urbanization

S.N. Eisenstadt
A. Shachar

SAGE PUBLICATIONS
The Publishers of Professional Social Science
Newbury Park Beverly Hills London New Delhi

For information address:

SAGE Publications, Inc.
2111 West Hillcrest Drive
Newbury Park, California 91320

SAGE Publications Inc.
275 South Beverly Drive
Beverly Hills
California 90212

SAGE Publications Ltd.
28 Banner Street
London EC1Y 8QE
England

SAGE PUBLICATIONS India Pvt. Ltd.
M-32 Market
Greater Kailash I
New Delhi 110 048 India

Printed in the United States of America

Library of Congress Cataloging-in-Publication Data

Main entry under title:

Eisenstadt, S. N. (Shmuel Noah), 1923-
Society, culture, and urbanization.

Bibliography: p.
Includes index.
1. Urbanization—Cross-cultural studies. 2. Cities and urban systems—Cross-cultural studies. 3. Sociology, Urban
I. Shachar, A., 1935- II. Title.
HT151.E366 1986 307.7′6 85-26249
ISBN 0-8039-2478-X

FIRST PRINTING

Contents

IN MEMORY

OF

STEIN ROKKAN

ANALYTICAL TABLE OF CONTENTS

Acknowledgments

Throughout our work we have had the help of several colleagues, assistants, and sources. In the earlier stages of this work, Martha Ramon and Atalie Gitter were both instrumental in compiling the basic sources for many of the chapters of this book.

Critical comments by several colleagues have been of great assistance, either on the theoretical framework of the book, on specific chapters, or both. K. Lambert Karlowsky, Paul Wheatley, and Robert Adams have commented on the general scope of the book. In addition, Wheatley has given us tremendous assistance in allowing us to draw from the results of his research on urbanization in the Islamic world and on Southeast Asia.

Jan Heesterman read and commented critically on the chapter on Indian urbanization; Fred Bronner on urbanization in Latin America; Nehemiah Levtzion on urbanization in the Islamic world; Ben Ami Shiloni on Japanese urbanization; Zvi Schiffrin on Chinese urbanization; and Erik Cohen on urbanization in Southeast Asia. Michael Toch wrote the original version of the chapter on European medieval urbanization, which provided the framework for the present chapter, and we gratefully acknowledge his important contribution. Mrs. Ruth Kark wrote some parts of the original version of the chapter on urbanization in the early periods of Islam, and her contribution is gratefully acknowledged.

Finally, we are deeply indebted to Mrs. Esther Sass, who was most instrumental in bringing this project to fruition. Without her relentless devotion, the editing and typing of the book would not have been completed. Our thanks as well to Lois Smith for her meticulous editing, which greatly clarified the text.

We would also like to thank the Truman Research Institute and the Institute of Urban and Regional Studies of the Hebrew University for providing a large part of the funds needed for this research, which was also partly supported by the Volkswagen Foundation. Of course, we accept full responsibility for any remaining flaws in the present volume.

To all those who extended us their much appreciated assistance, we wish to express here our very deep gratitude.

Preface

For more than a decade, the authors of this volume have wrestled with a series of complex, interrelated challenges in an attempt to achieve new insights about the ways in which the process of urbanization has taken place over long periods of time and across many different cultures and societies.

The origins of this book lie in a seminar on comparative urbanization conducted by the two authors in 1973-1974—one of the seminars in comparative civilizations initiated at the Department of Sociology of the Faculty of Social Sciences and Humanities and at the Truman Research Institute of the Hebrew University in Jerusalem—as well as in a series of international seminars.[1] During the decade since the conclusion of that seminar, we have pursued the subject systematically, far beyond our initial plan. The result has been a fusion of analytical approaches that offer exciting possibilities for further study. We are sharing the fruits of our intellectual labors thus far, in the hopes of stimulating further work by others who will utilize and enhance this framework for urban analyses in the future.

This book, we believe, exhibits some rather unique features. To the best of our knowledge, it is the first study on cities and urban systems in general, and on their comparative analysis in particular, that has been written jointly by a sociologist and a geographer. This collaboration has enabled us to take up the basic problems and analytical tools of urban life from both disciplines and combine them in a single framework. The problems of urban autonomy, of the nature of urban life, and the like are thus treated here in a common analytical framework with those of the analysis of central places, urban hierarchies, or internal ecological differentiation of urban space.

We hope to show that it is not only possible but, indeed, very necessary to combine these different approaches and tools of analysis in a common analytical framework. We believe that it is only within such a framework that the full contribution of each of these approaches can be attained.

The second specific feature of this book is its attempt to present a broad comparative study within a framework of explicit analytical assumptions and hypotheses, based on a combination of cultural, institutional, and ecological analyses. These premises and hypotheses are spelled out in detail in Chapter 2, following the review (in Chapter 1) of the major sociological and ecological approaches to the analysis of the urban phenomenon, and they are brought together again in the conclusions.

These analytical orientations follow, to a great extent, those developed in the other seminars on comparative civilizations mentioned above, but they have been enriched here by the incorporation of different modes of spatial analysis and by their application to the urban phenomenon.

On the basis of these orientations, in Part II we present, through a reexamination of the secondary literature, an analysis of some of the major characteristics of the structure of cities and urban systems in several traditional civilizations and societies. Our work will range between two limits: At one end, the fascinating problem of the so-called origins of cities in societies undergoing primary urbanization, which has been abundantly studied in the Near East, Mesopotamia and Meso-America, will be dealt with only tangentially. At the other limit, we shall not deal with modern civilizations and their cities—cities and urban hierarchies that developed in the context of industrial technology and systems of political modernization. Both of these problems have been amply researched, although we believe this research should also be reanalyzed in view of the different theoretical developments and analytical orientations analyzed in the first part of the book.

Instead we have concentrated on the analysis of that vast area that has been least systematically and comparatively researched: the problem of cities and processes of urbanization in several of major historical civilizations—namely, those of China, Islam, India, Japan, Southeast Asia, Medieval Europe, Imperial Russia, and Spanish Latin America.

These civilizations, as we shall indicate in greater detail in Parts II and III, shared several common traits but also varied greatly among themselves. The combination of the common traits and such variations provides a good background for a systematic comparative analysis not available easily in the literature. Such analysis will also provide, we hope, a good starting point for the reexamination of the origins of cities as well as of modern cities.

This approach to the analysis of cities will be based on two very simple but basic assumptions, which will be spelled out in greater detail in Chapter 2, and illustrated throughout the analysis of the case studies. One is that the various organizational, institutional, symbolic, and other "traits," identified with city life or with urban phenomena, may combine in various ways at different times and places. They probably stand—or, rather, are perceived as standing—in contrast to the "rural" sections of the society, but the exact relationship will differ in each specific case.

The other assumption is that the various constellations of traits, which attain their full meaning only within the context of specific civilizations or historical frameworks, should be studied not so much in relation to the general *Geist* (spirit) of particular civilizations but to the specific ways in which the social and cultural forces operate within them. In each case, it is the particular crystallization of these forces that indicates the historical context within which modes of spatial organization and urbanization develop.

Such historical contexts in a society or civilization are not necessarily homogeneous or unchanging. The degree of their homegeneity or continuity is empirically varied and, again, depends on the concrete operation of these forces.

But even within this context, we are not interested in listing such traits but shall rather, as indicated above, stress the major starting point of many of the studies of cities, historical and classical alike—namely, the analysis of different types of social and cultural creativity specific to cities or of cities as loci of such creativity, and the ways in which they are shaped by the forces specified above.

As in the case of other such comparative research, this one has also, necessarily, been based on secondary sources and shares the limitations inherent in such an approach. We hope that even if we have not overcome them, we have at least minimized some of these limitations by using a wide range of secondary sources and, above all, by taking into account the central controversies and disputes in these areas.

Nevertheless, we recognize such limitations and look on the analyses presented here as preliminary, hopeful, and fruitful hypotheses for the continued exploration of the urban phenomenon.

Jerusalem, Israel

S. N. Eisenstadt
A. Shachar

NOTE

1. The publications that grew out of these seminars are listed in the reference section below. In addition, three of the publications were the result of international seminars connected with these topics: Eisenstadt (1972), Eisenstadt and Graubard (1973), and Eisenstadt, Kahane, and Shulman (1984).

REFERENCES

S. N. Eisenstadt and Yael Azmon [eds.] (1975) Socialism and Tradition. The Van Leer Jerusalem Foundation. (German edition in J.C.B. Mohr, Tubingen, 1977).

The papers in the series of studies in Comparative Modernization (1973-1975) Sage Research Papers in the Social Sciences (Series Number 90-003). Beverly Hills: Sage.

S. N. Eisenstadt and L. Roniger (1984) Patrons, Clients and Friends. Cambridge: Cambridge University Press.

S. N. Eisenstadt, M. Abitbol, and N. Chazan [eds.] (1985) The Origins of the State in Africa. Philadelphia: ISHI Press.

S. N. Eisenstadt and I. F. Silber [eds.] (1986) Cultural Traditions and Worlds of Knowledge: Explorations in the Sociology of Knowledge. Philadelphia: ISHI Press.

Comparative Liminality and Dynamics of Civilizations (1984) Special issue of *Religion,* Summer.

S. N. Eisenstadt and S. R. Graubard [eds.] (1973) Intellectuals and Tradition. New York: Humanities Press.

S. N. Eisenstadt [ed.] (1972) Post-Traditional Societies. New York: Norton.

(The above two were initially published as special issues of *Daedalus.*)

S. N. Eisenstadt, R. Kahane, and D. Shulman [eds.] (1984) Orthodoxy, Heterodoxy and Dissent in India. Berlin and New York: Mouton Publishers.

PART I

THEORETICAL APPROACHES

CHAPTER 1

Theories of Urbanization

A
THE STUDY OF THE CITY IN HISTORY

Fascination with the city, with the urban phenomenon, has existed throughout history. It is probably as ancient as the origin of the city itself and can be found in the folk wisdom as well as the more sophisticated social and political speculations of the majority of civilizations.

This preoccupation with the city—with its singularity, its strengths and weaknesses, its distinction from the countryside, and a strong predilection to moral evaluation of the city—can be found in civilizations as diverse as the Jewish, Hellenistic, Roman, Christian, Indian, Chinese, and Islamic. In all of them a highly ambivalent attitude is also found: on one hand appreciation of all the power, wealth, and potential creativity stored up within the city, and on the other hand fear of its corrupting influence contrasting with the supposedly simple virtues of the countryside. In consequence, all these cultures searched for some formula of the ideal city that would compensate for the negative aspects of urban life.

As is well known, concern with the *polis* (the city-state) stood at the center of political philosophy in ancient Greece. Greek philosophers conceived the polis as the highest form of social life and the one best suited for the realization of an ideal society. Both Plato and Aristotle sought a formula for achieving the perfect city-state, Plato by means of a rigid system of laws and Aristotle by analyzing different types of city-states, their institutions, and social structure.

Many of the social philosophers, however, went beyond this search and attempted (in a somewhat Aristotelian way but in many respects going further than Aristotle) to analyze the actual changes that took place in cities and their functions in various civilizations. Among these, probably the most profound systematic observer was the great Arab philosopher and historian Ibn Khaldun, who lived in North Africa and Egypt in the 14th century and who has often been called the forerunner of modern history and sociology. Although his work was not primarily concerned with the city, the city was the focal point of his comprehensive analysis of the rise and fall of Islamic regimes. He compared tribal and rural life to life in the city and examined their changing interrelationship, which influenced the course of Islamic civilization. He maintained that in the early phases of their development, when uncorrupted tribal elements held sway, Islamic polities possessed great strength and creativity. Tempting and pleasure-filled city life eventually weakened tribal morality and corrupted the tribesmen, leading to disintegration of the Islamic political structure.

In the early modern period of Europe (17th-18th centuries), one finds this same preoccupation with the city and ambivalent attitude toward it. In the 18th century philosophers of the Enlightenment saw the city as a school of virtue, especially civic virtue and education. In the wake of industrialization, the rise of proletariat and city slums, and predominance of a commercial utilitarian attitude, a far more negative attitude emerged. From the mid-19th century, a new, rather negative conception and moral evaluation of the city developed under the influence of the Romantic Movement (Handlin and Burchard, 1963; Rodwin and Hollister, 1984).

Whatever the differences between these historical approaches, most of their authors shared the belief that there is a specifically "urban" form of social life as distinct from the "rural" forms. Above all, they sought to understand either the special types of social and cultural creativity characteristic of cities or the cities as loci of such creativity. This search has also constituted a very strong implicit assumption in the modern analysis of the urban phenomenon.

With the rise of modern social science, interest in the city and in its place in civilization became a central part of social science analysis and played an important role in macrosocietal analysis: the comparative analysis of total societies and their evolution.

The study of cities was a subject that had already appeared in the second part of the 19th century in early classical sociology with its celebrated dichotomies, such as Sir Henry Maine's distinction between status and contract (1931, originally published in 1885) and L. H. Morgan's (1877) contrast between savagery, barbarism, and civilization. It was further developed by F. Tonnies (1957), who contrasted "Gemeinschaft" and "Gesellschaft," and by Durkheim (1964a), who, in *Division of Labour in Society,* distinguished between "mechanical" and "organic" solidarity. Tonnies and Durkheim believed that the Gemeinschaft type of social organization, or mechanical solidarity, is, by implication at least, fully developed in cities, particularly in modern cities. Max Weber was also influenced by this approach, although through the abundance of its material and analysis, his work went far beyond the analysis of cities in these classical works. In a parallel way, the analysis of city life—of its origins, character, and development—was closely related to a search for the origin of civilizations, to an analysis of their evolution, and to attempts to define the special character of modern civilization.

Thus Fustel de Coulanges, in his famous work on the ancient city (1980, originally published in 1880), regarded it as a crucial stage in the development of all civilizations and particularly of Western civilization. He believed that religion was the primary motive in founding cities; that cities had come into being originally through a union of several families that, owing to religious symbolism, had been transformed into a religious community and the nucleus of a society.

Later, Weber, in his classical essay on the city (1921) and in many other works, provided a comparative study of the city in several civilizations, with special reference to the Western city, in which he sought a key to understanding European culture. Weber listed the major characteristics of European urban communities as including a fortress, a market, a court, and at least partial autonomy, with an administration in whose election the burghers participated (Weber, 1958: 81). He also noted the cities' political autonomy, originating with the community of burghers, the social distinctiveness of the city, and its importance as a center of production. His basic analytical approach to the study of social formations and historical processes provides some of the most important leads and insights into the comparative study of cities—insights which, as will be elaborated later in greater detail, also constitute the starting point for overcoming many of the weaknesses of the later approaches.

At the same time, a major contribution was made by historians such as Henri Pirenne (1946) who focused on the connection between the growth of medieval European cities and the "economic renaissance" brought about by the development of new economic institutions and the rise of a new merchant class, as against the old feudal order whose power resided in the countryside. These merchants were fighting for new codes of laws, the right to own property, and for an autonomous communal organization. Pirenne pointed out that the essential link between this merchant class and the new economic order was the market. "To Henri Pirenne," wrote Lopez, "the supreme test was whether or not a locality acted as a center of distribution of wealth. Without a market," he said, "one could not speak of a city" (Lopez, 1963: 28).

B
THE CLASSICAL SOCIOLOGICAL AND HISTORICAL STUDIES OF CITIES: CITIES AND ORIGINS OF CIVILIZATION

"Die Scheidung von Stadt u.Land". Mann kann sagen, dass die ganze ökonomische Geschichte der Gesellschaft sich in der Bewegung dieses gegensatzes resumiert.
—Karl Marx, *Das Kapital* I 12

1 Introduction

Subsequent, more specialized development of the study of cities exhibits two characteristics: close connections with the basic general concerns of the classical period of modern sociology and deep influences by the more implicit but strong analytical assumptions prevalent in this classical period. This perspective was bequeathed to later generations of scholars—social scientists and historians—who developed the study of urban structures and forms as a specialization within their respective disciplines.

The most important of these assumptions was the combination of the *ideal-typical* dichotomous approach with a strong evolutionary bias or orientation. Thus in the sociological literature, cities have often been characterized, as has been persuasively demonstrated by Paul Wheatley (1972) in his masterful survey of the literature, by "*trait complexes*" (groups of characteristics), which were usually contrasted with those of the countryside. At the same time, the analysis of cities has often been closely related to a comparative evolutionary study of civilizations. This study postulates continuous development of the city from a simple community, in which the urban element is weak, to more advanced societies, in which this element is much stronger and differentiated from the rural element.

Such evolutionary approaches have usually assumed that different modes of social division of labor provide the best explanation for variations in the institutional structure of different societies. These modes of social division of labor are defined in terms of social differentiation and specialization developing from Durkheim's mechanical to organic solidarity and are mainly shaped by the development of technology and demographic processes.

In various ways, all these assumptions have influenced most areas of study of cities and their relationship with society as a whole. These areas of study have been (1) the role played by cities in the emergence of civilization; (2) analysis of the sociological characteristics of the city; (3) the internal ecological structure of cities; and (4) the study (mainly by anthropologists and geographers) of the spatial organization of society and especially of systems of settlements.

We shall analyze the development of research in all these major areas of study and endeavor to show how, in all these studies, forceful paradigms initially evolved that were greatly influenced by a combination of trait complexes and ideal-typical dichotomous approaches, together with a strong evolutionary orientation. This was followed by the creation of rich and fascinating research agendas—which, however, could no longer be contained within the framework of the approaches and assumptions that gave rise to them.

In all these areas of the study of cities, a far greater number of components had been identified than could be anticipated from most of the classical approaches. It has been demonstrated that these components can combine in many more configurations and are influenced by many more factors than allowed for by the classical approaches.

2 Origins of Cities and the Evolution of Civilization

Childe's studies of the urban revolution (1947, 1950, 1954) focused systematically on the city's role in the emergence of civilization. In these works the relationship between the city and aspects of civilization were analyzed within an evolutionary framework. Childe made a significant contribution toward formulating a comprehensive theory of primary urbanization (the origin of the city). He identified the major turning point in the evolution of mankind as the "neolithic revolution"—that is, that period in which man progressed from hunting and gathering to sedentariness and food production. According to Childe, the emergence of urban settlements depended on the establishment of food production on a permanent basis, and an evolutionary sequence of sedentariness, agriculture, and urbanization was thus created.

Childe's earlier works contained a strong element of "environmental determinism." This approach claimed that food production began as a result of major climatic changes that caused gradual dessication of most of the Near East. He argued that this dessication provided the stimulus for a food-producing economy, which developed when the physical conditions were conducive to a symbiotic relationship among plants, animals, and human beings. In his later works, Childe placed less emphasis on environmental determinism and more on the development of the means of production as the most convincing explanation for historical processes. Lamberg-Karlovsky and Sabloff (1979: 31) stated that Childe argued that the means of production directly affected the legal, political, and religious aspects of a society, and cultures or civilizations that had reached similar technological levels possessed similar political, economic, and social institutions. Thus the stage referred to as the "urban revolution" (which was preceded by the "neolithic revolution") was marked by the appearance of the first literate communities living in densely populated settlements with complex bureaucracies. These indicate political organization and a highly developed division of labor. Childe claimed that the "urban revolution" has several important features, which are as follows (Childe, 1950): (1) urban centers comprising a population of 7,000 to 20,000; (2) a class of full-time specialists (craftsmen, merchants, officials, priests) residing in the cities; (3) a surplus of food production that could be appropriated by the government; (4) monumental public buildings symbolizing the concentration of the surplus; (5) a ruling class of senior priests, civil and military leaders, and officials; (6) numerical notation and writing; (7) the beginnings of arithmetic, geometry, and astronomy; (8) sophisticated art styles; and (9) the existence of long-distance trade.

Certain elements of Childe's approach were taken up and further developed by Robert Redfield, especially in his initial analysis of the differences between folk and urban societies (1947), which were conceived chiefly in antithetical terms. Redfield described the ideal-type "folk" (rural) society as a small, isolated, nonliterate, homogeneous community with a strong sense of group (family) solidarity, without mechanized manufacture or the use of natural power, little division of labor, a common experience of life among its members, and conventional ways of problem solving. Opposite characteristics are found in cities. Later Redfield attributed importance to the cities of a country or civilization as the home of "great traditions," and the development of the major symbols and creative works of a civilization, as opposed to "little traditions" that tend to evolve in the countryside or in provincial towns.

In the 1950s, this same notion of the city as the cradle and home of civilization was the central focus of the Chicago Symposium on the "City Invincible." Scholars (especially historians and anthropologists but also urbanologists and philosophers of culture such as Lewis Mumford) analyzed, within a comparative evolutionary approach, urbanization and cultural development in the ancient Near East.

Later Robert Adams, one of the editors of the symposium, in a work that became a landmark (1966), analyzed the societal conditions that gave rise to the emergence of cities in Central America and Mesopotamia. Adams found great similarities between these two early societies, especially in their development from theocratic to political control, culminating in the establishment of trade and tributary states that extended their domination over vast areas. The relation between urban origins and the emergence of civilizations became a central focus of research embracing various parts of the world (Sanders and Price, 1968).

3 Criticism of the Evolutionary Approaches

The early evolutionist view, represented by Gordon Childe, has been undermined by recent studies of the origins of civilization. Childe's theory of primary urbanization has begun to be challenged by the present generation of scholars because the empirical evidence from recent archaeological excavations appeared to disprove the neat sequence of sedentariness, agriculture, and primary urbanization. R. J. Braidwood's excavations at Jarmo (1974) in particular were a major setback for this theory. Following these excavations, Braidwood rejected the notion of catastrophic climatic changes as the main catalyst of the "neolithic revolution." He argued that a food-producing economy was the outcome of increasing differentiation and specialization in human society. Braidwood discovered several levels in the development of food production, which were unrelated in a sequential order, leading in a straightforward progression from hunting and gathering to sedentariness and agriculture. According to him, it is possible to assume a continuous interdependence between the village agriculturalist and the nomad herder, the two groups often interchanging their respective roles.

The clearest disproof of Childe's theory resulted from the excavation of the most ancient of all urban settlements, Jericho and Çatal Huyuk, dated to a far earlier period than the urban revolution. It appeared that Jericho's inhabitants had hardly begun to domesticate grains, yet they had the civic organization that enabled them to construct massive defense walls (Kenyon, 1957). Similarly, Çatal Huyuk already had pottery by 6500 B.C. and an elaborate set of religious shrines at about the same time that cattle began to be domesticated (Mellaart, 1967). These findings demolished the argument that urban development must be preceded by sedentarized agriculture.

4 Alternative Theories

In consequence, alternative theories of primary urbanization began to evolve. According to these theories, urban settlements do not have to develop out of villages of farmers or gatherers. They can be established as cities from the very beginning and exist side by side with rural settlements, subsisting through trade of natural resources or through the provision of services to the surrounding areas. Demographic pressure rather than agriculture is now thought to be the key factor in the emergence of cities (Binford, 1968). This pressure, it is claimed, disturbed the balance between population and resources, causing some people to move to marginal areas. Finding themselves in a region where agricultural conditions were unfavorable, these people either had to devise new techniques of food production and storage or had to establish a new form of economy based on services such as trade, religion, or defense. The pressure brought to bear by an expanding population and the resulting new type of economy necessitated a concentration of the population and led to the emergence of urban settlements.

These new theories suggest that an increase in population preceded large-scale food production, that food production began in the marginal areas of the mountain zones in the Middle East, and that from the very beginning there were many centers of food production (Braidwood and Willey, 1962; Flannery, 1969). This new interpretation of the "neolithic revolution," derived from Flannery, allows for many more variations in the path of civilization, the main elements of which are sedentariness, food production, domestication, growth of population, and the establishment of urban settlements (Lamberg-Karlovsky and Sabloff, 1979). Moreover, the recent theory, unlike Childe's, did not determine a set order for the developments that are said to have taken place. Thus in contradiction to

Childe's assertions, it is now claimed that cities could have emerged without evolving from villages and that the development of agriculture did not necessarily precede the invention of pottery and writing.

This point of view was reinforced by many studies made in other parts of the world on the origin of the state, civilization, and city (Flannery, 1972, 1976; Sanders and Price, 1968; Service, 1975; Whitehouse, 1977). These various studies have shown that in the Middle East and Central America, many different combinations are to be found of such elements as concentration of population, irrigation enterprises, as well as political and religious centralization, which the earlier evolutionists regarded as inseparable from the emergence of cities and civilizations. The new research, in contrast with the earlier, mostly evolutionary view, indicated that these elements combined differently in different places (Service, 1975). This evolved under the influence of other combinations of social and cultural forces as well as of the diverse levels and types of economic development and processes of political and religious expansion.

C
THE SOCIOLOGICAL ANALYSIS OF URBAN TRAIT COMPLEXES

1 The Ideal-Type Approach

The search for the specifically urban trait complex developed in the context of sociological analysis of city life. It was also influenced by a dichotomous conception of city as opposed to country life, as well as by some latent evolutionary orientations. We have already mentioned Weber's approach, which has often given rise to a tendency to analyze any city or groups of cities in terms of an approximation to an ideal type of Western city.

Later this search for specific trait complexes was connected with the study of internal urban ecology and of the social and moral quality of modern urban life. One of the most powerful expressions of this search can be found in the work of Louis Wirth, who indicated that size, density, and heterogeneity—regarded as the principal traits defining cities—are conducive to specific behavioral patterns and moral attitudes (Wirth, 1938). As mentioned by Wheatley, Wirth maintained that

> concentration in limited space . . . has certain consequences of relevance in the sociological analysis of the city . . . and an increase in density tends to produce differentiation and specialization. . . . Diversifying men and their activities and

increasing the complexity of the social structure, and a way of life characterized by such traits is impersonal, superficial, transitory, segmental [Wheatley, 1972].

Wirth (1938: 3) also stated,

> The city and the country may be regarded as two poles, in reference to one or another of which, all human settlements tend to arrange themselves. In viewing urban-industrial and rural-folk society as ideal types of communities, we may obtain a perspective for the analysis of the basic models of human association as they appear in contemporary civilization.

Gideon Sjoberg's study of the preindustrial city (Sjoberg, 1960) combined this trait complex approach with a comparative evolutionary approach. Sjoberg distinguished among three different types of societies, each with its own relationship to city culture: preliterate "folk" societies in which there are no cities, literate preindustrial city cultures, and industrial, urban ones. According to Sjoberg, preindustrial cities dependent on animate sources of power are subsystems of feudal societies or bureaucracies and do not possess the impersonality, secularism, and large size that are characteristic of cities in the folk-urban dichotomy. Conversely, industrial societies that possess a developed technology, derived from inanimate sources of power, have fully developed cities with independent economic resources.

Many of these analytical themes, together with a search for the ideal city, universal or historically specific, can also be found in the works of historians or philosophers of culture such as Oswald Spengler (1928), Lewis Mumford (1938, 1961), or Arnold Toynbee (1967). In these scholars' works, the city as the focus of a distinctive moral order and of cultural creativity or decay—a theme that was latent in the works of sociologists and anthropologists—became prominent, and different types of cities were often seen as epitomizing the spirit of whole civilizations, particularly those of the West.

2 Criticism of the Dichotomous and Trait Complex Approaches

Criticisms similar to those that arose with respect to the analysis of the relations between cities and origins of civilization were also voiced by sociologists, anthropologists, historians, and economists against the ideal-typical distinction between city and countryside and between the preindustrial and industrial cities, as well as against the search for the trait complexes that characterize cities. Richard Morse (1972), among others, stressed that this dichotomy between the two kinds of cities does not stand up under scrutiny of the variations in urban systems of various societies. Morse also maintained that the evolutionary approach, when applied to the study of cities in the Third World, has concentrated excessively on Western influences—especially the impact of industrialization—and has paid only cursory attention to indigenous features. He claimed that to place all cities in a single "preindustrial" or "industrial" category meant regarding them as artifacts rather than as integral parts of a larger society with a specific social structure, criteria for the allocation of political power, social mobility, or division of labor.

Similarly, Hoselitz (1955) distinguished between parasitic and generative cities and emphasized the different functions of cities that cut across relatively similar levels of economic development, thus going against many of the simple evolutionary assumptions of several former studies. Later Fox (1977) also offered a more diversified typology of preindustrial cities, distinguishing among regal, ritual, commercial, and administrative cities.

The search for the trait complex of cities has also come under fire. In his masterful survey, Wheatley (1972: 609) pointed out that this approach suffered from two weaknesses. The first is that the different traits by which a particular scholar might characterize the city are not always found in the same combination. Thus no decisive empirical evidence has been offered to support the assertion that "size and density of population necessarily induce variations in behavioral patterns." Similarly, psychological traits attributed in earlier research to the urban environment are contradicted by evidence from many parts of the world (Lewis, 1965).

The second weakness of the trait complex model is that any one characterization of the difference between city and country, which must be central to any "ideal" type of trait complex, does not hold for all times and places. In most times and places some distinction was made between them, but the way in which this distinction was defined varied greatly from place to place. Finley (1977) has shown that the ancients, whether directly concerned with cities, as were Plato and Aristotle, or only indirectly, as was Pausanius, had different views about their characteristic traits and functions. Unlike the moderns, they saw the city and its hinterland as a single economic unit, and they placed particular emphasis on the political distinctiveness of the city.

Perhaps paradoxically, similar indications can be found in Weber's vast program of comparative research into ancient civilizations, including China, India, and ancient Israel, as well as the Greek and Roman civilizations, which contained a rich analysis of their cities. In this analysis an ideal type of European city usually served as a starting point, but the abundance of material provided by the research indicated many intricate and varied relations between religious traditions and social and economic structures of these civilizations and the characteristics of their cities.

D
THE DIMENSIONS OF URBAN AUTONOMY

Weber's analysis shows that such diversity exists with respect to all the characteristics of the Western city he described, particularly with respect to the crucial concept of autonomy that, Weber maintained, was the major distinguishing feature of the Western city.

Indeed, both Weber's analysis and the rich comparative material collected since then show that many possible aspects of autonomy do not always appear in combination and do not always go hand in hand with the development of a particular kind of urban society. The entire set of elements combines only in specific circumstances, which occurred in the Middle Ages in Europe and, to a lesser degree, in the period of antiquity in which the city-states came into being.

Thus research indicates that it is important to distinguish the autonomy of urban groups (craft and merchant guilds, neighborhoods, and the like) from that exercised by the entire urban community. These too, in turn, are related to two other aspects: autonomy in purely municipal matters and a more political form of communal autonomy. The political form of autonomy involves the relative independence of the city with regard to its external affairs and to broader political entities of which, theoretically, the city forms a part. This political autonomy may be of various kinds, three of which occurred frequently throughout history.

One kind, to be found particularly in the ancient city-states and to a lesser degree in the medieval city (the Italian, Flemish, North German Hanseatic, and Indian towns), consists of recognition of a city's relative independence in accordance with existing definitions of sovereignty. Needless to say, such cities were often dependent on the political power of the larger entities of which they were a part, in some cases ultimately losing sovereignty and becoming incorporated into them. A second type of political autonomy, which in Medieval Europe often merged with the first, consists in regarding the city as a distinct corporate unit, recognizing its urban character and a degree of relative independence from the prevailing sovereign power, whose sovereignty is acknowledged and from whom, generally after a struggle, a certain freedom is gained in running its internal and even external affairs. A third type of relative autonomy—found in Latin America, India, and some Islamic countries—is characterized by the de facto usurpation of power by various local magnates and potentates, a process that usually accompanies a more general process of decomposition or weakening of central authority.

Just as with respect to the explanation of the origins of cities, so also the different kinds of autonomy tend to occur in far more diversified combinations than has been assumed by the classical approaches. They are also influenced by a great variety of social, political, and cultural forces that must be analyzed in greater detail. An important indicator of the nature of the forces that shape different patterns of urban life, different combinations of their different components, can be found in the development of the two additional areas of study mentioned above: the internal structure of cities and the analysis of urban systems.

E
THE INTERNAL STRUCTURE OF CITIES

The search for the trait complex specific to the city and the evolutionary approach have also greatly influenced the study of the internal structure of cities.

Many works—such as those of Bacon (1967), Lavedan (1926, 1941), Morris (1972), Mumford (1961), and Vance (1977)—deal with various aspects of the internal structure of cities. Some of these works have studied individual cities. Others—Gutkind's (1964-1972) monumental work on urban development and structure in various parts of the world is a good example—have examined the subject within a regional context. A comprehensive review of contributions in the field dealing exclusively with the modern city of the Western world can be found in *The Internal Structure of the City* (Bourne, 1982).

In principle several approaches can be distinguished in the study of the internal structure of cities. The first approach—that of urban ecology—has been concerned mainly with the spatial distribution of population in the city and changes occurring in this distribution. The second, closely related approach, in terms of land distribution and utilization within the city, has focused on analysis of the efficient utilization of urban land by the various activities, such as residential, commercial, and industrial land uses. The last approach deals with different aspects of design of the various elements of the city—the buildings, streets, walls, open spaces, skyline, and the like—and with the aesthetic and functional relations among the elements of urban design.

The urban ecological approach has developed a theoretical framework that is closely related to the evolutionary approaches previously described. This is the urban ecological approach that emerged in the 1920s and 1930s in Chicago, establishing human ecology as the basis for analyzing the internal structure of cities, especially the spatial differentiation of the urban population (Burgess and Bogue, 1964).

The Chicago School developed a general framework that explained the distribution of population according to socioeconomic status. The basic scheme of city structure was one of successive rings around the central business district. This distribution was brought about by a continuous stream of migrants entering the city, first settling on the fringes of the central business district and, with the constant demand for housing, pressuring the population to move from the inner to the outer rings. Thus a pattern was established whereby the outer rings housed people of higher status and the inner rings and the city center housed those of lower status (Burgess, 1925). This scheme, based on basic principles of biological ecology such as competition, invasion, and domination, was believed to provide a general framework for the analysis of the internal structure of cities. This was an essentially sociological interpretation of spatial behavior: Various groups of the urban population utilized their place of residence as a means of staying near the people of their own socioeconomic status or ethnic origin and of moving as far as possible from the groups with which they did not wish to be associated. The conceptual framework of this approach was based on several assumptions: that all parts of the city were clearly identified with regard to their respective status and prestige; that the prestige map was well known to the population of the city; and that there was increasing spatial differentiation between the functions of the city—work, residence, business, leisure, and so forth. The entire process of change in the urban ecology thus depended to a large extent on a continuous flow of migrants into the city. The human ecology approach gave tremendous impetus to the study of cities' internal structure, which culminated in the 1950s and 1960s in the form of the social area analysis and factorial ecology (multivariate techniques that identify the major features of the population in various quarters of a city) embracing many cities worldwide (Herbert and Johnston, 1978; Timms, 1971).

A different approach can be defined as the "design approach." It deals with the plan of the city and its various components and has deep roots in architecture. It has provided an extensive literature but almost no paradigm. It traces the aesthetic principles that guided the layout of the street patterns (Bacon, 1967) and is concerned with the relationship between building and transportation technology and the size, shape, and functions of buildings and public facilities (Giedion, 1963; Spreiregen, 1965). Particular attention is devoted to the influence of such natural features as topography, lithology, hydrological conditions, and drainage patterns of the site and to the physical development of the city (Dickinson, 1951). Public buildings, their architectural styles, and their relation to the main road network and other major buildings are also analyzed in great detail (Rasmussen, 1969).

Many of these architectural works on urban design, with their aim of identifying the unique features of the city's structure at a given period, are ideographic in their approach. Some of them trace the spread of architectural styles and city plans from region to region and the modification of these styles and plans caused by local conditions (Morris, 1972). The relations among societal processes and transformations of the physical fabric of cities—"urban morphogenesis"—have been explored more recently in Western civilization (Vance, 1977). Although the various architectural approaches have not developed a complete conceptual framework, the valuable information they have provided on architecture and urban forms (based on archaeological remains, maps, sketches, and descriptions of buildings, streets, and neighborhoods) has aided understanding of urban design and the basic elements of the internal structure of cities.

A field of study related to the design approach is the analysis of "ideal cities" (Rosenau, 1972). Although the vast majority of ideal cities never materialized but were merely imagined by philosophers such as Plato, theologians such as

Thomas Aquinas, architects such as Virtuvius, and more recently by social reformers such as Robert Owen or Ebenezer Howard, they constituted valuable conceptual tools by means of which social and cultural goals might be attained. With their pure forms and imaginative designs unhampered by material considerations, ideal cities reflected the spiritual climate of their age. Few cities were actually built in accordance with an ideal type, but in many cities it is possible to trace the strong influence of abstract ideas on their structural development.

1 Criticism of the Prevalent Approaches

With the development of research it was found that the urban ecological model, like the previous dichotomous and evolutionary approaches, was too rigid and narrow. It became clear that it could be applied mainly to North American cities at certain periods of their development, but it was not valid for most European cities (McElrath, 1962), for those of the developing world (Abu-Lughod, 1969; Berry and Rees, 1969), or for preindustrial cities, as indicated by Sjobert (1960: 91-105) despite being heavily criticized.

One of the main reasons the urban ecological approach of the Chicago School cannot explain the internal structure of traditional cities is because in these cities the level of prestige accorded an area was determined only to a small degree by the socioeconomic status of its inhabitants. More important was its distance from the center: The nearer an area to the center, the greater its prestige. Moreover, the premechanized transportation technology of the traditional city rendered the accessibility pattern highest at the city center. Accessibility declined rapidly at the fringes of the built-up area, thus emphasizing the importance of the city center in daily life. Similarly, in most traditional cities, residential and nonresidential activities were intermixed, the ground floor being used for commerce or handicrafts, the back rooms or other floors serving as living quarters. In most traditional cities urban space was shared by the various functions and land uses and was not differentiated and separated clearly in various areas.

2 Symbolic Approach to the Study of the Internal Structure of Cities

Important attempts to deal with some of the complex problems of the internal structure of cities were made within the framework of the symbolic approach to cities, which began as a corrective to the urban ecological and functional-economic approaches but developed far beyond them.

This was first elaborated in Firey's (1947) analysis of land uses in central Boston, which revealed the importance of symbolism as an ecological variable.

He emphasized the independent role played by social symbolism as a major force in shaping the internal structure of cities, particularly with respect to historically meaningful public buildings and open spaces.

The most forceful development of this approach is found in Wheatley's works on ancient Chinese cities (1971) and on Japanese cities (Wheatley and See, 1978), which analyzed the features of cities in terms of their religious or cosmological meaning. Wheatley demonstrates that in many cases the site of a city or particular structure within it has symbolic meaning in itself. It can be regarded as the center of the world, an "omphalos," the point of contact between this world and the world beyond. The alignment of walls, gates, and major road axes is determined by the degree of sacredness of their relative positions (Wheatley, 1971: 411-451). The same considerations apply to the allocation of different areas in the city to various groups in the population in accordance with their degree of social prestige, or of desirable as opposed to undesirable land.

In general, according to Wheatley, many of the ancient ceremonial centers were planned and built according to a design that was intended to be an earthly representation of the image of the cosmos. The plan of the city, the location and form of its shrines, temples, and many of its public buildings were seen as physical representations of the cosmological conceptions that presided over the city's foundation. A religious-symbolic interpretation has also been incorporated in the analysis of the internal structure of ceremonial centers in Central America (Hardoy, 1972), in Southeast Asia (Coe, 1961), and in Rome and the ancient world (Rykwert, 1976). In all cases it has greatly increased understanding of the founding of cities and of their internal structure.

The corrective of the symbolic approach, however, was not sufficient to eliminate two basic weaknesses of the prevalent approaches to the internal structure of cities. The first weakness was that the different approaches are generally unrelated to one another, and that the abundant material produced by the design approach was seldom integrated into, or even related to, the ecological and economic approach. The second weakness was that comparison of the various approaches clearly reveals that the internal structure of cities comprised a greater variety of elements than had first been considered. Hence these approaches proved to be incapable of providing a systematic analysis of the variety of material brought to light by research, and they were not sufficient to provide a satisfactory explanation for such variability.

The preceding discussion has outlined some of the major approaches to the analysis of the internal spatial organization of cities, most of which attempted to connect such analysis with some global macro-societal trends.

The current frontiers in studies of the internal structure of cities have been summed up by Tilly (1984: 120), especially in relation to the broader societal setting: "Cities are, above all, places whose analysis requires a sense of spatial and physical structure; analyses of broad historical processes rarely deal effectively with spatial and physical structure." What is called for is a better understanding of the interaction between large-scale political and social processes and the changing form of cities. Tilly identifies three possible ways to bridge the gap between large social changes and the evolving internal structure of cities:

(1) *Global-reach,* which consists of establishing a principle that pervades an entire civilization and then treats the spatial organization of cities as a direct expression of that principle. Mumford's *The City in History* (1961) exemplifies this approach; the structure of cities is analyzed in terms of the operation of two principles—accumulation and conquest—the balance between the two differing in various societies and periods. For Mumford, the extent and the particular combination of accumulation and conquest determined the internal structure of cities, and he thus created a direct link between large-scale political and social changes and the spatial organization of cities.

(2) *Space economy.* This approach emphasizes the powerful economic logic of costs and benefits that produces a distinctive spatial pattern for each system of production and distribution. The works of Pred (1973, 1977) and Skinner (1977) are reviewed by Tilly to illustrate this approach. It should be mentioned that these works deal with urbanization processes and with the evolution of urban systems and not with the internal structure of cities, but the economic principle remains valid and applicable.

(3) *The city-as-theater.* This approach conceives the city as the arena for human drama. The topics, actors, and struggles of the human drama have common sources and similarities that cut across civilizations and societies, but the development of the human drama is influenced to a large extent by the particular urban setting.

It is obvious that none of these three approaches provides a complete answer to the sought-after bridge between the large-scale social changes and the internal structure of cities. The global-reach approach lacks an explicit elaboration of the mechanisms that translate vast social processes into the form of cities. The space-economy approach says little about the relations of dominance, subordination, and solidarity that shape the city's social structure. The city-as-theater approach takes the stage setting for granted, without devoting much attention to tracing its evolution and changes (Tilly, 1984).

3 Basic Components of the Internal Structure of Cities

At the same time, the very abundance of material that has been collected in these studies has provided some important clues for undertaking a systematic analysis of some of the major components of the internal structure of cities, as well as of the forces influencing the development of the different constellations in such components.

The first among these components to consider is the extent to which cities are walled in by some physical girdle that emphasizes their distinction from the countryside and impedes their gradual spread into the rural area (Lopez, 1963). Second, the structure of the street pattern should be noted—whether it is orderly, allowing for a high degree of connectivity between streets, or takes the form of a multitude of narrow winding alleys and cul-de-sacs—together with the extent to which public space was used or neglected. Third, the degree of spatial differentiation between the urban functions should be studied—that is, the spatial separation between residential and nonresidential functions and the spatial differentiation between the nonresidential functions of commerce, industry, handicrafts, storage, accommodation for nonresidents, public services, and so forth. With regard to residential functions, the internal structure of the city may be characterized by the degree of segregation practiced by different groups of the urban society and the type of collectivity on which it is based (e.g., family or clan, ethnic group, cultural or religious affiliation, common geographical origins, occupation, and socioeconomic status). The interrelations among these organizing principles are usually variable and multifaceted. The relations, the combinations between the various societal bases, in shaping patterns of residential differentiation are a major characteristic in the analysis of the internal structure of cities and in the identification of various types of cities, both traditional and modern (Abu-Lughod, 1969; McElrath, 1968; Timms, 1971: 138-151). Fourth, cities can be distinguished by the existence or absence of a well-defined city center and by the various functions of that center. In many respects the city center determines the organization of land use and the distribution of population in the entire city (Bird, 1977).

Similarly, the structure of the city center can be distinguished according to the level of overlapping or differentiation of the various functions of the center—that is, of the buildings in which the functions were fulfilled: political administrative (the palace, castle, town hall, law courts), religious (temple, church, monastery, school, *madrassah*, etc.), or commercial ones (markets and shops) forming the hub of the entire city's economy. The differences in spatial overlapping of the various functions in the center or even throughout the city usually reflect the basic relations between the major functions of the city—that is, whether it is mainly religious, political, administrative, or economic (Bird, 1977).

These components, which shape the internal structure of cities, are often combined in ways that fail to conform to any of the patterns enunciated by the classical approaches. Although a satisfactory explanation of these variations has still to be devised systematically, it is possible, on the basis of the existing literature, to point out some explanatory principles.

The crucial element in shaping the internal structure of cities is the control of land and the mechanisms of land allocation. Control of land allocation is all-important in determining lot size, street patterns, major public building locations, the extent of open spaces, and their use. The urban map can be looked upon as "a mosaic of competing land interests capable of strategic coalition and action" (Molotch, 1976: 312). The structure of the city is effectively determined by whoever controls the allocation of land, whether it is the ruler, various elites, or social groups such as clergy, professional, or ethnic groups, or those who are in a position to resolve conflicts among the groups.

This control, however, may be influenced by various orientations. Two such orientations—the economic and the religious—have already been stressed in the approaches described. These approaches, however, fail to take account of other factors, such as aesthetics, symbolism, and the distribution of prestige, and to differentiate sufficiently between the different aspects of the orientations.

A systematic analysis of the approaches controlling the internal structure of cities was presented by Cohen (1976), who identified the following major orientations:

(a) *The instrumental orientation,* through which urban space is organized to achieve the greatest economic benefit. Accordingly, every plot of land in the city is occupied by the entrepreneur and owners, who utilize it to derive maximum profit. The form and internal structure of the city will be determined by the distribution of land values in accordance with the real estate market. The pattern of land values, which is determined to a large extent by the accessibility pattern, is the decisive factor in the allocation of land uses. Thus the people who determine the internal structure of cities to the greatest extent are the owners of the land and of the various economic enterprises. The instrumental orientation plays a decisive role in shaping the internal structure of the modern city.
(b) *The territorial orientation,* through which urban space is organized in order to achieve tight internal security and more complete political domination of the city. Many city sites were originally chosen because of their topographical situation. In dominating surrounding areas they were best suited for defense against nomads and external enemies. Within the city the alignment of walls, the position of gates (and approaches to them), and street patterns were determined chiefly by security and control considerations. The citadel is the prime example of this approach. Its size, shape, and relations to the rest of the city reflect not only considerations of external defense but the ruler's need to dominate the city itself.
(c) *The symbolic orientation,* which has two major aspects: the application of aesthetic criteria to the shape of the city and the individual sections and buildings within it, and the application of moral and religious criteria to the design of a city and its various components. Thus an entire city could be regarded as a holy place, within which various sites possess different degrees of sanctity.

The concrete form of a city is influenced by all these orientations. Any city is a mosaic, each of whose parts is the outcome of different orientations. Various orientations shape the different parts of a city at any given time and can conflict with the utilization of space in the same area (Cohen, 1976).

Accordingly, the most important task to be accomplished in studying the internal structure of cities—and in this way to take up the challenge posed by Tilly—is to analyze the different combinations of these factors as they determine the diverse components of the internal structure of cities and their particular combination in various cities, thus going far beyond the basic assumptions of initial studies.

All the approaches presented above, despite their criticism, are useful in advancing the quest for the links between large political and social processes and the changing internal structure of cities. This quest is at the base of the present work (and will be studied within the scope of various traditional societies and civilizations in the following chapters) and, while founded on the evolutionary approaches, will attempt to go beyond them.

F
URBAN SYSTEMS

1 City-Size Distribution of Urban Systems

While sociologists, historians, and, to some degree, anthropologists have searched for the specific characteristic trait complexes of the city or developed a general typology of cities, geographers and some anthropologists have combined a strong evolutionary orientation, similar to the one found in studies of modernization, with the analysis of urban systems and of the spatial organization of society.

The study of urban systems has concentrated on analyses of the relations among different cities within a given territorial framework (Carter, 1983: chap. 5). An urban system consists of a number of towns and cities that are interrelated by common societal, economic, and cultural links. Each unit of an urban system is characterized by the size of its population, its types and magnitude of economic activities, and its level of political authority and cultural influence. The settlements within an urban system are linked by a network of flows of people, money, commodities, regulations, and ideas. The volume and directions of the flows among the units of an urban system are strongly related to the size of the political and economic levels of the various settlements making up the system (Barker, 1978).

Because estimates or census figures of population size are more easily obtainable over long periods of time, the population size of a particular city or of an entire urban system is widely used in urban analysis (Chandler and Fox, 1974). An urban system can be characterized by the size distribution of its cities. The type of distribution has been determined by relating the population size of the largest city to that of the city next in rank (Jefferson, 1939), to the combined population of the four next in rank to the largest cities (Berry, 1961), or to the population size of each of the five cities at the top of the rank order of the city-size distribution (Stewart, 1958).

The ideal types of city-size distribution are identified in the geographical literature on the subject as "primacy" and "rank size." Primacy distribution occurs when one city dominates the others in population and economic power, thus controlling the flow of resources between itself and all other settlements in the system. Rank-size distribution occurs when the cities within a system are of various sizes, with a gradual decline in size along the rank order of the cities' sizes. The precise relations between the size of a city and its place in the rank order differ from country to country but, common to all cases and rank-size distribution, are systematic variations in the sizes of cities, their rank within the urban system, and the absence of any significant gaps in the distribution of their sizes.

Two additional types of city-size distribution have lately been identified in the literature (Carter, 1983:97-100). The first is an equal-sized distribution, in which all the urban places within a given territory are of similar size. A variation of this type is the "convex distribution" of city sizes, where the largest cities are of almost the same size, the size distribution falling sharply to urban places of small size (Johnson, 1980). Urban systems approaching an equal size or convex distribution were identified in situations of early colonization on the shores of a new land or in areas of highly decentralized polities. They reflect such a low level of integration and interaction among the urban places in a given territory that it is almost impossible to define an "urban system" there because, by definition, this term implies that interaction takes place between its constituent elements.

The second additional type of city-size distribution is a stepwise one, where urban settlements develop in a series of well-defined size steps, with a larger number of settlements in each step than in the overlying one. In the city-size distribution, the steps reflect the various hierarchical levels of a central place system. A hierarchical, stepwise distribution has been identified in urban systems whose main economic base was the provision of commercial services to the settlements' surroundings, the spatial system evolving out of the competition and interaction among the centers. It is thus evident that full-fledged hierarchical systems are found only in economies possessing a well-developed transportation network and a meaningful level of geographical mobility (Richardson, 1973).

The degree to which such full-fledged hierarchical systems have developed in traditional historical societies varied greatly, depending on the level of transportation, mobility, and commercial interaction that existed within them. As only few—if any—such societies have reached that level, it is perhaps wrong to assume that a full-scale city-size distribution of a stepwise hierarchical order developed. In this book, therefore, the term "urban hierarchy" is not used in its strict sense, as it cannot be applied to traditional economies in which complete central place systems could not evolve. This term is used here in a wider, nontechnical meaning to describe any urban system with an array of city sizes, the cities interacting and interrelated, and the analysis presented is based on the assumption that spatial characteristics and the size distribution of the urban system are determined not only by an efficient distribution of commerce and services but also by a combination of political, administrative, and cultural factors. Various combinations of these factors may be very important in determining the degree to which these hierarchies approach a full-scale stepwise hierarchical order. Thus the term "urban system" is used in this book to describe both the hierarchical and the nonhierarchical structures.

It is essential at this point to examine some additional aspects of the urban systems.

2 Urban Hierarchies

Another aspect of urban systems, closely related to the one analyzed above but not identical with it, gave rise to Christaller's "central place theory" (1966;

originally published in German in 1933), an elaborate theory of spatial and functional organization of urban settlements that is concerned with the size of centers, the extension and shape of the areas served by them, and the distances and spatial arrangements of the centers in a given territory (Berry and Pred, 1961; Christaller, 1966). According to this theory, a central place system is composed of several hierarchical levels of settlements, each level defined by the size and type of the centers and the variety of goods and services provided by them to their respective zones of service. A central place system is the most efficient one for distribution of any item over space. In societies in which the main functions of urban settlements are commercial and administrative, central place theory provides a useful tool to explain the number, size, and spatial distribution of urban settlements and of the flows of resources among the central places themselves and between them and the rural areas related to them.

Within a hierarchy of central places there develop flows of people, capital, commodities, information, and regulations. The volume and intensity of the flows increase in accordance with the level of a place in the hierarchy of settlements. All central place systems are similar: They have a hierarchical structure and consist of numerous settlements of small size and of a decreasing number of larger-sized settlements, as one rises in the hierarchy, and a direct relation exists between the size of an urban settlement and that of its dependent area. Central place systems vary, however, with regard to such factors as the number of levels in the urban hierarchy, the ratio of the number of settlements in consecutive hierarchical levels—the "K ratio"—the quantitative relations between the size of the centers and their zones of influence, and so on. These factors are influenced by the type and level of transportation technology, the density of population in a given area, the marketing system, the supply and demand for goods and services, and, above all, the organizing principle underlying the system of settlements and interactions among them.

In his original formulation of the theory, Christaller enunciated three such organizing principles: the "marketing principle," the "transportation principle," and the "administrative principle" (Berry and Pred, 1961). The ratio of the numbers of urban settlements between consecutive levels of the urban hierarchy (the K ratio) varies in accordance with the organizing principle of the settlement system; the ratio is lowest in regions organized according to the marketing principle and highest in regions organized according to the administrative principle. Central place systems based on the marketing principle evolve over a period of time and are the outcome of a large number of decisions by individuals and firms, producers and distributors, and buyers and sellers relating to the most effective pattern of location for efficient economic transactions. Spatial competition is the dominant element of central place systems based on the marketing principle. The gradual way in which this form of system—which is the most common—evolves gives it a great deal of internal cohesion, stability, and potential of survival. By contrast, central place systems based on the administrative principle are established by decree. The administrative centers are defined by the central authorities, the boundaries of their service areas are clearly demarcated, and the entire population living within these boundaries is compelled to receive services from the center and to be under its control. For this reason, central place systems based on the administrative principle tend to be much better defined but much less stable than those based on the marketing principle and are subject to rapid changes in the wake of political upheavals.

3 Developmental Models of Urban Systems

In the works of the first scholars to deal with the size distribution of cities—G. K. Zipf (1949), who studied the rank order of cities in the United States and in other developed countries, and M. Jefferson (1939), who studied the primacy patterns in countries such as Austria, Denmark, Thailand, and Uruguay—the various types of city-size distribution were related to the levels of social or economic development or types of political regimes. Thus Zipf related the development of rank-order distribution to a highly diversified yet integrated economic system, whereas Jefferson related a situation of primacy to a centralized political system mobilizing economic resources. This pattern is especially characteristic of colonial systems.

A more detailed examination of the relations between the various types of city-size distribution and the economic and demographic characteristics of the respective countries was undertaken by Berry (1961). His analysis, based on data from 38 countries relating types of city-size distribution to certain demographic and economic characteristics, disproved two central hypotheses based on evolutionary assumptions. The first hypothesis related the type of city-size distribution to the degree of urbanization in a given society. It maintained that the lower the level of urbanization, the greater the tendency to primacy, whereas the higher the level of urbanization, the greater the tendency to rank-size distribution. The second hypothesis related the same phenomenon to the level of economic development. Thus the higher the level of development, the greater the tendency to rank-size distribution, whereas low levels of development are related to primacy.

The rejection of these two hypotheses led Berry to propose a far more flexible "developmental model" of city-size distribution in a country. According to this model, the major factor affecting the type of distribution is the degree of complexity—that is, the number of different forces participating in the shaping of an urban system. Berry believed that this factor was related to the following variables: size of the country, length of its urban history, and level of its economic development. The weaker each of these variables is, the less complex will be the play of forces affecting the urban pattern. Primacy in this model is regarded as the simplest form of city-size distribution, affected by few strong forces. Rank-size distribution, on the other hand, is regarded as an outcome of a complex economy, affected by the operation of many forces.

The developmental model proposed by Berry assumes an evolutionary sequence in the development of urban systems. According to this model, the initial distribution in any given country is one of primacy followed by an intermediate stage marked by the growth of small- or medium-sized towns, and the final stage is one of rank-size distribution. Implicit in Berry's work is the conviction that in a primacy situation, the primate city will embrace several functions—political, economic, and cultural—whereas in a rank-size situation, the urban system will be much more differentiated, with functional specialization of the various cities and of many centers of decision making throughout the country.

The combination of the analysis of urban systems, of central place hierarchies, and of an evolutionary approach to the analysis of the organization of space came to full fruition in the work of Gilbert Rozman (1973, 1976, 1978). The aim of his analysis was to identify successive stages in the development of urban systems in premodern societies and to relate different central place systems to specific stages of societal development.

Rozman's comparative study of the structure and characteristics of urban systems in five countries (Japan, China, Russia, England, and France) identifies several components of an urban system, the combination of which establishes a particular national urban system. The first component is the number of hierarchical levels of cities in a system. The number of levels varies in any given period, beginning with no levels at all in preurban society and culminating in a full-fledged seven-level urban hierarchy. Rozman maintains that the chronological sequence by means of which new levels are added to an urban system was similar in all the countries he studied. It began with the emergence of towns of low level (level 5—the lowest administrative centers) simultaneously with towns of high level (levels 1 and 2—national and regional centers). The system of urban hierarchies then gradually came into being through the emergence of levels 3 and 4—towns and cities of intermediate size. The regularity of this occurrence would appear to support an evolutionary interpretation of the development of urban hierarchies, beginning with a primacy-like distribution in the early stages and evolving into a more rank-size distribution at a mature stage.

In the early stages of urban development, the main force behind the evolution of the urban network was "administrative centralization" (Rozman, 1976: 263). Commerce played only a secondary role and was confined almost entirely to limited sections of cities that were administrative centers. In later stages of urban development, however, the major influence was the centralization of commercial activities, which took place first through the appearance of periodic markets in settlements removed from administrative centers, then through the emergence of intermediate marketing centers, and, finally, of a complete commercial hierarchy.

Two other closely related components of urban systems are the average size of an "urban place" at each level of the urban hierarchy and the average size of the population served by a central place at each level of the urban hierarchy. Rozman (1976: 247) sees these components as a measure of the "urban efficiency," which is estimated by the average size of population served by one central place. The average population size of central places and of their hinterlands is generally related by Rozman to the type of rural society and economy involved. They are especially related to the rate and system of taxation—whether in kind or in money payments—to the system of landownership, to the form of manpower—whether free peasantry or serfdom—to the size of markets for agricultural products, and to the means of transportation. All these social and economic factors influenced the flow of resources from the rural areas into towns, which, in turn, influenced the size of the towns and of their hinterlands.

After examining a great deal of evidence from China, Japan, Russia, England, and France, Rozman came to the conclusion that the urban systems of these countries had passed through a similar course of evolution, and he stressed the existence of a strong connection between this sequential transformation of the urban system and the main facets of societal development. This interrelation, which, according to him, remained a constant factor despite their variations in the different countries, led Rozman (1976: 203) to suggest that the "basic pattern of social change in premodern countries is directional, cumulative and immanent within existing conditions." Thus the maturation of urban networks marks the "inexorable course" of social change, spanning a period of 1,000 to 3,500 years.

4 Criticism of the Foregoing Theories

The strong evolutionary assumptions of the theories analyzed earlier, which are closely related to the theories of modernization and development prevalent in the 1950s, have since been considerably revised and subject to criticism not unlike that raised in relation to the trait complex and dichotomous approaches to urbanization processes (Barker, 1978; Carter, 1956, 1983).

Thus Berry's work refuted some hypotheses derived from over-simplistic evolutionary assumptions (Berry, 1961). Berry's developmental model, however, was in turn modified by El Shaks (1972), who suggested that the relation of city-size distribution to economic development is not a linear one, and that primacy does not necessarily develop at an early stage of economic development. Rather, primacy tends to occur in the intermediate stages of economic development, not in the early stages. According to El Shaks, the interaction and competition among cities are at a low level in the early stages of economic development. The prevailing principle of spatial organization is that of separation, in which each settlement served the surrounding area but in which strong interaction between the different centers was prevented by lack of communications systems. At this early stage most towns essentially assumed the same functions, whether political and administrative or rudimentary economic activities of production or provision of services. The forces shaping the structure of the system were varied and mainly local in character.

A major shift in the pattern of economic activity may occur with the emergence of some major economic force or process, such as specialization in the export of a commodity or the specialized cultivation of a staple crop. Such specialization may have the effect of encouraging a particular city to take the lead and become the major center of the urban system. In this way a primate city and an urban system characterized by a primacy structure can come into being. At this stage, because the prosperity of the primate city depends on a single economic activity that is affected by fluctuations in world markets, and because the structure of the urban system is not very stable, rapid changes may occur in the city-size distribution within the country (Carter, 1972: 41).

El Shaks maintains that as the national economy matures, the emergence of many kinds of production and distribution renders the urban structure more diversified and complex. This diversity is reflected in the establishment of secondary centers and the improvement of channels of communication between all parts of the urban systems, and ultimately results in complex spatial and functional integration of the urban system. All these factors constitute a change from a primacy to a rank-size structure, with a complete hierarchy of cities of various sizes and fully developed mechanisms for distribution and provision of services to the entire population of the country. Because urban systems of this kind are based on a variegated, differentiated economy, they are less dependent on one particular activity and consequently much more stable than those characterized by a primacy structure. Thus El Shaks's work considerably expands the scope of the evolutionary model by postulating an early stage of small, equal-sized urban settlements preceding a primacy stage and then evolving into a rank-sized urban system.

Several other researchers, such as Linsky (1965), McGreevey (1971), and Morse (1971), have also indicated the existence of a "disappointingly small" degree of association between primacy and a relatively low level of economic and social development. Morse and McGreevey state that, on the contrary, in many Latin American countries primacy is a recent 20th-century phenomenon. Morse's work (1971) on city-size distribution in Latin America is most significant in this respect, as it clearly negates the one-directional developmental model whereby urban systems evolve from primacy to rank size. Morse argues that in many of the Latin American countries it was only after national unification had been achieved, political systems centralized, and export earnings increased that the primate city asserted itself as "a locus where political clientele and economic concessions were dispersed, financial intermediaries thrived, elites and foreigners tasted urban pleasures and basic industries appeared" (Morse, 1971: 6). He concludes that the evidence from Latin America indicates the existence of an opposite sequence of change from rank size to primacy.

But although he abandons the one-way sequence of Berry's model, Morse retains the same basic explanation for the causes of change in an urban system—namely, that growing complexity and specialization in economic activities and changes in the political power structure within a territory are the main factors shaping the urban system and the distribution of city-sizes within it. And McGreevey (1971: 126) maintains that the change from rank-size distribution of cities to one of primacy is related to the growing importance of the export sector in the national economy.

Paralleling these interpretations, Rozman's analysis indicates important differences in the structure of urban systems in various countries, differences which, to some degree, go beyond the basic evolutionary assumptions that have guided his work. He demonstrates that the rate of change in urban development can differ from one society to another. These differences depend on local geographical, economic, and political circumstances; in addition, the number of levels and "central places" at each level of the urban hierarchy may vary in each case. Thus he has shown that England and Japan have an "overrepresentation" of cities at the highest levels—levels 1 and 2—China and France have a larger proportion of cities at intermediate levels—levels 3 and 4—whereas England and France are overrepresented in the number of towns at the lowest level—level 7. Rozman has related these findings to different constellations of the three basic societal processes: (1) centralization, (2) commercialization, and (3) urbanization. Accordingly, he explains that the overrepresentation of large cities (levels 1 and 2) in England and Japan is due to "local administrative centralization," and the large number of "central places" of the lowest level (7) in England and France is due to "individual commercial participation" (Rozman, 1976: 254).

One may draw, in more general terms, the following conclusions from Rozman's analysis:

(1) In small countries with a strong central government, cities of the highest level will dominate the urban system.
(2) In countries with a strong central bureaucracy organized on a regional basis, regional capitals that are medium-sized cities will be overrepresented in the urban system.
(3) The more elaborate the commercial network, the more developed will be the lower and middle levels of the urban hierarchy. The greater the flow of resources and commodities between rural areas and cities and among urban centers themselves, the more integrated will be the national economy and society.
(4) A high degree of spatial integration will be achieved by the overlapping of an administrative and commercial hierarchy within a single urban system.

Thus Rozman's analysis constitutes probably the most important and systematic attempt to combine the study of urban systems, the "central place" theory, and a social evolutionary approach. It suggests a far greater degree of variability and diversity in the developmental patterns of urban systems than has previously been assumed—even by Rozman—and indicates several new dimensions of the political, economic, and social factors influencing the structure and dynamics of urban systems.

A similar emphasis can be found in Blanton's (1976: 250) work. He advocates adopting a regional approach to the study of cities, involving a far-reaching modification of the central place theory. He identifies the dual role of central places, which are centers for both the distribution of commodities and services and the processing and distribution of information, and has shown that the relationship between the two can vary greatly. He explores the implications of different patterns in the relations between these two functions for understanding variability in the size and functions of central places and different types of hierarchy that they create. Blanton analyzes two types of urban hierarchies in detail. (1) The primate pattern is one that occurs when the major central place of a region is also its capital. This kind of center will tend to dominate the region in such a way that the development of secondary centers will be delayed. (2) The other pattern is one in which the commercial central place hierarchy is more developed. In this case, primacy does not evolve because the political capital is usually disembedded in the marketing hierarchy. Blanton suggests that this variable relationship between the administrative and marketing hierarchies, which is both cross-cultural and diachronic, can be explained by three factors: the power of the central authority, the power of the state relative to that of the marketing institutions, and the capacity of administrative institutions to process information.

Another important contribution to the study of urban systems and their interpretation can be found in Carol Smith's work (1982). She differentiates, far more than in other approaches, among the types of city-size distributions identified, deals explicitly with premodern urban systems as well as with modern ones, and does not assume a priori any evolutionary process that leads in a single direction from one type of city-size distribution to the next. To the well-known primacy, rank size and convex types of distribution, she adds a composite distribution of primacy on the urban system's upper level and any of the three following distributions beneath it: rank size, convex, and concave. She argues that primacy occurs in both traditional and modern societies, but their societal interpretation differs according to the specific type of distribution at the lower level of their urban systems. Rank-size distributions are found mainly in the developed economic systems of the modern world, whereas convex distributions prevail in traditional societies. Interpretation of the occurrence of different city-size distributions in premodern and modern urban systems alike rests on the basic forms of the respective society's social and economic organization. In premodern economies, labor mobility was restricted, either because people were tied to the land or because urban institutions (guilds, town citizenship) inhibited settling in the cities. Under these conditions, urban migration was limited and resulted in a convex or equal city-size distribution. The limited urban migration could have been selective in its destination if the ruling elite had for some reason oriented the flow of migrants toward one particular city, thus creating a primacy situation, overlying the convex or equal size distribution. Smith's work thus clearly allows for greater differentiation in the types of city-size distributions and multidirectional paths in their evolution.

The urban system approach has recently been applied in a most thorough analysis of urbanization processes in Europe by Jan de Vries (1984). De Vries studied the postmedieval, preindustrial period (1500-1800), and the major methodology used established the rank-size distribution for the European urban systems, defined according to three scales: the whole of Europe, the subcontinental units (Northern and Southern Europe), and individual countries. Based on de Vries's important work, a developmental model of the European urban systems can be formulated. (See Chapter 11 on urbanization in Medieval Europe.)

De Vries's detailed and meticulous analysis proves the validity and usefulness of the "urban system approach" in the study of urbanization processes and, even more important, highlights the multipath trajectories of change that occurred in urban systems over time, reflecting the specific combination of political, economic, and social forces operating in each period. The concept of the city-size distribution of supranational urban systems was applied recently within a global context (Chase-Dunn, 1985). By constructing an index that measures the deviation of a city-size distribution from a log-normal one, Chase-Dunn analyzed the dynamics of the global urban system in the period 800-1975, interpreting the changes in the city-size distribution toward a more or less hierarchical (rank-size) one as an outcome of both the existence of a "world empire"—a major hegemonic political power—and the level of spatial integration of the "world economy."

5 Basic Components of Urban Systems

By critically reviewing various analyses of urban systems, subsequent generations of scholars have modified approaches and typologies. Beyond this, critical reviews over time led to far more serious consideration of these approaches. They prove that urban systems are composed of different components and that these components do not always combine in the same way—that is, they may occur in varying permutations and combinations.

The most important components of urban systems are the following:

(1) *Function:* What is the main function fulfilled by an urban system—administrative, economic, or religious? What degree of overlap exists between urban systems fulfilling these different functions? For instance, how much overlap exists between an economic and an administrative urban hierarchy?

(2) *Hierarchical levels:* How many hierarchical levels are contained in any given urban system? (Each hierarchical level consists of a number of towns of roughly equal population size.)

(3) *Sequence:* In what order are new hierarchical levels added to the system? New levels may be added in various sequences, beginning at the lowest level and ascending; or low and high levels may appear simultaneously, with the intermediate levels emerging at a later stage.

(4) *Size:* What is the average size of an urban settlement in terms of population, territory, and places served, and the zones of influence at each level of the hierarchy?

(5) *Patterns of resource and information flows:* Are these flows between the parts of the urban system unicentered or multicentered? To what degree? In unicentered systems all flows are oriented toward or emanate from a single major urban center; in a multicentered system the flows have many origins and destinations.

(6) *Relations among levels:* What degrees of strength, self-sufficiency, and overrepresentation exist at each of the levels? How much is each level dependent on flows from higher or lower levels?

(7) *Autonomy:* To what degree is the urban system enclosed within itself? Some urban systems have a strong external orientation, which is reflected in the amount of interaction with areas outside the system. Others have a greater level of self-sufficiency.

The above list of component elements is not necessarily exhaustive. However, the significance of this approach to our study is that contrary to the assumptions of the major approaches analyzed previously, these components can be combined in a number of ways, thereby enlarging our range of inquiry and enriching our understanding.

Moreover, as we have indicated above, although some of these combinations can be related to an evolutionary model (Rozman, 1976) or to the increasing complexity of an economic system (Berry, 1961), these hypotheses are not sufficient to explain all the possible variations of urban systems. And just as with studies of the origins of civilizations, trait complexes of cities, and internal structure, they indicate that it is necessary to undertake a more differentiated analysis of the conditions that shape different variations of the urban system.

G
THE IMPACT OF SOCIOLOGICAL ANALYSIS: TOWARD A NEW APPROACH

1 Major Sociological Controversies

The preceding analysis has indicated that in all areas of the study of cities—whether analyzing the relation between the development of cities and the growth of civilizations, the internal structure of cities and the scope of their autonomy, or the urban systems, we find that far more components have been identified than could be assumed in most classical approaches. These components appear in many more configurations and are influenced by more factors than allowed for by previous approaches in general and by the evolutionary approach, which stresses the different modes of division of labor, defined in terms of degrees of social differentiation and specialization as the major explanation of different patterns of city life. Thus developments in all these areas of urban analysis call for a far more differentiated approach.

The chance that such an approach will be developed is increased by the fact that criticisms of the major classical studies of urbanization and the new approaches we have analyzed have followed closely or paralleled—and sometimes were even interconnected with—some of the major controversies in sociological theory that emerged from the 1960s onward. Those directed against the structural-functional school and the closely related studies of modernization with their neo-evolutionary emphasis were especially powerful (Eisenstadt, 1973, 1981; Eisenstadt and Curelaru, 1976, 1977).

These controversies have focused on what seemed to be a major implicit assumption of both the structural-functional school and these related studies: that identifying the mechanisms of social division of labor and the systemic nature of societies, in terms of such mechanisms, does explain the variability and dynamics of institutional formations.

Such an assumption was often considered by critics of these approaches to neglect one of the major insights of the sociological tradition, as crystallized by its founding fathers—above all, Marx (1965), Durkheim (1964a, 1964b), and Weber (1951, 1952). This was the nonacceptance of the assumption, implicit in utilitarian ethics and classical economics, of the predominance and sufficiency of the social division of labor in general, and of the market in particular, as the regulators of social order, as a mechanism that assures the maintenance of the social division of labor.

The founding fathers did not, of course, deny the importance of social division of labor and of the market as such a mechanism. Indeed, in many ways they elaborated some aspects of market analysis, the processes of exchange in social life, as well as the impact of different aspects of market structure on the behavior of individuals and on the crystallization of forms of social life. But they all questioned whether such mechanisms were sufficient to explain the operation of any concrete social division of labor and of any concrete social order. In different ways all showed how such mechanisms in general, and the market in particular, cannot assure such operation. Indeed, they stressed that the organization of the social division of labor—social exchange in general and the market in particular—generates problems that render problematic the working of any concrete social division of labor.

The most important of such problems were the construction of trust or solidarity (stressed above all by Durkheim), the regulation of power (emphasized by Marx and Weber), and the provision of both meaning and legitimation for social activities so prized by Weber. These theorists emphasized the fact that the mere construction of social division of labor generates uncertainties with respect to each of these dimensions of social order: the construction of trust and solidarity, the regulation of power, and the provision of meaning and legitimation. Simultaneously, because of this fact, no concrete social division of labor can be maintained without solving these problems.

Despite their stress on the importance of analyzing the regulation and legitimation of power relationships and the construction of trust and meaning, the founding fathers (with the exception of Weber in his analysis of charisma) did not systematically analyze the institutional mechanisms of these dimensions. They pointed to some of the most important areas of social life—especially legitimation, ideology, and ritual—that bear on such construction. However, their analysis of the institutional structure of these dimensions of social order was, on the whole, much weaker than that of the operation of the market or of direct power relations.

This situation was due partly to the relatively low level of appropriate analytical and conceptual tools; significant developments along these lines came much later. There was probably also a linkage to a strong awareness on the part of the founding fathers of the great tension existing between the organizational division of labor on one hand and the regulation and legitimation of power and the construction of trust and meaning on the other (Eisenstadt, 1981).

The structural-functional school has further elaborated a new, systematic conceptual and analytical apparatus for analyzing social relations, behavior, and organization, connecting it to the classical problems of sociological analysis. As a result, this school addressed itself directly to the problem of how the dimension of solidarity (trust), meaning, and, to a smaller degree, power is institutionalized in the construction (or production) of social order. These dimensions of social life were defined as needs that every social system (and, in a different way, also personalities and cultures) has to cope with, or as prerequisites of the inherent working of such systems. In one version of this approach—probably the best known—these dimensions were defined as follows: the need for solidarity (integration), which can be seen as equivalent to trust; the need for pattern maintenance (meaning); the need for maintenance of instrumental orientations (closely related to but not entirely identical with regulation of power); and the adaptive need—that is, the need for adaptation to the respective systems' environment, which is accommodated through the organization of social division of labor. In this conception pattern maintenance and integration were seen by Parsons as having a higher cybernetic role in the regulation of social activities than the instrumental or adaptive needs. Thus using the terminology of cybernetic theory, Parsons indicated that the last two provide the actors or the systems with energy, whereas the first two supply them with the information that "molds" such energy. This higher cybernetic role of the construction of meaning (and, to a smaller degree, of trust) was evident in the central place in the social systems that Parsons allocated to value consensus—a point that very early became the subject of continual criticism.

Thus by defining in such a way the construction of trust, the provision of meaning, and the regulation of power, and by analyzing them systematically, Parsons (as well as Merton [1963] and many others) was able to achieve what the founding fathers had done only tangentially: namely, to specify the institutional processes through which these dimensions of construction of social order are interwoven in the structure of society.

This specification was greatly facilitated by the important achievement of this school or approach: its restructuring of certain central concepts that hitherto had been widely accepted in sociology and social anthropology. Many concepts—such as role, status, role sets and institutions, prestige, power, and solidarity, which were common to both sociological and anthropological research—underwent a far-reaching analytical transformation in the structural-functional school. Rather than being used in a descriptive or classificatory manner, they became conceptual tools specifying how different systems of action and different social settings are related to one another and to the broader macrosocietal settings. New conceptual refinements such as role sets and status sets were also developed. All such analytical refinements of concepts were closely connected—although certainly not fully identical—to the analysis of the structuring of power, trust, and meaning discussed earlier.

Research was increasingly influenced by these conceptions, and hardly an area was left untouched. The structural-functional approach provided, for almost all fields of sociological research, a general view, image, or map of the social system as well as analytic guidelines that led to new, far-reaching studies, the likes of which had not been seen hitherto in the history of sociology.

The single most distinctive feature of many of these studies, which used these new conceptual and analytical tools, was the indication that the concrete institutional structure researched—be it a hospital, factory, professional organization, or broader macrosocietal settings, or the structure of roles and status that is prevalent within them—could be explained by the manner in which the basic needs of any social system were defined and met.

All these developments had many impacts on research and have given rise to a variegated research agenda in the fields of both micro- and macrosociology. At the macro level, especially in studies of modernization and comparative institutional analysis, we see a rich variety of research strata in which the above analyses of cities and urban hierarchies constitute an important part.

As already mentioned in our analysis of studies of both cities and urban systems, most comparative studies were infused by a strong evolutionary vision. This vision assumed that human societies grow through different stages that can be distinguished, above all, by the degree of complexity of the division of labor, of structural differentiation, developing as it were from mechanical to organic solidarity (to use Durkheim's terminology).

But in all areas of research—just as in the studies of urbanization described earlier—these assumptions were inadequate in explaining the variety of actual institutional forms and led to increased criticism of the researchers. The most sweeping criticism of such approaches argued that they tended to take for granted the emergence and crystallization of different institutions. By explaining institutions only in terms of growing differentiation, or of their functional contribution to the maintenance of the overall systems, the creative autonomy of groups or individuals was negated and the tension minimized between the organization of social division of labor and the regulation of power and construction of trust and meaning.

2 New Sociological Approaches

Criticism of these studies was closely related to a more general and theoretical critique—that of the structural-functional school—and to the development of new research perspectives and problematics, new theoretical approaches, and attempts to revive alternative theoretical models. As a result, these sociological controversies generated a variety of new approaches and models of human society, such as the conflict model espoused by Ralf Dahrendorf (1964), the exchange model of George C. Homans (1961) and Peter Blau (1964), and the symbolic-structuralist models of Claude Levi-Strauss (1963). This trend also led to the reaffirmation and further development of older models, such as symbolic-interactionism (from which ethnomethodology developed) or Marxism (and its variant forms). These models, and especially those contrasted with the structural-functionalist model, were a focal point of sociological debate and particularly of macrosociological analysis during the last three decades.

The common factor in all of these models and approaches was an unwillingness, in the face of assumptions of the structural-functionalist school and "classical" sociological studies of modernization, to accept the supposition that a particular type of institutional arrangement (in the present study, a certain type of city or urban society) could be explained by the organizational needs of a specific social system or of a particular stage in the evolution of a social system. In other words, they did not accept the assumption that any concrete institutional pattern could be explained in terms of the pattern of social division of labor prevailing in it, its needs and exigencies (Eisenstadt, 1981).

The models differed in their proposals for coping with this problem and for explaining any concrete institutional order. One such approach, found in the individualistic and conflict models as well as among symbolic-interactionist models, stressed that any such institutional order develops, is maintained, and is changed through a process of continuous interaction, negotiation, and struggle among those who participate in it. In this approach it was stressed that explaining any institutional arrangement has to be attempted in terms of power relations and negotiations, power struggles and conflicts, and the formation of coalitions during the processes. Concomitantly, a strong emphasis was laid on (a) the autonomy of any subsetting, subgroup, or system that could find expression in the definitions of goals that differed from those of the broader organizational or institutional setting and of the groups dominant in it; (b) on the environments within which the social setting operates; and above all (c) on the international system for the analysis of "total" societies or macrosocietal orders.

The second, seemingly contradictory approach is found among structuralists and Marxists. As we have discussed, that approach explained the nature of any given institutional order and especially its dynamics in terms of some principles of "deep" or "hidden" structure, akin to those that, according to linguists such as Chomsky, provide the deep structure of language. In attempting to identify the principles of this framework, structuralists stressed the importance of the symbolic dimensions of human activity, of some inherent rules of production and reproduction of different social formations, and of the relations between modes and relations of production as carried by different classes.

Cutting across these two approaches to explain the crystallization of concrete institutional patterns was the growing emphasis on power, on symbolic orientation ("culture"), or on some combination of the two as the major principles influencing the shape of such patterns beyond the structure of social division of labor.

3 New Directions of Research on Cities and Urbanization

In connection with some of these theoretical models, numerous but relatively segregated new lines of research developed (Eisenstadt, 1981; Eisenstadt and Curelaru, 1976) in many fields of study, including analyses of cities and urban systems (Elliot and McCrone, 1982; Gale and Ollson, 1979; Pickvance, 1976). The first such relatively new research orientation emphasized the distinctive characteristic of cities within a specific civilization. This is evident, for instance, in the tendency to assume, in regard to the Islamic city (Von Grunebaum, 1976), that special characteristics (or lack of characteristics) of the "real" or Western city were inherent in the Islamic city—or in the cities of other civilizations—without comparing it systematically with those of other civilizations (except, perhaps, the Weberian ideal type of the Western city) and usually without allowing for systematic analysis of the variations within it.

A second spate of research on urbanization processes and dynamics of urban systems developed within a rather broad framework of Marxist approaches (Castells, 1977; Castells and Godard, 1978; Harloe, 1977; Harvey, 1973, 1978, 1982; Lefebre, 1972; Peet, 1977). These studies emphasized such variables as class struggle, class domination, and international dependency as the major forces explaining the processes of urbanization and the structure of urban systems and cities. While referring to Marx and Engels, these studies have in fact presented a rather different mode of interpretation of cities.

Marx and Engels regarded the historic division between town and country, which characterized all human societies from antiquity to the period of modern capitalism, as the basis for the major division of labor in society. According to Marx, the first fully developed class society was that of the ancient city-state, and foremost the Roman society. This society was based on a slavery mode of production, the wealth of the ruling class accumulating through ownership of agricultural land. The rich landowners lived in the city, but the mode of production remained rural. "Ancient classical history is the history of cities, but cities based on landownership and agriculture" (Marx, 1964:77). Until the Middle Ages, the history of human societies is thus the history of the countryside, as the city did not become the locus of a new mode of production. In the feudal period a new mode of production emerged for the first time in Europe. The growth of a merchant class in the towns during the Middle Ages had the effect of extending trading links among different areas, thus stimulating specialization in different cities and regions and encouraging the growth of new industries. The accumulation of wealth in the feudal city resulting from flourishing commerce created the essential contradiction between capital and feudal landownership, expressed in the two distinct forms of settlement: town and country. Thus in Saunders's exposition of Marx, he concluded that "in the feudal period, the division between town and country not only reflected the growing division of labor between manufacture and agriculture but was also the phenomenal expression of the antithesis between conflicting modes of production" (Saunders, 1981:21).

Once capitalism was established, the division between town and country lost much of its significance. This stemmed from the fact that the essential contradiction existing within the capitalist mode of production—that between capital and labor—was no longer represented by the forms of settlement of town and country. With the breakdown of the feudal system and the establishment of the capitalist mode of production, agriculture, in the same way as industry, became characterized by capitalist social relations. The city, then, is no longer the real subject of analysis, because the conflicts between proletariat and bourgeoisie cut across the urban-rural boundaries, and workers in town and country were drawn into the social relations of the capitalist mode of production.

Although the study of the city plays a secondary role in Marx's analysis of capitalism, both Marx and Engels regarded the process of urbanization as a condition necessary to the transition to socialism. This was not because the city was considered as the locus of a new mode of production (as was the case in Medieval Europe) but because it was in the capitalist city that the proletariat fully matured and conditions were provided there for an effective class struggle. The concentration of capital in the city, the polarization among the classes, and the appalling deprivation of the proletariat led to the emergence and formulation of class consciousness and revolutionary organization (Saunders, 1981).

Marx and Engels condemned the consequences of urbanization under capitalism, as observed in Manchester and elsewhere, but viewed the concentration and pauperization of the mass of workers in new urban metropolises as a necessary stage in the creation of a revolutionary force. Although misery and degradation constituted a major aspect of urbanization, equally important was the destruction of the social nexus of the traditional community and its replacement by the rational world of a city. Man, as a rational being, was a product of towns, not of countryside. Workers had to be freed from the limitations of rural life, the conditions of rural confinement. Urban industrial civilization was the necessary basis for the new higher relations of production. Urbanization was ultimately liberative (Mellor, 1977: 170).

A different assessment of the urbanization process was made in the neo-Marxist writings. Common in the works of Castells (1977), Fanon (1965), Frank (1969), Harvey (1982), and Marcuse (1964) is the belief that the conditions of capitalist urbanization inhibit community formation, destroy social involvement, distort personality, and encourage apathy and alienation. The assumptions of orthodox Marxism are overturned, as urbanization is no longer seen as the basic condition for a socialist transformation of society.

Given that urbanization is regarded, in human terms, as mutilating, the cities themselves are viewed as instruments of capitalist or imperialist domination. In their function as residence and centers of power of the regional, national, and even international bourgeoisie, cities are links in the chain of expropriation from peripheral territory to dominant metropolis. The growth of cities is due not to the accumulated surplus of the honest toil of their citizens but to the labors of the rural, provincial, and colonial masses. The development and wealth of cities depend on the effective exploitation of the material and human resources of their peripheries. The effectiveness of this exploitation rests with the political control exercised by the major urban centers over the regional and national territories. Thus the characteristic feature of cities is seen not in their economic specialization but in their role as centers of dominance. They represent the sphere of social domination by the bourgeoisie (Castells, 1977).

Control of the metropolis over national and regional territory and economy is not only exercised by political power and economic dominance but is supplemented, following Gramsci, by social authority, whose ultimate sanction is a profound cultural supremacy or hegemony (Gramsci, 1971: 103). The Gramscian concept of hegemony must be added to exploitation and domination when explaining urbanization processes. Pervasive exercise of authority is derived from the cultural and social prestige of the metropolitan bourgeoisie; the culture of the cities' ruling elite is diffused throughout society. Thus metropolitan culture comes to legitimate control. Hegemony plays as important a role as dominance in understanding the metropolitan control over the entire society.

Some of the Marxist approaches have tended to deny the existence of any universal urban trait characterizing the "city" or a particular type of city. It is worthwhile, in this respect, to quote from Castells's contribution titled "Is There an Urban Sociology?":

> Urbanism is not a concept. It is a myth in the strictest sense since it recounts, ideologically, the history of mankind. An urban sociology founded on urbanism is an ideology of modernity ethnocentrically identified with the crystallization of the social forms of liberal capitalism [1976: 70].

This approach argues that the only scientific way to study and understand urbanization processes and urban phenomena is by placing them into the much wider perspective of whole economic systems and relating urban development to particular stages in the evolution of the economic systems in general and of class relations in particular. Class struggle is the dynamic force carrying the economy and society from one stage to the next, and it is this struggle that structures the entire range of social relations and institutions. Most of these Marxist scholars contend that the important urban institutions and practices are the outcome of class struggle and conflict and firmly argue that national and urban institutions in capitalist societies operate in a way that maintains the status quo, appease class conflicts, and ensure the reproduction of the existing systems of labor, capital, and production. In classical Marxism, the reproduction of the existing economic system and class conflicts were related to urban issues of employment and housing. In neo-Marxist approaches, a strong emphasis is put on conflicts in the urban arena revolving around new issues of consumption of collective goods such as education, health, public transportation, and other services provided by the state and local authorities. This provision of collective goods by the state and its agencies opens up the eventuality of new inequalities and new class struggle. Marxist approaches called for the construction of a "political economy" of the city for studying urban processes and institutions, not as a distinct phenomenon but just as the very product of general economic forces, of sweeping political and social changes, and continually evolving class struggles and conflicts.

Without going into a detailed critique of the neo-Marxist approaches to the study of urbanization processes and urban phenomena, it is most important to point out two major drawbacks of most of these approaches. The first is the almost total rejection of history, of detailed historical analysis in structural Marxism, despite paying lip service to the importance of historical contexts (Hindess and Hirst, 1975), and the consequent lack of specification of the components of historical context within it. This was stressed by Marxist historian E.P. Thompson (1975), who severely denounced "the gross confusion which runs through the structuralist critique of empirical procedures, empirical control" as reported by Elliot and McCrone (1982: 14). Thus one finds in structural Marxist analysis of urban phenomena that historical contexts are defined in very general terms, often derived in rather simplistic ways. This trait, paradoxically enough, is quite similar to one found in historicist approaches. The second drawback, in contrast to these approaches, is the lack of specificity in studying problems of urban life, whether general ones, common to many urban forms, or those related to different societies or civilizations.

In none of these approaches was it thus possible either to compare systematically between the contexts or to explain the great variety of components of urban life and urban systems and the patterns in which they coalesced. In most of these researches, as well as in the basic theoretical approaches to which they were connected, it was not well specified how and through which social mechanisms and activities of the different social actions, the dimensions of social life emphasized by each—be they power or class relations, symbolic orientations, or the social division of labor—were related to one another in the crystallization of concrete institutional patterns.

These indications moved—even if in an irregular and often halting way—in what may be called a Weberian or neo-Weberian direction. It would therefore be worthwhile to spell out here in greater detail the implications of Weber's analysis for the comparative study of cities (Bourdin and Hirshhorn, 1985).

Two aspects of such a neo-Weberian approach are of special importance. The first, stressed recently by Elliot and McCrone (1982), focused on the structuring of conflict and on the identification of the ways in which various social groups—be they classes, status groups, or ethnic collectivities—struggled to control and dominate the city, to establish urban institutions, to form coalitions, all in order to gain and maintain economic and political power in the cities. In this context, Weber emphasized the complexity, rich variety, and numerous constellations, all historically rooted, of the political, economic, and social relations existing in cities. His analyses of different civilizations highlighted the specific historical constellations of forces shaping the political and social structure of society as a whole and of its cities in particular (Spencer, 1977). In this respect, being quite close to Marx's own work but not to that of the neo-Marxists, Weber's extensive comparative studies of societies and periods bear witness to his commitment to in-depth historical analysis.

In his analysis of the internal structure of cities, Weber stressed three facets—lately emphasized strongly by Elliot and McCrone (1982)—that are closely related to different dimensions of the autonomy of cities: first, the nature of internal collectivities constituted in the cities, and second, the pattern of life developed by various class and status groups that evolved in cities, and, third, the movements of protest and social conflict. Thus in his analysis of the Medieval European city, he outlined the various modes of collective action that developed within it and stressed the formation of oath-bound fraternities of burghers and guilds of merchants, professionals, and artisans. Throughout his work, Weber provided numerous comparative indications with respect to cities of other civilizations.

With respect to the pattern of life of different groups in the city, Weber went, as is well known, beyond a simple economically determined conception of stratification and stressed the autonomy of the political and status dimensions. He identified many more bases for the formation of status groups and status conflicts, such as race, ethnicity, and religion, any of which could constitute an avenue for differential allocation of material resources and of social honor as well as for the emergence of distinctive patterns of consumption and life-styles. The numerous bases for the formation of status groups enrich most significantly the theoretical framework of social stratification to encompass a larger number of classes and a more complex structure of the social hierarchy. This less rigid and more differentiated approach to the definition of class structure enables a thorough examination of many political conflicts that take place within the economic dimension of class formation and of struggles for power and privileges among particular constellations of the various status groups.

Finally, in studying urban collective action, special attention was devoted by Weber to the analysis of the evolution and operation of social movements that aimed at expressing discontent with domination by the elite groups, with political oppression, and with economic deprivation. The struggles of the social movements that were internally differentiated by class, religion, ethnicity, or neighborhood assumed many forms of collective action, from civil disobedience to strikes and urban riots: "As forms of domination have changed, so too have the collective responses" (Elliot and McCrone, 1982:23). The study of protest movements constitutes a central focus of the sociological analysis of the city, as most strikingly representing the power structure in the city, the evolving tensions within the political and social systems, the competing economic interests of various groups, and, above all, the patterns of domination and revolt that characterized urban society throughout history. Thus Weber's analysis indicates—even if it does not always fully explicate—that the cities can be studied as the major arena for political struggle and social conflict, as a full-scale laboratory in which can be observed and analyzed the integration and disintegration of political regimes, the rise and fall of dynasties, the changing forms of domination and revolts, the multifaceted interactions among classes, status groups, and ethnic and cultural collectivities.

The second major aspect of Weber's comparative analysis, even if sometimes rather implicit, was the cultural or religious focus of society, most fully elaborated in his "Gesammelte Aufsätze zur Religions-soziologie" (1920-1921). In this work he investigated how the major religious or cultural orientations and traditions of the major civilizations, as well as their carriers as they evolved in different societies, have influenced the various patterns of urban social relations and movements mentioned above.

Weber applied all these considerations, sometimes even implicitly, in his analysis of the different historical types and complexes of cities. At the same time, however, his various comparative historical studies did not lead him to an evolutionary approach to history. Instead of arguing about phases and stages in the march of history toward an ultimate stage of society, Weber placed much more value on detailed historical studies of particular cities and urban systems in well-defined historical and cultural frameworks. It was this more differentiated approach to class structure and cultural orientations that Weber applied to the analysis of the Western European medieval cities as loci of many of the most fundamental processes of change that shaped the transition from feudalism to modern society. In the medieval cities, Weber observed and analyzed by meticulous historical study the struggles among the different classes and status groups, the emergence of new institutions, and the construction of new forms of social relations—all of these processes taking place within a well-defined cultural context.

As indicated above, at least some of the new developments in the study of cities have indeed moved in one way or another in such a neo-Weberian direction, but they already take into account many of the more recent theoretical and methodological advances as well as historical research. Important indications in this direction have emerged slowly. Thus some later works that stressed the uniqueness of the urban pattern of different civilizations—especially Morse's (1972) study of Latin America—already gave, beyond relatively generalized concepts, a more explicit analysis of the particular features of a society. These were, among others, concepts of authority and political orientations that differed considerably in the various civilizations and had a major influence on the cities and urban systems within those civilizations.

A major contribution toward the establishment of a new paradigm for the study of urbanization was made by John Friedmann (1973). In the chapter entitled "Urbanization and National Development," this scholar proposed a core-periphery framework comprising the basic structural components of the spatial organization of society. The autonomy-dependency relations between core and periphery are fundamental to the paradigm and are the prime driving force behind the spatial and social dynamics of society. It is very important to emphasize that Friedmann did not equate the core and periphery with the urban-rural dichotomy: "Because urban settlements can occur in the periphery and rural settlements in the core, it is possible for the periphery to become urbanized . . . without losing its dependency status" (Friedmann, 1973:68). The paradigm identified four basic urbanization processes that relate core and periphery in a national spatial system: diffusion of innovations, control of decisions, migration movements, and flows of capital. These four processes are asymmetric, a condition that stems from the fundamental imbalance in authority-dependency relations between core and periphery. The proposed paradigm of urbanization is helpful in perceiving societal development in terms of the operation of the above basic processes, which lead through a series of structural transformations to successively higher levels of spatial integration within a national framework.

Friedmann called for new studies of urbanization that will be interdisciplinary in nature, will focus on open national systems instead of on cities, will make extensive use of models based on the core-periphery paradigm, will apply a comparative perspective, especially a cross-cultural one, and, finally, will emphasize dynamic processes interaction. Friedmann's review of the existing urbanization studies demonstrated their relative paucity and the lack of a sound comparative approach.

Such a search for a more differentiated approach appears, in a more fully articulated way, in some recent works on center-periphery relations having a strong political economy approach, especially in that by Stein Rokkan (1980) as well as in three collections, one edited by Abrams and Wrigley (1978), the second by Gottmann and Laponce (1980), and the third by Agnew, Mercer, and Sopher (1984).

As a consequence of these theoretical debates and controversies and the researches they spawned, it is now possible to develop a more differentiated approach, based on a neo-Weberian starting point. Such a perspective can combine the analysis of the interaction between cultural orientations and political institutions and conflicts, international political, and ecological patterns as they influence the shaping of both cities and urban hierarchies. We shall now attempt to suggest such a new approach to the theoretical analysis of some of the major aspects of cities and urban life.

CHAPTER 2

Cities and Urban Systems: Toward a New Comparative Civilizational Approach

A
ASSUMPTIONS OF THE NEW APPROACH

In this chapter we shall delineate a new approach to the study of cities and urban systems. This approach is based on developments in sociological theory and on studies of urbanization through which, despite certain weaknesses, the basic insights of these approaches can be preserved.

The present approach to the study of the urban phenomenon is rooted in a broader framework of the analysis of institutional patterns, indicating how the combination of political and cultural forces with those that structure the social division of labor shapes the institutional patterns and contours of a society and influences its evolution and maintenance. This approach can best be designated as a comparative civilizational approach.

Based on a critical examination of the major controversies in the social sciences, this approach recognizes that the establishment of any institutional setting is affected by the combination of the following components: (1) the level and distribution of resources among the different groups in a society—that is, the type of division of labor dominant in that society; (2) the institutional entrepreneurs or elites available or competing to mobilize and structure such resources and to organize and articulate the major groups formed by the social division of labor; and (3) the nature of the conceptions or "visions" informing the activities of these elites, from which they combine the structuring of trust, provision of meaning, and regulation of power with the division of labor in society.

The structure of such elites is closely related to the basic cultural orientations prevalent in a society. In other words, different types of elites carry different orientations that shape the basic premises of their respective civilizations or societies and that are institutionalized through the exercise of different modes of control over the allocation of basic resources in society. These elites or coalitions of elites wield control primarily over the access to the major institutional markets (economic, political, cultural, etc.), over conversion of the main resources between these markets, and over production and distribution of information that is central for structuring cognitive maps of the members of a society, their perception of the nature of their society, its basic premises, and their reference orientations and reference groups. This control is applied by combining organizational and coercive measures with the structuring of the cognitive maps of the social order and the social groups' major reference orientations.

The concretization of these tendencies takes place in different political-sociological settings. Two aspects of such settings are of special importance. One, much stressed in recent research, is the structure of the international political and economic systems, the place held by societies within them, and the different types of relations of hegemony and dependency. The second aspect is the recognition of the existence of a great variety of political-ecological settings of societies, the differences between small and large societies, and their respective dependence on internal or external markets.

B
THE SHAPING OF INSTITUTIONAL STRUCTURES

Our basic assumption is that the continuous interaction between the forces shaping the social division of labor and the regulation of power and construction of trust and meaning designs the contours of institutional structures, particularly the structure of cities and urban systems.

This interaction shapes the contours of the institutional structures (a) through the generation of different levels and types of resources, including information and positions, and (b) through the regulation and control of the flow of resources between different positions and individuals by the major dominant ruling groups of elites. These are the political elites who deal directly with the regulation of power in society by applying the models of the cultural order, whose activities are oriented toward the construction of meaning. They are also the articulators of solidarity of the major groups involved in the construction of trust; in this way they shape the basic premises of these civilizations or societies, their basic conceptions about the norms of the social, political, and economic orders.

The structure of these elites is closely related to the basic cultural orientations or codes prevalent in a society. In connection with the types of cultural orientations and the premises of societies connected with them, these elites or coalitions of elites tend to exercise different modes of control over the allocation of basic resources in the society, primarily over the access to the major institutional markets, over exchange of major resources among these markets, and over production and distribution of information. This latter function is central in the structuring of cognitive maps of the members of their society, perception of the nature of their society, and their reference orientations and reference groups (Eisenstadt, 1963; Merton, 1963). Such control is wielded by applying a combination of organizational and coercive measures, by constructing cognitive maps of the social order, and by structuring the major symbolic and institutional reference orientations of the social groups.

Such regulation affects the flow of resources and hence of urban life, urban forms, and urban hierarchies by structuring the following institutional frameworks essential to the organization of this flow:

(1) major institutional markets, particularly the economic ones;
(2) major collectivities, especially the political ones;
(3) center and center-periphery relations.

Each of these institutional frameworks greatly influences the flow of resources in a given society. The structure of institutional markets determines the possible scope of the flow of resources. The structure of the boundaries of major collectivities influences the degree to which resources flow within or beyond those limits. The structure of center-periphery relations molds the nature of the institutional control of the flow.

C
MODES OF CONTROL OVER RESOURCES

The different forms of control over the flow of resources in a society can be distinguished according to the following criteria:

(1) The degree to which the scope of the major institutional markets is affected and, especially, the degree to which such markets cut across different ascriptive communities (Eisenstadt, 1965: chap. 7).

(2) The degree to which they limit or encourage the flow of resources between the markets and the convertibility of various resources (for instance, from economic into political, and so on).
(3) The manner in which they dictate the direction of the flow of resources between various markets.
(4) The breadth and diversity of reference orientations constructed by the various elites.

The forms of control, influenced by the different structures of the elites themselves and by their coalitions, are closely related to the cultural orientations they articulate, the societal premises connected with them, and the conditions under which they act.

It is the different coalitions of such elites and the modes of control they exercise that construct and shape the major characteristics and boundaries of the respective social systems—political and economic, those of social stratification and class formation, and the overall macrosocietal system—or the structure of the economic, religious, and political institutional markets. This structure can be distinguished according to the following features:

(1) Scope of the markets, nature of the ascriptive and territorial units affected, degree of overlap among the different institutional markets, their autonomy and predominance relative to one another;
(2) Nature and definition of the major collectivities (ethnic, religious, regional, and especially political), which can be distinguished according to their territorial compactness and the degree of coalescence between their boundaries;
(3) Nature and definition of the center and center-periphery relations, which can be distinguished according to the number of centers (political, religious, or administrative) that exist in a given society, to their degree of coalescence, to the form of the center's distinctiveness in relation to the periphery, and to the mode of permeation of the periphery by the center.

D
PROTEST, CONFLICT, AND CHANGE

The different coalitions of these elites construct the boundaries of social systems, collectivities, and organizations. Yet no such construction can be continuously stable.

The crystallization and reproduction of any social order—with its systemic tendencies or qualities—of any collectivity, organization, political system, or civilizational framework, are shaped by the different forces and factors analyzed earlier. In order to maintain and reproduce themselves, such collectivities require the special mechanisms of control and integration alluded to above.

The more complex the societies, the more autonomous such mechanisms are, and it has been Herbert Simon's great contribution (1977) to point out that the mechanisms of control are autonomous analytical entities. Each such mechanism has a second set of stability and instability built in, and thus in every social order there are at least two sets of stability and instability, the instabilities built in the very construction of the system and in the mechanism of control.

For all these reasons, the crystallization and reproduction of any social order, of any collectivity, organization, political system, or civilizational framework, shaped by the different forces and factors analyzed above generate processes of conflict, change, and possible transformation.

The ubiquity of such changes is rooted in the fact that the processes of construction of collectivities, social systems, and civilizational frameworks are those of continuous struggles in which ideological, "material," and power elements are continually interwoven. These processes are structured and articulated by the different social actors and carriers already mentioned—that is, by those groups structuring the division of labor in a society, the level of its economic and ecological differentiation, and by carriers articulating ideologies and political control that are extremely important in the study of construction of boundaries of collectivities.

Between these different carriers a very complex interaction develops, going beyond what has been assumed in sociological, anthropological, and historical analysis, which is of crucial importance for understanding the crystallization of institutional formations.

Conflict is inherent in any setting of social interaction for two basic reasons: (1) because of the plurality of actors in any such setting and (2) because of the multiplicity of principles inherent in the institutionalization of any such setting and the multiplicity of institutional principles, cultural orientations, and power struggle and conflicts between the different groups and movements, which such institutionalization entails.

Any setting of social interaction, particularly the macrosocietal order, thus involves a plurality of actors—elites, movements, and groups—having differential control over the natural and social resources and continuously struggling over such control, ownership, and the possibility of using those resources, which all generate the ubiquity of conflicts on all levels of social interaction. These conflicts are intensified when the plurality of actors merges with basic characteristics of the organization of social division of labor and the construction of institutional formations. Different social actors may also have different cultural orientations, leading to conflicts of principles, premises, and prerequisites. The actors may stress the centrality of their respective spheres as distinct from others and develop their own autonomous dynamics at the expense of others, generating systemic contradictions.

The very processes of institutionalization of any social order thus entail heterogeneity and pluralism. Accordingly, whatever the success of attempts made by any coalition of elites to establish and legitimize common norms, these are probably never fully accepted by all the participants in a given order. Most groups tend to exhibit some autonomy and differences in their attitudes toward these norms and in their willingness or ability to provide the resources demanded by the institutional system.

In any social order a strong element of dissension always exists in respect to the distribution of power and values. An institutional system is hence never fully homogeneous in the sense of being fully accepted, or accepted in the same degree, by all the participants. Even if for long periods of time a majority of the members of a given society may be identified to some degree with the values and norms of that system and are willing to provide it with the resources needed, other tendencies develop that may give rise to changes in the initial attitudes of any given group to the basic premises of the institutional system, as discussed above.

Thus the possibility exists that "anti-systems" may develop within any society. Although the anti-systems may often remain latent for long periods, under propitious conditions they can activate and become important foci of systemic change. The existence of potential anti-systems is evident in all societies in the development of themes and orientations of protest and of social movements and heterodoxies that potentially exist and are often generated by secondary elites. Such latent anti-systems may be activated and transformed into processes of change by several methods connected with continuity and maintenance or reproduction of the different settings of social interaction and the macrosocietal order.

The most important of these processes are (1) shifts in the groups' relative power positions and aspirations; (2) activation, among members of the new generations of the upper classes and elites, of the potential rebelliousness and antinomian orientations inherent in any process of socialization; (3) sociomorphological or sociodemographic processes through which the biological reproduction of settings of social interaction is effected; and (4) the interaction between such settings and their natural and intersocietal environments—movements of population, conquest, and the like.

The crystallization of these potentialities of change usually takes place through the activities of secondary elites who attempt to mobilize groups and resources in order to change some aspects of the social order as shaped by the ruling coalition of elites.

The possibility of failure of the integrative and regulative mechanisms is thus inherent in a society. Every civilization or type of political and economic system constructs specific systemic boundaries within which it operates, although the very construction of such civilizations of social (economic or political) systems also generates conflicts and contradictions that *may* lead to change, transformation, or decline (that is, to different modes of restructuring their boundaries).

Although these potentialities of conflict and change are inherent in all human societies, their concrete development, their intensity, and the directions of change they engender differ greatly from society to society, according to the specific constellation of forces they contain, discussed earlier. Such forces shape the patterns of social conflict and movements, rebellions, and heterodoxy developing within the society and the relation of these movements to processes of institution building, as well as the direction of institutional order and the consequent pattern of transformation of such order.

These constellations also mold the development of sectors of or entire civilizations, of different patterns of cities and urban systems, and of the different modes of institutional and cultural creativity of cities.

E
CONSTRUCTION OF BOUNDARIES OF SOCIAL SYSTEMS

In the preceding analysis, several implications were described for understanding the construction processes of the boundaries of human collectivities and their systemic qualities:

(1) The construction of systemic boundaries is a basic component or aspect of human life.
(2) As has often been assumed in sociological, anthropological, and historical analyses, such systems and boundaries do not exist as some sort of natural closed systems, but rather are continuously constructed and are open and very fragile.
(3) No human population is confined within any single such system, but rather in a multiplicity of only partly coalescing organizations, collectivities, and systems.
(4) Such systems and the division of labor entailed are constructed by special social actors, by different carriers; ideological, power, and material components are always closely interwoven in the process of such construction.
(5) Such construction of boundaries delineates the relations of the various collectivities or systems with their respective environments, but such environments are not given in "nature"; they are contrived by social actors through construction of the respective boundaries and social systems.
(6) The restructuring or reconstruction of the boundaries of collectivities and social and political systems is a continuously changing process.
(7) In the construction and maintenance of such systems, different integrative mechanisms are of central importance and acquire autonomy of their own; their continued operation is of crucial importance in the maintenance and change of societies or civilizations.
(8) Such integrative mechanisms become more important and autonomous—and hence also more fragile—the more complex the social systems, political systems, and civilizational frameworks become.
(9) The construction of any such systemic qualities of patterns of human interaction in general and such mechanisms in particular generates the possibility of conflicts, contradictions, and changes in these systems.
(10) The nature of such changes, ranging from their total demise to far-reaching transformations, varies greatly among social systems.

F
SPATIAL CONCENTRATION AND CREATIVITY OF CITIES

Our basic assumption is that the combination of the processes we have analyzed shapes the contours of cities and urban systems as well as the patterns of social and cultural creativity specific to cities in different societies. The major dimensions of creativity in cities are generated by some of the basic characteristics of the process of urbanization that distinguish it from other institutional frameworks. Spatial centralization or the concentration of activities, control, and information in a society or in sectors of a society is at the core of the urban phenomenon. A tendency to spatial concentration has even been identified in relatively primitive societies in which full-fledged urban concentrations hardly existed. Full spatial concentration can emerge only in more differentiated or developed societies, mainly in the so-called archaic societies that were built on the basis of concentrations that appeared to have existed previously (Adams, 1960).

The emergence of spatial and urban concentrations depends on the availability of surplus free resources that can be processed and distributed in a continuous and concentrated way beyond self-enclosed and self-sufficient dispersed socio-economic units. The development of any type of urban system—the network of relations between several such units of concentration—requires constant availability of resources and the organization of their flow in relatively wide and diversified societal ecological frameworks through mutually reinforcing activities and mechanisms. It is such spatial centralization and concentration that generated the potential for special modes of institutional creativity within cities, the most important of which are perhaps storage and control of information and of flow of resources.

Analysis of the developments in urban studies reviewed in Chapter 1 provides us with systematic indications of the major dimensions of institutional and cultural creativity. The potential of such creativity lay, of course, in the fact that the spatial concentration of resources and population in defined modules made them distinct from their surroundings and gave rise to several problems. These problems are related both to the internal organization and structuring of such concentrations of population and resources and to the molding of their relations with other sectors of society.

The very concentration of resources and population created distinctiveness from both the centers and the peripheries of their respective societies, expressed in the ways the cities related to the various sectors of their societies. The distinctiveness of cities generated the specific patterns of social and cultural creativity, the special mode of generation and control of resources and information that evolved within them.

Several dimensions of distinctiveness of cities may be distinguished. The first, made up of two components, related to the distinctiveness of the urban structure. The first component is the extent to which special types of urban, social, and economic structures and political organization evolved within the cities, differentiating them from other sectors of the society—especially from the symbolic institutional center that, although usually located in the cities, must not always be identified with the overall population and habitat of the city (Eisenstadt, 1978; Shils, 1978) or with its rural sector. This distinctiveness of the social organization was closely related to the second component—the specific internal ecological structure of cities, the major elements of which were mentioned in Chapter 1.

The development of distinctive social organization and ecological structure necessarily gave rise to several crucial problems concerning the relations of this specific organization with its surroundings and other sectors of society—especially with its symbolic center and the rural sectors of the periphery. This problem constituted the second dimension of the distinctiveness of cities: the autonomy of their relations with other sectors, whose central focus lay in the extent of the center's regulation of the internal and external political autonomy and the confrontation between such regulation and the tendency of urban groups toward self-regulation or toward relatively active participation in such regulation.

Three aspects of urban life evolved, bearing directly on cities as loci of distinct types of social and cultural creativity, out of the continuous interaction and confrontation between the social organization and ecological structure of cities, the tendencies toward control and self-regulation. These three aspects are (1) the nature of the social struggles and conflicts that emerged within cities, the relation of these conflicts and movements of protest in the society, and their impact on the centers of society; (2) the definition of an urban identity, prevalent in a society and expressed in the self-conception of cities and its active groups, in relation to other sectors of society; and (3) the conception of the city and its moral evaluation prevalent in a society. Finally, the distinctiveness of cities found expression not only in their internal structure or organization but also in the interrelations between them as well as the different types of city systems.

It is our contention that these different aspects or dimensions of the distinctiveness of cities, and of the special types of social and cultural creativity that developed within them, were influenced by (a) the combination of the forces we have mentioned, (b) the levels and types of technology and economic development and the modes of control over the flow of resources exercised by different elites in the different types of geopolitical settings, and (c) the ways in which these forces have shaped the structure of markets, collectivities, and center-periphery relations. This contention requires elaboration.

G
CONCENTRATION AND CENTRALITY

It is the combination of these forces, as they influenced patterns of spatial concentration, that shaped the basic contours of cities and urban hierarchies.

From the point of view of spatial concentration, these forces coalesced according to two main types of processes that may be called "concentration" and "centrality."[1] In most societies the trend toward spatial concentration can be identified in varying degrees by the internal structure of these processes and by their different constellations and convergences.

Concentration is above all the concentration of population, usually as a result of demographic and economic processes that lead a growing number of people to concentrate in specific areas, generating processes of social differentiation and division of labor or growing interaction between various groups and the emergence of crafts and services. Needless to say, the concentration of population and its socioeconomic manifestations vary in extent and intensity from one place to another and often constitute intermittent processes. They are always determined by the specific conditions of the place and period in which they occur, but whenever they do occur, they display certain general tendencies of spatial organization and concentration that we will discuss shortly.

Centrality is the process—often intermittent—whereby the symbolic and political centers of a society, through which it transcends its daily routine of existence, are constructed and crystallized. Centrality implies the crystallization and symbolization, in a specifically defined space and ecological setting, of the cultural, political, and moral order of a society and the domination of society by such a center (Eisenstadt, 1971, 1978; Shils, 1978). The two most important characteristics of the forces of centrality, above all in early history, were the religious and political administrative manifestations that appeared, sometimes simultaneously.

The nature of the production and control of resources and information and the spatial concentration generated by each of these two processes, with regard to both their general institutional implications and the different components of urban life and urban systems, varied a great deal, and major differences existed in their internal structuring. Each of them generated specific tendencies to crystallization of the patterns of spatial and internal organization and structure of cities and their relations with their environment and the urban system as a whole. The crystallization of these patterns was a continuous process whose expression in each society changed constantly, often receding but sometimes proceeding in a progressive direction.

Concentration and centrality are thus spatial manifestations of the basic social forces that, as we have indicated, shaped the crystallization of different institutional formations—that is, generation of resources and the social division of labor, and the control and direction of such resources, through the activity of the various elites. The different constellations of these forces generated the distinct patterns of cities and urban systems, different patterns of institutional and cultural creativity of cities, and their impact on social transformation.

H
URBAN PATTERNS GENERATED BY CONCENTRATION AND CENTRALITY

We have shown that each of these trends displayed specific but general tendencies in the construction of cities. Concentration, largely influenced by demographic and economic forces, tended to produce cities or urban sectors with a relatively high level of internal economic division of labor and specialization. These cities, which resembled Redfield and Singer's (1954) "heterogenetic" cities, were distinguished from the rural hinterland chiefly by the different level and mode of their social and economic organization. The single most important manifestation of spatial organization generated by the forces of concentration were various kinds of fairs and markets.

Urban places that were market centers and centers of information and controlled the flow of economic resources, were mainly sustained by the continuous exchange of these resources among spatially dispersed economic units, both within the national boundaries and beyond them. Their continued existence depended on the availability of these resources in the hinterlands and their "processing" by the specialized urban social organizations. Their economy was based chiefly on commerce and services to the neighboring areas or to distant centers. The most prominent groups in such cities were merchants and providers of economic services.

In such cities the atmosphere was characterized by mobility and openness, influenced by a fluctuating population, and possessed those characteristics that at different times have been derided by moralists of various civilizations. A social ambiance favorable to the personal freedom of the citizens may have been fostered, within which an individual's achievements defined his or her position.

The urban systems produced by concentration were based mainly on marketing, usually strongly related to the rural economy; they tended to originate in the periphery or in secondary centers and to expand gradually toward the major center of the emerging urban system or toward international trade.

The actual form of a system was determined first and foremost by the type of economy prevailing in a certain area, by its level of development, and by the mode of international economic relations. The higher the level of development of the rural economy, as influenced by ecological conditions and the cities themselves, the higher the level of technology. The more varied the types of cultivation, the greater the probability of emergence of marketing centers in which the surpluses of agricultural products could be exchanged for manufactured goods and services. Similarly, the stronger and more continuous the international trade, the greater the possibility that such urban concentrations would emerge.

The contours of these types of cities were, of course, greatly determined by the relative strength of the different demographic and economic forces that shaped them and by the balance between the demographic pressures and availability, on the one hand, of population and, on the other, of economic outlets.

Centrality generated urban complexes resembling Redfield and Singer's (1954) "orthogenetic" cities (see also Wheatley, 1972) and produced urban patterns that, through different constellations of political and religious processes that shaped the center in a mold symbolic of the Great Traditions of a society and to impose them on the rest of the country, differed distinctly from those generated by concentration. Dominance and coercion and their legitimation constituted a crucial element in these processes.

In ancient times one of the main agents in the construction of a system based on centrality was a strong military-political authority (Adams, 1966), a God-king who dominated an area and, in exchange for providing security, taxed the population, representing at the same time the moral order and assuring its maintenance. The internal ecological pattern produced by centrality was characterized by hierarchical, sacerdotal, patrimonial, or administrative organization, not by specialization and division of labor. In this kind of system the rulers, bureaucrats, clergy, and army formed the backbone of the urban society; the people who assumed more purely economic functions (such as merchants or artisans) served mainly as appendages.

Such cities displayed a relatively strong emphasis on kinship and tribal ties and tended to stress the essential differences between core and periphery as a means of maintaining the efficient control of the center over the entire space. The sociocultural ambiance of such a city was characterized by an attempt to uphold the social, moral, and cosmic orders of a particular civilization through the legitimation of political power by means of religious sanctions, and not by the "looseness," mobility, and seeming frivolity distinctive of the "economic" city. The latter characteristics were limited mostly to the outskirts of such cities and occurred only during periods of pilgrimage or important political events.

The spatial expression of such political-military or religious dominance was usually the emergence of a large city—generally the capital—coexisting with many small villages and usually incorporating the centers of political and religious power: the palace and the ritual center, the archives, and the large storage areas. Such capital cities were the seat of the political-religious aristocracy, who were granted land as a reward for services rendered to the ruler, so that tenant-landowner relationships were common among most of the population. Because of the compulsory transfer of food products from the villages to the capital city, a greater flow of resources took place than would be the case in a completely spatially restricted society, thus leading to a larger degree of spatial integration. The relations among villages, however, remained very restricted spatially. The exchange of resources in such an embryonic spatial system was almost solely unidirectional, proceeding from the villages to the capital.

The relation of such a city to its hinterland usually was based primarily on the exacting of tribute as a means of political or religious control. The main form of control exercised by the city was domination—control of access to markets and the interrelations among them—contrasting with the free flow of resources in relatively open markets. The economic interaction of the city with its hinterland was mainly of a mobilizatory-redistributive kind (Polanyi et al., 1957). The degree to which the social composition and structure of the city differed from those of the hinterland was chiefly determined by the center-periphery relations, not by the degree of economic specialization existing within the society.

Concentration produced cities and urban hierarchies that depended on the continuity of internal markets that mediated between a relatively high level of productivity and a potentially widespread consumer demand. The scope of this type of market was influenced by three factors: (1) the productivity and structure of units of production (i.e., the degree to which they were independent of closed ascriptive units and generated a high level of free resources); (2) the existence of a high level of relatively widespread consumer demand; and (3) the existence of adequate technologies of communication and transport.

The combination of these factors enabled resources to be continually processed at various points of concentration and to serve as the focus of a relatively high level of manufacture and services. The more pronounced these factors were and the higher the level of the rural economy, the greater was the likelihood of emergence of a marketing system, enabling agricultural surpluses to be exchanged for manufactured goods and central services, and of a commercial central place system. Urban hierarchies of this kind tended to change in accordance with the alterations in the regional rural economies on which they depended. The continuity of urban centers based on international marketing depended chiefly on the maintenance of mutual economic interests among different international communities and on assurances of safety of transport among them.

Centrality produced urban hierarchies that depended on the continuous availability of a surplus that could be mobilized for use by a center, without this necessarily implying the existence of a high level of widespread consumer demand within the units creating the surplus. Generally the religious or political center constituted the major consumers here, and very often its resources had to be obtained through trade or tributary relations with external groups beyond its own political limits.

However distinct these two processes of centrality and concentration may have been in practice, they frequently overlapped from an analytical point of view. They complemented each other, and every society contained elements of both. Even the most commercial society required a measure of political control or religious authority, and these were at least partly provided through the processes of centrality. In the same way, the uninterrupted activity of well-organized markets was needed to some degree for the continuous exercise of control by central authorities.

Elements of both processes existed in most situations, but they combined in different ways and in different proportions at each specific time and place and together generated specific patterns of creativity. These forces, moreover, were never homogeneous, usually being composed of different components that, as illustrated by the analysis presented in Chapter 1, were very diversified and did not necessarily always combine in the same manner. Hence the concrete patterns of institutional creativity of cities and urban hierarchies varied greatly, far beyond the two general tendencies generated by the forces of centrality and concentration we have analyzed.

The processes of concentration and centrality were thus of crucial importance in shaping the many patterns of concentration of activities, control, and information—that is, the different dimensions of cities and urban systems. It would be wrong, however, to suggest that cities and urban hierarchies were only a kind of epiphenomenal result of these processes. Cities did become generators of these processes, creating new levels of generation of resources, activities, and information and changing their momentum and direction. But here, as in the construction of the cities themselves, important differences can be discerned between the two processes. These differences are influenced mainly by the constellations of the different social forces, the different components of the processes that we have analyzed, and by the different components of construction of civilization.

I
THE SCOPE OF THIS WORK

The number of possible combinations among the social forces we have analyzed and among the different aspects of the three institutional frameworks and the corresponding number of elites is very great. Some such combinations have been more frequent than others, however, and have generated some of the most important types of political-ecological structures and civilizational frameworks in human history and in the so-called traditional historical civilizations.

We shall analyze in the following chapters, through reexamination of secondary literature, some of the major characteristics of the structure of cities and urban hierarchies in several traditional civilizations and societies. Our work will cover the field extending between two limits. At one end, the fascinating problem of the so-called origins of cities of primary urbanization, which has indeed been abundantly studied in the Near East, Mesopotamia (Adams, 1966), and Meso-America, will be dealt with only tangentially, and we shall make only brief comparative remarks on this problem in our analysis of Southeast Asian urbanization. At the other limit, we shall not deal with modern civilizations and their cities and urban hierarchies that developed in the context of industrial technology and systems of political modernization.

Our analysis will focus on some of the traditional historical civilizations—China, Islam, India, Japan, Southeast Asia, Medieval Europe, Imperial Russia, and Spanish Latin America. Most of the civilizations that we will analyze, with the partial exception of those in Southeast Asia and Japan, belonged to the so-called Axial Age civilizations as designated by the German-Swiss philosopher Karl Jaspers, and evolved in the first millennium B.C.—namely, Ancient Israel and, later, Christianity in its great variety, Ancient Greece, Persia (partially with the development of Zoroastrianism), China of the early imperial period, Hinduism and Buddhism, and, much later, beyond the Axial Age proper, Islam. The development and institutionalization of a basic tension between the transcendental and the mundane orders were common to all these civilizations (Eisenstadt, 1982).

Several common characteristics distinguished these civilizations from other pre-Axial ones and were of interest from the point of view of our analysis. They were thus characterized by the emergence of the following: (1) specific civilizational or religious frameworks and collectivities, which to some degree were distinct from other ethnic, regional, national, or political ones; (2) relatively wider institutional markets; and (3) autonomous religious as well as political elites, which were not entirely embedded within ascriptive collectivities but recruited and organized according to relatively autonomous criteria.

Beyond these common traits, however, these civilizations varied greatly both by the relative importance of those characteristics and by their economic and political-ecological features. Within these civilizations evolved various groups of markets, boundaries between the major collectivities and center-periphery relations typifying these imperial, patrimonial, and complex decentralized forms of society, and several variations within each of them (Eisenstadt, 1978), which we shall analyze in greater detail in Parts II and III.

NOTE

1. The terms *concentration* and *centrality* derive, in Hebrew, from the same root word and were thus presented, as applied to cities, in Eisenstadt (1967, in Hebrew).

PART II

CASE STUDIES

CHAPTER 3

Urbanization in Southeast Asia

Our analysis of urbanization in Southeast Asia deals geographically with the area extending from Burma in the west to Java in the east. There were numerous variations in the civilizational and historical processes within this vast area. On the whole, however, the prevalent basic political structure was a fragmented configuration of many small kingdoms, loosely related to one another and strongly influenced by India and China, the two major cultural and political powers. The entire area was characterized by low levels of urbanization and development of the urban systems. Apparently the concentration and centrality forces had not been very effective in building urban settlements and urban hierarchies in this area.

A
MAJOR HISTORICAL STAGES IN THE DEVELOPMENT OF SOUTHEAST ASIA

1 The Formative Period: Transformation from Tribes to City-States and Kingdoms

Three stages can be distinguished in the historical development of the region. The first, the "formative stage," from the 1st century B.C. to the 9th century

A.D., was characterized by tribal organization and the emergence of city-states. During this period the process of Indianization influenced the local culture, religion, and sociopolitical orientations. The second stage, from the 9th to the 13th century, is referred to as the "classical period" and was characterized by the development of important inland and island kingdoms, based on syncretic cultural orientations that combined Indian concepts (both Hindu and Buddhist) with local traditions. Some of these kingdoms left splendid monuments of the highest architectural and artistic level. The third stage lasted from the 13th century until the penetration of European colonial powers in the 18th century and was characterized by the almost complete dominance of Theravada Buddhism (with the partial exception of Vietnam) and the spread of the Thai-speaking people. It was also characterized by the region's fragmentation into numerous kingdoms with shifting capitals and boundaries.

The evolution of civilization in Southeast Asia was based on two economic foundations: (1) intensive irrigated agriculture and (2) maritime coastal trade. By the time Indian influence became noticeable (around the 1st century A.D.), the mainland and island realms could be characterized as a fairly common technological-economic basis in the region. Among these elements it is possible to identify irrigated rice cultivation, domestication of the water buffalo, limited use of bronze and iron, and boat building and navigation (Farmer et al., 1977: 321).

This formative period, during which the classic civilization of mainland Southeast Asia crystallized, lasted from about the 1st century B.C. to about the 9th century A.D. At that time the tribal societies incorporated concepts borrowed from India into their cultural traditions, and the eventual synthesis of local and Indian cultures forged the region's civilizational framework. This process of Indianization was enhanced by strong trading links between India and Southeast Asia. During this formative period Indian influence on Southeast Asian civilization was decisive, gradually shaping (as various archaeological finds testify) the religious and political premises of the region's various polities. The Indianization process was initiated by the rulers of Southeast Asian societies who realized that the Indian concepts could provide strong legitimation to their rule (Keyes, 1977:66). Contrary to former views that regarded Indianization as a simple eastward expansion of Indian society and culture, the process involved continuous and differential selection.

The polities and people of Southeast Asia used certain elements of Indian culture, altering them according to their needs and orientations (Coedes, 1969; Hall, 1968; Spencer, 1983; Wheatley, 1983). Contact with Indian traders and local merchants returning from India made the elites aware of the ideological and symbolic sources of Indian governance and control. Southeast Asian rulers found it useful to invite and patronize Indian Brahmins and monks to assist them in organizing the administration and court ceremonials. Some basic aspects of Indian civilization, however (above all the institution of castes), were never adopted in Southeast Asia. On the other hand, various rulers invested heavily in the construction and maintenance of Hindu or Buddhist cult centers, thus establishing nuclei for their new capital cities. These cult centers, courts, and capitals, with their strong Indian influence, were culturally remote from the vast rural population, which was still influenced by the indigenous pre-Indian culture. However, they exerted a steady influence on the Indianized urban elites.

The impact of this new Indianized civilization gave rise to a syncretic blend of concepts and symbols either from Hinduism or Buddhism but strongly interwoven with motifs from the indigenous traditions. This syncretism was not identical regionwide. Hindu concepts dominated the culture that developed among the Khmers and Chams occupying the eastern part of the region, whereas Buddhist concepts dominated the Mons and the Burmans living in the central and western parts of the region. This syncretic culture predominated in the classical period of the great kingdoms of Angkor and Pagan, from the 9th to the 13th century.

The first important Southeast Asian polity that emerged in the formative period was the kingdom of Funan. Founded in the 1st century A.D. and enduring until the 6th century, its core area was in the Lower Mekong Valley. The rulers of this kingdom were called "Kings of the Mountain" after the local cult of the Sacred Mountain, which marked the center of the universe. Funan dominated the coastal trade of the region and the interregional trade with India and China for about five centuries. Chinese visitors were deeply impressed by the size and splendor of this capital city and by the dense network of canals, which were used for drainage and irrigation, as well as for linking the kingdom's principal towns.

2 Transition to the Classical Period

The decline of the Funan Kingdom brought about the emergence of several rivaling small kingdoms in the area of Cambodia, which were in existence for about two centuries, as well as the rise of the Buddhist kingdom of Srivijayya. Srivijayya became the main maritime power in Southeast Asia and provides the best illustration of the second major type of Southeast Asian polity: the coastal state (Spencer, 1983). This kingdom, which remained the major maritime power in Southeast Asia from the 8th to the 12th century, was situated along the northeastern coast of Sumatra, and its capital was located near the present city of Palembang. Srivijayya built up its political power by subjugating and incorporating small polities in Sumatra and the Malay peninsula, making the local rulers part of the governing system of the kingdom. The Srivijayyan rulers promoted international trade and thus brought economic prosperity, principally to the capital and the port towns. The kings were followers and patrons of Mahayana Buddhism and established important Buddhist centers of learning. Except for a brief period of subordination to the maritime empire of Cholas in India during the 11th century, Srivijayya maintained its political and economic supremacy among all the other maritime states of Southeast Asia until the mid-12th century, after which its political power and territorial extension gradually disintegrated.

The Khmer political unity was restored at the beginning of the 9th century, and the royal cult (mostly Shaivit Buddhism) was then reinstituted. The Khmer Kingdom was characterized by tremendous efforts to build an extensive canal and irrigation network. Of the greatest importance was the technical capability to preserve the run-off from monsoon rains in large man-made reservoirs, thus allowing year-round rice cultivation and huge increases in the annual yields.

The Khmer Kingdom's political power and territorial extent grew steadily. At the height of its power in the 12th century, the kingdom stretched from the coast of Annam in the east to the Salween River in Eastern Burma in the west, and dominated almost the entire northern part of the Malay peninsula. The core area of this kingdom centered on the Lake of Tonle Sap, and its capital cities were often moved to short distances within this area. The population of the kingdom was subjected to compulsory labor and military service. The governance system, at least in the core area and its immediate surroundings, was regimented closely and the material resources and human capital were mobilized fully to construct the tremendous projects of capital cities and the most elaborate temples undertaken during the 12th and 13th centuries. The Vaisnavite Temple of Angkor Vat, the Mahayana Bayon Temple, and the capital city of Angkor Thom were all built within the span of only about 100 years in the 12th century. This achievement indicates the very high level of resource mobilization by the state, in terms of both manpower and foodstuffs provided to the building sites by the rural areas. This is a prime example of the operation of forces of centrality in establishing the capital city of the kingdom that, both politically and symbolically, overwhelmingly dominated the entire state.

The two major classical civilizations that existed from the 9th to the 13th century exemplify the variations in the syncretic tradition: The Khmer Empire of Angkor had a stronger element of Hinduism, whereas the Burman Empire of Pagan drew more from Buddhist concepts. Both civilizations, however, were structurally similar and were based on the concept of a cosmic order that was well defined, unchanging, and absolute. The rulers' role was to harmonize the human world with the cosmic order, and they attempted to fulfill this role by constructing, with tremendous effort, architectural monuments as models of the cosmos. These were the famous cult centers of Angkor and Pagan, whose internal structure will be analyzed later in detail. These syncretic civilizations disintegrated in the 13th century and were followed by the Thai-speaking societies, which became dominated by the Theravada Buddhism orthodoxy.

3 Predominance of Theravada Buddhism and the Spread of the Thai People

The 13th century represents a turning point in the history of Southeast Asia (Keyes, 1977:262). This was due to two processes: successive Mongol attacks, culminating in capture of the kingdom of Pagan in 1287, and the spread of Thai-speaking people in the region, resulting in reshuffling of the political system. The Thai ascendency was dominant between the mid-13th and mid-14th centuries, during which period the Thai founded a number of new kingdoms and extended their rule over most of the region. This ascendancy brought about the demise of the Khmer Empire.

The major change that occurred in the political system in the wake of Thai ascendency was establishment of the Thai kingdom of Sukhothai, which came into existence in the early part of the 13th century and evolved to differ structurally from the classic Indianized states. The religion was no longer the syncretism that characterized the Angkor kingdom but was based on orthodox Theravada Buddhism. The kings of Sukhothai actively supported the efforts of missionary monks to spread orthodox Theravada Buddhism among the people of the kingdom and make it the popular creed. Sukhothai enjoyed a brief period of dominance over much of the area of today's Thailand. However, in mid-14th century Sukhothai was challenged by another Thai kingdom, called Ayutthaya after its capital, which was located in lower central Thailand and succeeded the Mon Kingdom that had occupied that area since the 11th century. The two Thai states fought for supremacy in the region, and Ayutthaya ultimately gained the upper hand and absorbed the kingdom of Sukhothai. The rise to power of the Ayutthaya Kingdom engendered a fierce conflict with the Angkor Empire. This external political clash was exacerbated by internal contradictions in the Angkor Empire, which began to undermine the legitimacy and authority of its rulers. At the center of these contradictions stood heavy demands made on the population to support the construction and maintenance of enormous shrines and monuments, which created great unrest. The attacks on Angkor by Ayutthaya in the 14th and 15th centuries resulted in disintegration of the Angkor Empire, abandonment of the sacred capital of Angkor, and emergence of a new Khmer culture based on Theravada Buddhism.

Upon the earlier collapse of Pagan and the later one of Angkor, both Burmans and Khmers adopted Theravada Buddhism. In the beginning of the 15th century at the very latest, the vast majority of the region's inhabitants (which today comprises Burma, Thailand, Cambodia, and Laos) had become adherents of Theravada Buddhism (Keyes, 1977:82).

B
BASIC CHARACTERISTICS AND TRANSFORMATIONS OF SOUTHEAST ASIAN PATRIMONIAL SOCIETIES

Throughout their history the Southeast Asian kingdoms evinced some of the classical features of patrimonial systems. The relatively embedded elites adopted orientations characterized by a limited perception of tension between the transcendental and the mundane orders, limited modes of control, relatively narrow and embedded markets, little distinction between center and periphery, and very narrow status segments in the society (Eisenstadt, 1973).

The first crystallization of these patrimonial kingdoms took place in the formative and classical periods, under the impact of Indian civilization, with the arrival of Indian merchants in the region at the beginning of the Christian era. Within a century or two new kingdoms emerged whose governance was based on the Indian, Hindu, or Buddhist conceptions of social order (Wheatley, 1983).

The polities that emerged in the earlier centuries of the Christian era were kingdoms of conquest ruled by Indianized dynasties and expanding territorially along the great South Asian trade routes. They transformed the weak hierarchical (political and juridical) organizations that had existed in some parts of Southeast Asia before the arrival of the Indian merchants into much more compact political units. Hinduism and Buddhism dominated in the area, shaped its cults, and engendered the numerous temple-cities that formed the backbone of the region's urban development.

With the spread of Buddhism and its establishment as the official religion of the more centralized Southeast Asian kingdoms, and with the consequent incorporation of these polities into the framework of the Buddhist civilization, transformations occurred in the structure of the elities and collectivities. The most important of these changes were the emergence of an autonomous religious elite—the Buddhist clergy, the *Sangha*—and a new, wide civilizational framework carried by this elite and its major institutions—the monasteries—as well as the concomitant development of broad religious markets. Yet at the same time, the basic nature of the political regimes remained patrimonial, albeit with more compact boundaries, stronger centers, and some additional characteristics that distinguished it from earlier patrimonial states. Insofar as their cultural or religious activities were concerned, these religious elites evinced a relatively large degree of symbolic and organizational autonomy from the major ascriptive groups and political rulers.

This autonomy, however, as well as the specific activities of these elites, was generally confined to the cultural or religious sphere. In the more mundane spheres they attained some organizational autonomy (usually limited), but on the whole these elites were much more dependent on the political authorities, both symbolically and organizationally. Their organizational autonomy was contingent on their acceptance of the basic rules of the political game directed by political elites. Although they were called upon to legitimize the political order and took part in forming new political regimes or restructuring the scope of ascriptive communities, the religious elites evinced very little autonomous, potentially critical participation in the political realm; indeed, they tended to legitimize any victorious ruler. Despite these limitations, the very activities and orientations of these elites had far-reaching (though specific) impacts on certain aspects of the social structure, above all on the formation of collectivities.

Through the activities of these elites, a new dimension in the construction and definition of the basic ascriptive collectivities crystallized. Given their strong religious orientations, the "civilizational" collectivities were those that developed as distinct frameworks, symbolically autonomous and highly articulated. These orientations also effected important changes in the construction and definition of the local, national, and political communities. To the "usual" primordial or territorial components of such a definition they added a broader orientation. This orientation was the basis of and framework for the crystallization of new symbols and boundaries of political collective identity, of national political communities expressed in the fact that the political realm was conceived as a reflection or representation of basic conceptions of the cosmic order (Eisenstadt, 1976).

This transformation of the Southeast Asian civilization under the impact of Buddhism did not directly change the basic patrimonial characteristics of these polities; rather, it infused them with a new social dimension. There was a greater tendency toward more compact frontiers and new dynamic initiatives, especially toward the expansion and multiplication of political units, creating what S. J. Tambiah has called the "galactic polity" (1976: 102-131). These tendencies had important repercussions on some aspects of the cities and hierarchies that developed in these societies.

C
URBAN DEVELOPMENT IN SOUTHEAST ASIAN SOCIETIES

1 The Formative Stage

The development of cities and urban hierarchies within the civilizational framework of Southeast Asia was shaped, during the formative stage, mainly by the geographic-economic characteristics of the area and the basic features of its political regimes. It was a highly productive agricultural region with identical crops grown throughout the area; regional differentiation—the basic condition for exchange and trade—did not occur on any significant level (Coe, 1961:67). Its insular and peninsular character, combined with its rugged topography, greatly limited accessibility to parts of the area. The low level of specialization in agricultural production, the abundance of foodstuffs grown locally, and the poorly developed transportation system all contributed to the emergence and preservation of the cellular structure of the numerous self-contained rural units usually found in a tribal area. As consumer demand was almost nonexistent there, no internal markets of importance could arise; thus there was almost no stimulus in this civilization for the emergence of towns that could become commercial centers. The internal economic activities were insignificant, and the forces of concentration therefore did not operate in the construction of cities. This can be seen with respect to the two major types of cities (and states) that evolved in Southeast Asia.

2 Temple and Trade Cities

Two basic types of these cities can be distinguished: (1) the sacred or temple cities and (2) the "entrepôt" or trading cities (McGee, 1969:30-31; Wheatley in Sabloff and Lamberg-Karlovsky, 1975). The temple cities primarily contained buildings that fulfilled religious functions; these temples were geomantically oriented and served as symbolic expressions of the power of Buddhism, with which the rulers of the state were identified. The temple cities located within an overwhelmingly agrarian society were founded on the construction of religious symbols and commitments that were seen as necessary to ensure the success of the crops. The better known cities of the formative period (from the 1st to the 9th century A.D.) were Vyadhapura, capital of Funan; Panduranaga and Indrapura, capitals of Champa; and Srikshetra, capital of the Pyū Kingdom. These cities were good examples of the operation of the forces of centrality in shaping the emergence and structure of urban settlements. The religious buildings that formed the core and the *raison d'être* of these cities heralded the appearance of the monumental buildings that were later erected in the classic temple cities, such as Pagan and Angkor.

The sacred cities of the classic period began to appear at the beginning of the 11th century and reached their apex in the 13th century. The two most prominent sacred cities of this period were, as mentioned, Pagan and Angkor, whose internal structure will be analyzed later. Other examples of classical sacred cities, built on a smaller scale and attended by the population of smaller territories than those of the two prime centers and less affluent, are Vijaya, the capital of medieval Champa, Lavo and Haripunjaya, capitals of medieval kingdoms in central and northern Thailand, and Hamsavati and Suddhammavati, capitals of Mon Kingdoms in Lower Burma.

The major feature of the temple cities was that their internal structure was meant to symbolize a sacred cosmography, each city having a sacred center that represented Mount Meru, the center of the cosmos. The site of the capital city was regarded as an Axis Mundi, a meeting point of Heaven, Earth, and Hell (Wheatley, 1971). The geomantic meaning of the city structure required the selection of auspicious locations for these cities. As mentioned, the temple cities were the centers of royal administration and the places where the kings established their courts and lived with their entourage and chief ministers. The temples' priests constituted the major sector of the population of those cities. The building and maintenance of the monumental temples required a large force of architects, stone-cutters, masons, sculptors, wood carvers, and other artisans. Given the demand of the large temple cities for labor and resources, they were the main consumers of agricultural surpluses produced by the peasants of the entire region.

The second type of city found in mainland and insular Southeast Asia was the entrepôt or trading cities, almost all of which were located on the coast and often were centers of the maritime kingdoms. It is not clear when the earliest trading cities were founded in mainland Southeast Asia, but evidence indicates they were in existence by the 1st century A.D. One of the earliest trading or entrepôt cities is the port city of Oc-eo, in the ancient kingdom of Funan. In the classical period two cities of this type, Takkola and Nagara Sri Dhammaraja, were located on opposite sides of the peninsula in today's South Thailand (Keyes, 1977:261).

The major characteristic of the trading cities was their particular economic base; they drew their wealth from the maritime trade and fulfilled the role of trading emporium (McGee, 1969:30). The entrepôt cities rarely exercised control over large rural peasant populations, as their prosperity did not depend on such dominance. These cities were externally oriented and were mostly related to the international and regional maritime trade routes. Obtaining food and resources from their rural hinterland was their major link with the outlying regions.

3 Structure of Urban Systems

Given the weakness of the forces of concentration in Southeast Asia, it is not surprising that no full-fledged urban hierarchy was able to emerge. The sociopolitical organization of the region was not conducive to the operation of centrality forces in bolstering urban hierarchies. This civilization, in which each respective area was controlled and directly organized to support a chieftain or a religious cult through the payment of tribute or contribution of corvée labor, did not encourage the emergence of an administrative hierarchy. The fragmented nature of the political system, composed of small autonomous units, did not foster the formation of an administrative hierarchy that could give rise to an enduring urban system. Because of the fuzziness of the political boundaries between the regions, most of the rural settlements and small towns remained inward-looking and internally oriented, kept their limited size, and did not form a significant urban hierarchy. The coastal settlements, which were externally oriented and mainly based on external trade, were segmented and did not develop into urban systems. Because there was no export of agricultural commodities, these tiny towns, as well as the coastal communities of fishermen, did not evolve into outlets for the large rural hinterland. They remained isolated outposts of small-scale coastal trade and later became the guardians of major sea routes.

The process of transformation in the formative period took place progressively from tribal domains to isolated city-states to kingdoms that, at various periods, extended over parts of the region. Often this occurred when one chieftain subdued others and merged their territories into his own. This process generated forces of centrality that stimulated the establishment of towns throughout the unified territory. These towns fulfilled administrative-ceremonial functions, thus creating a rudimentary urban-administrative system tending toward a simple primacy structure. The unifying principle of the kingdom was the practice of a common religion emanating from the main temple complex.

This ceremonial center was the locus of a High Priest, and adjacent to it were built the sovereign's palace and the residences of his ministers and attendants. The major ceremonial center defined the apex of the kingdom's urban system, and the smaller administrative-ceremonial centers were subjected to the authority of the chief center.

There are two examples of such a process. The first is the Funan Kingdom, which encompassed the areas along the Mekong River, between Chou-Doc and Phnom-Penh. It gave rise to a discernible urban system comprising some 30 settlements that were walled, palissaded, and moated and were located at focal points in a hydraulic system that consolidated the heartland of the Mekong Delta into a functional and economic polity. The second is the Pyū Kingdom, which was established in Burma in the 6th century and brought about the evolution of a local urban hierarchy made up of military-administrative centers, all subjected and controlled by the Pyū overlordship (Wheatley, 1983).

The structure of the urban systems generated by polities created around coastal cities did not greatly differ from those created by the ceremonial centers; their commercial activities were oriented outward and were not based on any internal markets.

By the 5th century A.D., chiefdoms and city-states had made their appearance on the Isthmus, in Java, and in Sumatra. The process of transformation from tribe to city-state to kingdom can be followed here. There is evidence that in Sumatra, the Srivijayyan polity was transformed from tribe to city-state to kingdom toward the end of the 7th century. Vijayya apparently held sway during that period over the southern half of Sumatra, the island of Bangka, and parts of Western Java. Once the territorial base of the kingdom was laid, it could readily utilize the benefits of its nodality in relation to the maritime trade routes of Southeast Asia. In order to consolidate its position as the premier power in the archipelago, Srivijayya strove to achieve complete monopoly of the commerce routes in the area by subjugating the isthmian sector of the Malay Peninsula.

This is an illuminating example of the way in which forces of centrality directed the forces of concentration to transform a congregation of city-states into a kingdom. This process enabled the Srivijayyan Kingdom to attempt to monopolize the major sea routes of the area without creating any internal markets (Bronson, 1977; Bronson and Wisseman, 1978; Leur, 1955; Spencer, 1983). It is well summarized by Spencer (1983: 137):

> Although the picture of Srivijaya has become more complicated in recent years and the conventional view of it [as a maritime empire] more doubtful, we still have a very clear impression of a Southeast Asian world enjoying a high degree of commercial prosperity, sufficient to attract the attention of an aggressive king like Rājēndra.

4 Processes of Urbanization in the Classical Period

Compared with the formative stage, the processes of urbanization in the classical period evinced several significant changes. The two major directions of change were the construction of greater monumental capitals-temple cities and that of a hierarchy of such temple cities—one which, however, remained weak and intermittent. In the absence of a clear demarcation of territorial control, the classical state showed a consistent inability to incorporate regions beyond the core domain, which consisted of the area surrounding the capital and which was directly administered by the central authorities of the kingdom. The king was usually a major landowner in the core area, but the landholding rights of the other owners were also protected. The areas beyond the core maintained a tributory relationship with the state. Control was quite limited, and although they paid homage to the center, the local elites remained in power (Hall, 1976:4).

The mechanism of the center's control over the peripheral regions operated through the imposition by the center of a dependency relationship upon the population of these regions. These processes are well illustrated in the development of the Khmer Kingdom. The Khmer king incorporated tribes on the outer edges of the core area into the kingdom and appointed their chiefs as district rulers. In this manner the Khmer administration was able to absorb many of the indigenous elites and could control and protect the local interests of the periphery.

In order to integrate the subordinate population, the Khmer Kingdom utilized the elaborate Devaraja cult in which the personality of the king symbolized the interaction of the human and the divine. Traditional symbols of divinity and authority combined local ancestral cults with the cosmological symbolism of Hinduism to form the ideological base for the Khmer Kingdom. Under Khmer rule, ancestor worship and the traditional symbols—the *Linga* and the mountain, representing fertility and the abode of the dead—were integrated and became the major elements of the state religion. Deities of Hindu origin, such as Siva, Vishnu, and the Buddha, merged with the indigenous Devaraja gods and were installed in the capital city to represent the king's control over the protective forces of the ancestors (Aeusrivongse, 1976).

The royal court and its life-style recreated the world of the gods. Thus the king maintained world order by fulfilling the symbolic role of focus of all sanctity and power. The subordinate centers were ritually linked to the king's court by adopting its style and routine. This ritual unity was more important in maintaining the state's dominance over areas outside its core than the formal administrative control. By integrating indigenous folk traditions, symbols, and religious beliefs into the royal cult, which was visibly concentrated in the center, it was possible to sustain the kingdom and to cement the peripheral area to the core, under the king's rule (the divine authority).

Large estates, both in the core and in peripheral areas of the Khmer Kingdom, were integrated into the state's ritual system via a network of local temples that were brought into a structured relationship with the large, state-level, central temples. The local elites, many of whom had been given their estates by royal grant, often established local places of worship. In return for recognition as legitimate entities (the priests of the local temple were allowed ritual privileges in a central temple), the temples paid a percentage of the annual yield of the endowed lands to the central temple. Thus a hierarchical network of temples extended throughout the kingdom, strengthening royal control. At the same time, by keeping the temples' landholdings fragmented, the royal authority ensured that the religious network of central and local temples would not become the dominant power of the kingdom. This fragmentation was effected by forbidding the amalgamation of lands by the various local temples (Hall, 1976).

In the classical kingdoms of Southeast Asia, the settlement system was thus shaped by a religious hierarchy, from central to local temples, through which royal authority was manifested and applied to all parts of the kingdom. Given the instability of these kingdoms and their frontiers, however, and the lack of strong internal economic and agricultural bases for these cities, such control, and the ensuing urban hierarchies, were inherently unstable and weak.

D
INTERNAL STRUCTURE OF CITIES

1 Temple and Capital Cities

The internal structure of capital cities in the formative and classical periods reflects the gradual social and political transformations that brought about the emergence of classical states in many parts of Southeast Asia. These transformations, strongly affected by Indian religious principles and political patterns, replaced the tribe with the kingdom, the tribal chief with a divine king, the *shaman* with a Brahmin, and the tribesman (as an occasional cultivator) with a peasant. These institutional changes influenced in a cardinal way the internal structure of the urban settlements by converting the chief's hut into a palace, the spirit house into a temple, the spirit stone into the *Linga*—which came to symbolize the essence of royal authority and strength—and the village fence into the city wall.

The internal structure of the capital cities of classical states is a prime example of the almost exclusive operation of centrality forces. The source of all authority in these states was the archaic kingship that mediated between the mundane and the other-world, among the various planes of existence, and between macrocosmos and microcosmos. In a manner common to much of Southeast Asia, it was perceived to rule from the axis of the world. The Axis Mundi was endowed with the highest level of sanctity; the level decreased with movement away from the axis.

The basic principle that governed the urban structure was the concept that the capital city in general and the king's court in particular were shaped as an image of the divine order and served as an idealized structural model for society. The need to model the court and the capital on those principles required the implementation of certain cosmological rules that almost always incorporated the basic elements of the *Mandala*—a central image with surrounding and circling entities. Thus the temple, the palace, and the capital city constituted two basic structural elements: the center and its surrounding entities, which numbered four or multiples of four. The resulting pattern, including both the center and its perimeter, was likely to be composed of 5, 9, or 17 components (the central figure plus any multiple of four). In this precise geometric pattern the central entity always represented the ruler, whereas deities of lesser rank, in multiples of four, were in the perimeter.

The structure of the capital city was aimed at paralleling the macrocosmos with the microcosmos, without which prosperity for mankind could not be achieved. The capital city was conceived as a reduced version of the cosmos that, in many parts of Southeast Asia, meant the Indian cosmos. This principle of urban design assumed different forms in different areas at different periods, but most of them shared some common elements of spatial organization. Urban space was structured in the image of celestial space. The mirror image thus established had to be upheld carefully by a rigorous schedule of rites and worship. But the urban design—the microcosmos image—was not merely a reflection of the celestial order. Whatever happened on the terrestial level might induce a chain of reactions of the macrocosmos plane; therefore, strict adherence to royal authority and the rules of urban design was essential in order to preserve the cosmic order.

The celestial and the terrestrial spaces intersected at one point. This was common to the macrocosmos as well as to the microcosmos and therefore was the source of sanctity and creativity. The axis of the world passed through that point. It was the responsibility of the priests to guard the Axis Mundi and keep it from dislocating. The safest way to do so was by establishing the major temple, a monumental structure, to watch over the Axis Mundi, which symbolized the sacred pivot of the universe. The temple guarding the Axis Mundi was in the middle of an extensive sacred enclave; the wall enclosing it and its gates were built in accordance with the cardinal compass directions. The walls surrounding the temple defined the sanctified enclave within the continuum of profane space comprising the entire kingdom.

Although the kingdom could be organized only symbolically on cosmological-religious principles, the capital could be laid out in concrete form to incorporate all the symbolic, allegoric, and imitative modes of the other-world. The most rigorous manifestation of such concrete representations of heavenly prototypes were the great cult cities of Cambodia, especially those built between the 12th and

14th centuries. The successive state capitals—above all that known as Angkor Thom—were a complete replica in stone of the cosmological order. The Khmers spared no effort to reproduce in their capital city the counterpart of India's mythological capital on Mount Meru. The hydraulic technology, which was so essential to the wet rice civilization of Cambodia, was also incorporated into the symbolic layout of the capital. Large pools were built adjacent to the sacred enclave, and the water, conducted by canals to the moat, was then distributed to the rice fields. The water in the moat flowed under the bridges leading to the main gates, where stone figures performed churning operations. The churning was believed to bring eternal prosperity to the kingdom, transforming the water into an ocean of ambrosia (Pym, 1968:74). The Hindu myth of churning the ocean thus provided the link between the instrumental needs for irrigation in a rice civilization and the symbolic structure of the capital city.

The shaping of the internal structure of Southeast Asian cities by forces of centrality thus assumed its fullest expression in the Khmer capitals. Among them, the most elaborate and monumental was the city of Jayavarman II, built on the site of Angkor in the 12th century. The same principles of symbolic representation of the cosmic order were applied, on a smaller scale, to the buildings of the provincial district and to the lesser seats of the kingdom. The symbolic orientation became less pronounced in the lower levels of the urban hierarchy. The resources available to the district chiefs were very meagre, limiting their ability to display the grand symbolic designs so prominent in the great capital. Centrality forces were thus dominant in shaping the internal structure of the upper level of the capital cities' urban hierarchy, but efforts were made to meet the need of maintaining harmony between the macrocosmos and the microcosmos in establishing the basic structure of even the smaller towns on the lower levels of the urban hierarchy.

2 Coastal Cities

Unlike the inhabitants of the temple cities, those of the trading cities were much more diversified and included Indians, Chinese, Malays, and the indigenous population. This ethnic diversity was associated with an ethnic division of labor. The traders were mainly Indians or, more rarely, Chinese. Boat crews were mostly Malays. Rulers and servants were drawn from the local population. Trading cities did not represent a cosmographic symbol; thus their internal structure was much less defined and demarcated than the temple cities. The central focus of these cities was the harbor. Adjacent to it was the ruler's residence and between it and the harbor was the market with its shops and warehouses. The many ethnic groups occupied distinct neighborhoods in the city.

The trading cities of Southeast Asia provide a good example of the way in which forces of concentration influenced the evolution of urban centers (Bronson, 1977; Leur, 1955). The most important aspect of this process was that the trading cities did not produce a hierarchical structure; they were neither related to service areas of different size nor part of a hierarchical administrative framework. The size of their population was quite similar, and whatever differences could be observed reflected a different magnitude in the level of maritime trade.

Beginning in the 14th century the port towns in Java, as indicated by Geertz (1963), became independent political and economic centers. Geertz contrasted Javanese inland "castle towns" of the river valleys, which were ritual administrative centers, with the bazaar towns of the northern coast, which were mercantile trade centers. The inland towns were the capitals of kingdoms based on a rural rice civilization and served as the residence of the ruler, his court, and the sacred center's religious attendants. In contrast, the coastal towns were dependent on regional and international commerce and were not related to the agrarian economy. The coastal towns were basically outward-looking, oriented toward the sea and engaged in relations with far-away commercial centers (Bronson, 1977; Spencer, 1983). The population of the coastal towns was gradually augmented by numerous foreign merchants of diverse ethnic origin. As the power of the inland Javanese kingdoms declined, the coastal towns gained semi-independence from their rulers and constituted another, equal-level urban system, maintaining strong interaction with the other coastal commercial centers within a regional and international framework.

3 Impact of Theravada Buddhism on City Structure

The institutionalization of Theravada Buddhism as the dominant religion in most Southeast Asian polities had significant impact on their cities. The capitals of the Theravada Buddhist states of Southeast Asia remained the political and religious centers of their agrarian societies. In this respect, they continued the tradition of the sacred capitals in the classical period. Examples of those capitals are Sukhothai and Ayutthaya in Thailand, Pegu and Ava in Burma, Oudong and Phnom-Penh in Cambodia, and Luang Prabang and Vientiane in Laos. All these cities dominated the history of the Theravada Buddhist states until mid-19th century. Compared with the capitals of the classical period, however, Theravadian capitals underwent a significant change in their structure, which culminated in a spatial differentiation between the royal palace and the religious temples. Many of the temples moved out of the city and relocated at some distance from it.

The cities of Sukhothai and Ayutthaya in Thailand provide good examples of the typical Theravada capitals. Sukhothai was established in the 13th century as the capital of the Thai Kingdom. The first inscriptions found there from the period of King Ram Khamhaeng, dated 1292, described the major act of royal merit-making, which was the presentation of robes and other offerings to the monks. It is significant that the king traveled in a procession to the main monastery located outside the walls of Sukhothai (Griswold, 1971:212), which illustrates the separation between the royal palace, located in the center of the walled city, and the main monastery (reflecting the religious authority), located outside the walls of the city. This spatial differentiation did not exist in the cities of the classical period (Keyes, 1977:263).

Although the basic relationship between political and religious authorities greatly influenced the spatial differentiation between the royal palace and the religious monasteries and temples, the strong cosmological orientation in the construction of the cities persisted. The magical and spiritual center of the kingdom was in the main shrine within the walls of Sukhothai (Griswold, 1967:3). These two basic foundations of the Theravada capital—the Buddhist merit and the cosmological expression—were joined by a third feature, which was association with the most powerful spirit of the kingdom. The shrine to the deity of Sukhothai, protecting the city, was located outside the city walls, on a hill to the south of it. The combination of these three foundations in shaping the structure of capital cities can be found in various degrees in all Theravadian capitals.

A further significant feature of the differentiation between the political and religious functions of Theravadian kingdoms was the growing importance of markets and trading activities in their capital cities. Two examples of capital cities that acquired major trading functions are Pegu in Lower Burma and Ayutthaya in Thailand. The latter, especially, became a center of international trade, a focal point in the transshipment of goods between Europe, India, and the Far East. Pegu, which fulfilled a similar role, though to a much lesser extent, became the residence of numerous foreign traders (Indian, Chinese, and, later, European) who organized the international trade in which Burma was involved. It should be emphasized that this involvement in international trade proved to be a major factor in the growth and stabilization of cities in the Theravada Buddhist Southeast Asian kingdoms; domestic trade never played such a significant role.

The incessant conflicts among Theravada Buddhist polities led to the destruction of the existing capitals and establishment of new capital cities that characterized the urban system until mid-19th century, when European colonial powers achieved dominance over the region. Only those capitals possessing a strong base of international trade could sustain the constant upheavals and frequent changes occurring in the urban system throughout this period.

4 Chinese Influence—Vietnam

A different pattern of centrality of a political-administrative nature had its origin outside Southeast Asia. This foreign agent of change was the southward extension of the Chinese political authority. Chinese penetration into Southeast Asia began shortly before the spread of the Indian religious-cultural influence and was aimed at a territorial expansion of the Chinese Empire. Chinese political domination proceeded in stages, each phase associated with a particular mode of political organization. The process began by maintaining a tributary relationship with a Chinese-style kingdom; it proceeded by granting the dominated territory a protectorate status and culminated with its complete incorporation within the Chinese Empire.

The establishment of an elaborate system of urban centers fulfulling political-administrative-legal functions based on the Chinese model became one of the major instruments of Chinese rule. Into the areas of Southeast Asia under their control the Chinese authorities transplanted the administrative system of the *Hsien* centers, which had a well-defined hierarchical organization of functions and service areas.

In the areas under Chinese domination, the urban system was the outcome of a foreign colonial power's centrality forces. It emerged not in response to the needs of the indigenous society but as strongholds and control centers of the Chinese authority. Many of the urban centers were populated by Chinese metropolitan officials and by local people who had adopted the Chinese culture and way of life. Thus, the *Hsien* and prefectorial towns stood out as islands of governmental authority and foreign elements in a sea of indigenous rural population and were maintained for long periods to restrain a dissident population.

Chinese rule prevailed for almost 1000 years, and it took a long time for the urban centers to be accepted by the indigenous society. Urban development stimulated by centrality forces of foreign origin was not conducive, however, to the operation of forces of concentration. With limited internal trade and a low level of economic interaction among the regions, the urban system remained principally administrative in nature and did not encourage its utilization for economic activities.

CHAPTER 4

Urbanization in Colonial Latin America

A
GENERAL BACKGROUND

In this chapter we concentrate on the process of urbanization in Latin America during the colonial period, from the beginning of the 16th to the beginning of the 19th century. Throughout this period Central and South America were ruled by Spain and Portugal, which shaped their political institutions, social organization, and juridical system and, above all, imposed a single form of religion over the entire continent. Urbanization in the colonial period represented, in most respects, a complete break from the pre-Colombian civilizations and the structure and characteristics of the cities and urban systems of Latin America were overwhelmingly shaped by the orientations and policies of the Spaniards and Portuguese.

The origins of the settlement system in Latin America were unique in one respect: The establishment of towns was the outcome of an induced urbanization process, in which selection of the site, planning of the physical layout, allocation of land, determination of the political municipal boundaries, and shaping of the institutional framework were all initiated and closely controlled by the courts in Europe and by their representatives in the newly discovered continent. This purposeful and centrally controlled urbanization process, recalling the systematic colonization process in the Roman Empire, gives us a rare opportunity to trace the evolution of the urban system, from its inception in the Caribbean Islands in the last few years of the 15th century, and to pinpoint the historical and societal origins of the new urbanization. We will also be able to define the motives for establishing settlements based on the explicit orders and regulations issued by the court, the Council of the Indies, and the royal representatives in the new territories.

Urbanization in Latin America, initiated and carried out by the Spanish explorers and conquerors, had its roots in the medieval Iberian urban experience. That experience was shaped by the slow reconquest of Spain from the Moors and by the recapture of land, and was reinforced by the formation of well-protected urban strongholds throughout the newly reconquered areas.

The process of imperial colonization of Latin America, which was accomplished mainly through the establishment of towns in the new territories, may be considered a continuation of the reconquest of Moorish Spain applied overseas. The much faster pace of the conquest of America, the tremendous size of the territories involved, and the commercial revolution accompanying it accentuated even more the urban character of the colonization process in America. At the same time, the role of the Latin American town as an agent of colonization and as an outpost and control point of the imperial government shaped significantly the location of towns, their functions, their relationship with the hinterland, and even their physical layout. The urban nature of the colonization effort was also strongly influenced by the urban origin of arriving Spaniards (Gakenheimer, 1967:35).

As the towns were perceived as the instruments of colonization, their intended function, location, layout, and municipal organization were all specified by the Spanish court and its representatives in the New World. Almost no town was established without following the precise instructions of the viceroy's office. After the first stages of conquest and settlement, this planned, deliberate system of urbanization evolved into a large-scale urban development process in which the towns reflected the aims, needs, and basic political and cultural orientations of the imperial authorities, within a societal organization that gradually emerged as patrimonial.

Even in the first stages of the conquest of Mexico and, later, the discovery and overthrow of the Inca Empire, the king's prerogatives were nominally asserted (Haring, 1963). Initially this process was carried out mainly as private enterprises, motivated by the hope of acquiring power and legendary riches, and was given little financial or other assistance from the crown.

When Spanish rulers realized the immense potential riches and opportunities made available following the conquest of Mexico, the first phase of the colonization, characterized by private enterprise and the semi-autonomy of local administrations, came to an end. The territorial jurisdiction of the governors (*adelantados*) and the governor's position were now incorporated into the royal system of administration. From 1530 onward, the crown of Castile (the center of the imperial bureaucracy) established its dominance over the new territories by limiting or revoking the authority of the first governors, which threatened to encroach upon its power and sovereignty. The *audiencia* (the high juridical court, which was also endowed with legislative and administrative powers, such as reviewing the acts of the viceroy) was established in Mexico in 1528; in 1535, the first viceroy of New Spain was appointed. In 1544, an *audiencia* was established in Lima, Peru (Haring, 1963: 83). These actions symbolized the consolidation of the absolute power of the king, assisted by his Council of the Indies. Throughout the 16th century, a series of laws called the "Laws of the Indies" shaped the political and administrative system, each law representing a further step in controlling and curbing the political emergence of local elites striving to obtain a greater share of political power and grant of economic privileges. The "Laws of the Indies" epitomized a "paternalistic, punctilious regulation of the smallest details of colonial life and administration" (Sarfati, 1966:25).

B
STAGES OF URBANIZATION IN LATIN AMERICA

The initial process of urbanization in Latin America took place in the first 40 to 50 years after the discovery of the New World. Until the capture of Tenochtitlan (1521), the geographic extent of the colonization effort was limited to the Caribbean Islands and Panama. Town and fort sites established in the earliest phase of colonization were chosen on the basis of the natural advantages they offered. When the newly discovered areas became better known, however, settlements were often abandoned and reestablished at new sites.

Until the 1520s, Santo Domingo on Hispaniola Island (present-day Haiti and Dominican Republic) was the capital of Spain in America. It was built in a regular grid pattern; church and administrative buildings were grouped around the plaza. This pattern, which was quite innovative in medieval Spanish town planning, was elaborated and formalized in later years and became the physical symbol of the new urbanization of Spanish America. The city was the seat of the first governor, and the first archbishop and many of the political and administrative institutions of the colonial regime made their appearance in Santo Domingo. These included the *audiencia,* the *cabildo* (the city council), and the *visita* (the viceroy's inspection tour).

This early stage of colonization was characterized neither by the discovery of precious metals nor by a serious attempt at agricultural colonization, but rather by the establishment, fast removal, and relocation of small military centers and outposts.

Once the Aztec confederation had been subjugated and the conquest of the Inca Empire was under way, Santo Domingo lost its centrality. The newly reconstructed city of Mexico became the capital of Spanish America and the political center of New Spain. Mexico was joined in 1535 by the new city of Lima, capital of the newly established Viceroyalty of Peru, and these two cities remained the centers of power in the Spanish Empire throughout the entire colonial period. The main thrust of urbanization occurred between the capture of Tenochtitlan and the mid-16th century, the urban frontier moving south in the steps of the military expeditions of the *conquistadores*.

Most of the major centers of the continent were thus built within a short period in the first half of the 16th century. As stated by Morse, the Spanish town represented "the intrusion of formal, metropolitan bureaucracy into an empty continent" (1962a:332). These towns did not emerge and develop through economic and social interaction with their surroundings but were planted and imposed on the newly conquered territories as outposts of the metropolitan imperial and ecclesiastic bureaucracy. This provided one of the clearest illustrations of the operation of forces of centrality in stimulating an urbanization process.

The major factor in determining the cities' particular pattern of development was not so much the date of their foundation as the period when they achieved regional or continental dominance. The traditional viceregal capitals of Mexico City and Lima became dominant interregionally and reached maturity in the 16th and 17th centuries; thus their internal structure, social hierarchy, racial pattern, urban design and architecture, and the distribution of ownership of property in them bore the imprint of these periods. In contrast, Buenos Aires, Caracas, Santiago de Chile, and Montevideo reached regional dominance only in the 18th century. Regardless of the date of their founding, these cities reflected the new economic and administrative order of the Bourbon regime. Their coming of age at this late period relieved them from the rigid socioeconomic and racial hierarchies that marked such cities as Lima and Mexico City, which had reached their climax of development at an early date (Socolow and Johnson, 1981:32).

C
PATRIMONIAL-BUREAUCRATIC REGIMES IN LATIN AMERICA

The urbanization process in Latin America was shaped by the nature of the major forces that developed there and gave rise to the crystallization of a patrimonial regime (Eisenstadt, 1973; Sarfati, 1966), though evolving from an imperial regime.

Although Spain and, to a lesser degree, Portugal upheld strong imperial centers until the late 15th and 16th centuries, these were nevertheless changing in a patrimonial direction. A variety of historical causes contributed to this process: the reconquest of the Iberian peninsula from the Moors, the transformation of Iberian Catholicism by the counter-Reformation, and the decisive economic effects of the conquest of the Americas. These developments weakened the pluralistic elements in Spanish and Portuguese societies as well as the autonomy of their major institutional entrepreneurs and gave the centers control over those resources that were beyond the reach of the broader groups. The orientation of the centers toward mobilization of the broader groups was thus weakened. The centers turned instead to distributive policies through which they could buy off the principal autonomous groups, thereby increasing their dependence on the center, lessening their autonomy, and reinforcing the segregative tendencies in society (Eisenstadt, 1978:284).

These patrimonial characteristics became even more reinforced in Latin America through the crown's establishment there of sole control of the flow of resources. Thus some of the basic characteristics of patrimonial regimes indeed crystallized very early in Latin America.

This patrimonial regime displayed few symbolic and organizational differences between the center and the periphery. The comparative lack of center-periphery differentiation was especially evident in the nature of their relationship and, above all, in the small degree of restructuring of the periphery by the center. The center impinged on local communities mainly in order to administer law, preserve peace, collect taxes, and maintain the economic and religious ties linking them to the center. The conception of the center generally held by the patrimonial society stressed its role as the preserver of a static social and cosmic order. This attitude was strongly expressed in the policies adopted by patrimonial rulers. Of special interest for understanding the urbanization processes were those policies that were aimed at accumulating available resources in the centers and monopolizing the means of distribution of surpluses throughout society by the center.

The character of center-periphery relations in this, as in many other patrimonial regimes, greatly influenced the prevalent conception of the relationship between political boundaries and cultural, social, or ethnic boundaries. Seldom was explicit emphasis placed on the convergence of these boundaries in a single territorial framework, around a single geographic center or focus. Hence a relatively weak symbolic attachment to frontiers and territory existed.

The nature of structuring social hierarchies and class relations and consciousness was closely related to the characteristics of center-periphery relations in the patrimonial society. In such regimes lack of independent strata—especially of intermediate status—tended to emerge together with a preponderance of bureaucratic elements. The centers allowed various groups and strata to maintain autonomous though segregated status arrangements, but they attempted to preserve control of the access to macrosocietal status attributes. Small groups—

territorial, occupational, or local—became the major status units, very often organized in horizontal status sets. These groups exhibited the strong clientelistic tendencies of such strata but little leaning toward political autonomy or consciousness.

The upshot of all these characteristics of this system was that the forces of centrality, as shaped in the patrimonial-bureaucratic manner, have predominated over those of concentration—a tendency that was reinforced further by the specific circumstances of the colonial regime.

D
CITIES AND URBAN SYSTEMS IN THE PATRIMONIAL FRAMEWORK

These characteristics of patrimonial-bureaucratic regimes crystallized quite early in colonial Latin America and shaped some of its most distinctive institutional patterns, including, of course, cities and urban hierarchies.

The predominance of forces of centrality over concentration forces and the patrimonial features of the Spanish American regimes molded the major aspects of cities and urban hierarchies as they evolved in Latin America: the location of cities, the structuring of their functions (especially economic), and the major characteristics of the urban systems, their internal ecological and social structure, and the nature of their autonomy.

1 The Location of Cities

The location of towns was guided by considerations of political control, extraction of taxes, and mercantile control over trade, with almost no attention paid to possible internal economic development and the stimulation of internal markets.

In earlier stages of colonization two types of such locations were important. One was the establishment of towns on the sites of former indigenous cities. The building of a new town on the site of a former capital or a political-administrative center gave that city a "role transfer" and conferred upon it the authority and controlling power of the former seat of government. The transfer of roles offered the additional advantage of providing conquerors with the opportunity to utilize some of the existing urban infrastructures, such as major roads and water systems. A Spanish capital thus replaced an Indian one and a Christian church replaced a pagan temple.

The other most frequent location of towns at that period was in the midst of densely concentrated indigenous rural populations (Hardoy, 1975:20). Cities like Puebla, Guadalajara, Lima, Trujillo, and Arequipa are examples of this kind of urban site in densely populated areas. This location provided efficient means of control over taxation of the rural population; it thus served the chief aims of the conquest, the territorial control, and derived fast returns on colonization efforts.

With the implementation of the *Encomienda* system, central locations of this kind afforded Spaniards the opportunity to capitalize on the labors of the local Indian population, within the scope of the powers granted them through this system by the crown and its representative in America. The *Encomienda* was "the patronage conferred by royal favor over a portion of the natives concentrated in settlements near those of the Spaniards; the obligation to instruct them in the Christian religion . . . and to defend them in their persons and property, coupled with the right to demand tribute or labor in return for these privileges" (Sarfati, 1966:126). It should be understood that the *Encomienda* did not relate to land, only to people. Not being an allocation of land but a statement of reciprocal obligations and benefits between Spaniards and Indian laborers, the *Encomienda* could most efficiently be handled in towns situated amid a dense concentration of Indian population. In the 16th century, however, the establishment of towns in an area having a dense indigenous population enabled the unidirectional flow of resources into the towns from their respective hinterlands, in terms of tribute, taxes, and labor.

2 Role of the Church in the Establishment of Cities

Parallel to the colonization process, in which towns were military outposts and centers of patrimonial bureaucracy, another important motive for establishing urban centers can be identified. That is the evangelization of infidels, the spreading and teaching of Catholicism. Even the briefest royal directive for establishing towns rarely failed to mention the spreading of Catholicism and that God would be better served as the reasons for building towns (Gakenheimer, 1967:37). This Christian duty, however, could not be accomplished by the *conquistadores* themselves; it required the organization of numerous members of a well-trained clergy, recruited especially and dispatched to the new territories to fulfill this immense task. The prime responsibility for encouraging and organizing missions to the unbelievers lay with the Papacy, but at that time the Popes had neither adequate resources to undertake this task in America nor a body of trained men to accomplish it. The Popes could only implement their will by bringing moral pressure to bear upon the temporal rulers.

From the early days of discovery of the New World, the Papacy realized that an apostolic mission to America could be organized only by the Spanish crown.

The famous *Patronato* was then enacted (1508), conferring on the Spanish crown an immense and unique authority, unlike any then existing in Europe; the sweeping terms of the Papal concessions subordinated the entire clergy of a great empire to the royal will.

As a result of the constitutional agreements between the crown and the Papacy, therefore, the administrative arm of the patrimonial order was joined by the clergy, which constituted the ecclesiastic arm of government. The situation of the clergy differed somewhat from that of the bureaucracy because, even here, the Church enjoyed the special privileges and prerogatives of a "state within a state." In the early phases of conquest, at least, the clergy constituted "the representative and functional agents of the spiritual order, in which the sanctity of the king's rights and ultimate ideological justification of the conquest were rooted" (Sarfati, 1966:40).

Like the administrative bureaucracy that established towns and reconstructed the plundered centers of government, the clergy also fulfilled an important role in the urbanization process of the 16th century. The crown saw to it that friars or priests accompanied military leaders and explorers in their missions of conquest, and the conquerors had to follow the firm royal policy of creating a bishopric immediately upon colonizing any new region (Haring, 1963:140). The friars spread out quickly over most territories, founding churches, monasteries, and convents wherever they went. Initially the churches were modest temporary structures, but from mid-16th century, with almost unlimited labor available (including a large work force accustomed to building with stone, mortar, and adobe), the clergy began constructing permanent buildings for their churches, often of considerable size and splendor (Parry, 1973:151). Numerous churches were built on the sites of destroyed pagan temples to symbolize the victory of the "true" belief.

Many churches served as the nucleus of a village and, eventually, a town. Like the officials who chose sites for reasons of military control and political administration, the clergy sought to locate churches in areas with a dense indigenous population. In the course of time, when evolution of the administrative hierarchy was reflected in the size of cities, the ecclesiastic hierarchy followed a similar development and the hierarchical levels of the Church corresponded closely, in each case, to those of the administrative hierarchy (Hardoy and Aranowich, 1970:85). The administrative and religious hierarchies, therefore, reinforced each other in establishing the framework of the loosely structured urban hierarchy.

In many areas of the new colonies there were no dense populations because the Indians lived among their cultivated fields according to a settlement pattern. In those areas the missionaries devoted much energy to persuading or compelling the Indians to move into concentrated forms of settlement—large villages or towns built around a church and a school, with some sort of local government. These new settlements, established to facilitate evangelization, had an important impact on the urbanization process of thinly populated regions. The large-scale resettlement of scattered Indians into new villages or small towns (*reducciones*) created a particular type of spatial organization—that of the ecological segregation of different groups in society, quite characteristic of patrimonial regimes. Because of the need for cheap labor in the large cities, however, the Indians were concentrated in *reducciones* at the periphery of the large cities (Violich, 1944:29; Kubler, 1948:82). The policy of establishing Indian compounds was at the root of the ecological segregation, on both a regional and an urban scale.

3 Mercantile Economic Policies and Their Effects on Urban Development

Analysis of the location of towns in Spanish America indicates how the urban development was shaped and determined by the forces of centrality, the towns constituting either military outposts or political-administrative centers of the bureaucratic-patrimonial Spanish regime. The Church's activities in evangelizing the Indians added an ideological dimension to these forces of centrality, the Church in Latin America being the spiritual arm of the king, the *"patronato real."*

Against this strong impact of the forces of centrality shaped in a patrimonial mode, those of concentration and autonomous economic and social development were very weak. This can be observed most clearly in the impact of the mercantilist policies of the center on the location and function of cities—especially port and mining towns—as they evolved from about the 16th century onward.

The mercantilist policies represented an important element of the bureaucratic-patrimonial configuration, stressing that the power of the king would be strengthened by the increase of his revenue. In Spain, "the mercantilist structure derived its 'raison d'être' from the existence of a vast colonial empire which had to be kept under the economic and financial control of the crown" (Sarfati, 1966:17). Increase in revenue could, in principle, be effected either by means of a monopolistic mercantilism, where a purely fiscal orientation prevailed, or by encouraging local industries and enterprises and padding them by a strong protectionist system. These two approaches had completely different impacts on urban development; the first approach was applied to the colonies by the Spanish regime and gave rise to a rather special mode of urban growth.

The core of the Spanish mercantilist policy consisted in maintaining a monopoly on trade with the Indies that, by royal decree, was reserved to the powerful merchant guilds *(Consulados).* This restrictive trade policy with its extractive nature, together with the closure of the economic market, prevented the forces of concentration from playing a meaningful role in the urbanization process in Spanish America.

Beyond the many colonized cities established in the first two stages of urbanization in Latin America, the particular type of mercantile system practiced by the Spanish government gave rise to the development of two kinds of urban settlement: ports and mining towns. The growth of these two types of settlement characterized the third stage of the urbanization process. This took place in the second half of the 16th century, when "Spanish exploitation of the most important mineral resources began and the principal ports were definitely established" (Hardoy, 1975:27) and the fate of these two types of cities in different parts of colonial Latin America fully demonstrated the predominance of forces of patrimonial centrality over those of concentration.

Port cities fulfilled two important functions in Spanish America: communication and transfer. They acted as relay stations, major nodes of communication in the bureaucratic-administrative machinery that, in the Spanish Empire, was the main feature of the patrimonial government. The strong centralization of Spain's administrative apparatus, both secular and ecclesiastic, required a highly developed system of intercontinental maritime travel. (It should be pointed out that the administrative machinery had an extended "response time" as, even under the most favorable conditions, a round trip between Peru and Madrid took about a year.) The merchant fleet transporting passengers and cargo between Spain and America was definitely organized and established by law during the second half of the 16th century. The cities most strongly characterized by their function as communication and transfer centers were those located at the main convergence of transport routes in the Caribbean area, such as Panama, Veracruz, Cartagena, and Portobello.

In some cases, political-administrative centers were established in a location selected to ensure easy communication with Spain, thus combining the functions of a port with those of a political capital. A classic example of this phenomenon was the transfer of the capital of the Viceroyalty of Peru to Lima and its adjacent port of Callao, instead of allowing it to remain in its inland location at Cuzco, capital of the Inca Empire, which was too remote from the sea (Gakenheimer, 1967:33). The combination of its functions, as the political center of the most extensive viceroyalty of the empire and as the major node of exchange and distribution, allowed Lima to become the largest and most prosperous city in Spanish America.

The strength and prosperity of Lima exemplify the influence on urban development of the coalescence of political and economic markets, directed above all by political considerations. The economic advantages offered by Lima as a port and as a central point in the distribution system materialized only because of the town's political primacy as the administrative capital of a tremendous hinterland served exclusively by it. This imposed administrative hinterland included distant regions such as the La Plata Basin and prevented other towns—such as Buenos Aires, with better accessibility to markets—from developing into major ports and commercial centers (Davis, 1975). The case of Buenos Aires, which was unable to develop as an important economic center throughout the 16th century, illustrates the consequences of limitations and restrictions imposed by the patrimonial center on the periphery. The various provinces of Spanish America were denied the privilege of trading with one another, but each of them related directly to the center. The political and economic dominance of the center over all the distinct parts of the New World was thus ensured. "The case of Lima and Buenos Aires is an example of how New World cities had to strive, not merely for commercial advantage but for what might be called their very legitimation for commerce" (Morse, 1962a:325). This illustration shows how, under the impact of monopolistic mercantile policies that predominated in Latin America, the forces of concentration were insufficient to ensure urban growth. When they collided with forces of centrality, with administrative schemes and structures of control, they were bound to yield to them. Indeed, as Hardoy expressed it, "the progress achieved regionally by certain cities . . . over others was due to their pre-eminence as administrative centers of the crown and Church" (Hardoy, 1975:27), and to differences in the extent of their areas of commercial influence, defined not by market forces but by administrative regulations reflecting the interests of the metropolis in Spain or the local center in Spanish America.

The relative ineffectiveness of the forces of concentration in developing towns in Latin America is best demonstrated in the mining towns. The extraction of mineral resources and their shipment to Spain constituted one of the main stimuli to the discovery and conquest of the colonies in America and epitomized the particular monopolistic mercantilism of the Spanish Empire. Immediately after the discovery of ore, mining towns grew with unparalleled speed and, nearly without exception, almost none of them held the distinction of having been formally founded as the other towns had been (Gakenheimer, 1967:40). Due to their fast growth and to the major role played by private entrepreneurs and adventurers in their rapid development, mining towns lacked the systematic planning that characterized most towns in South America. Despite their large population and their economic strength and affluence accumulated in a short period, mining towns were frequently not recognized as cities by the patrimonial bureaucracy. The inferior political status of large, rich mining towns such as Potosi illustrates the methods used by the patrimonial center in the economic sphere to secure its dominance by limiting the autonomy of groups having potential power (Sarfati, 1966:52).

4 Urban Systems

The patrimonial-mercantile policies of the Spanish colonial center greatly affected the major characteristics of the urban systems in Latin America.

The general level of development of the urban system was quite low if measured by the percentage of urban population from the total population (Hardoy and Aronowich, 1970). Even more important, however, was the low level of interaction among the cities of various hierarchical levels of the urban system:

> Urban networks developed feebly; geographic barriers to regional transportation were often formidable, while the crown's mercantilist policies did little to encourage centers of complementary economic production. New World cities tended to be

> related directly to overseas metropolises and isolated from one another [Morse, 1962b:477].

This constituted a decisive factor in shaping the particular structure of the urban system. Rather than interacting with towns at other levels of the urban hierarchy, which would create a regular economic hierarchy tending toward a rank-size distribution, each of the urban settlements in Spanish America tended to interact separately with the center in Spain. Without economic interchange based on local specialization, no economic hierarchy of central places could evolve. The clearest evidence of this particular relationship between the colonial periphery and the imperial center in Spain is found in the shift in road patterns after the Spanish conquest.

> The magnificent Inca highway system, centering on Cuzco, was soon fragmented, not simply because the Spaniards abused and badly maintained it but because in contrast to the directional emphasis of the Inca state, the Spanish economy emphasized the lateral connecting roads as outlets from mining areas to the sea [Morse, 1962a:324].

The slicing of Spanish America's huge internal market into small, fragmented, unrelated submarkets, each oriented separately toward the external market, constituted a monopolistic pattern that was instrumental in shaping the urban system as a political-administrative hierarchy with a strong primacy tendency. This primacy structure and the concentration of all urban functions in the capital city (Hardoy and Aranowich, 1970:60) prevented any significant development of towns at lower levels of the urban hierarchy and reflected the dominance of the political elite over all other elite groups.

These basic characteristics of the location of cities, their functions and organization, as well as the structure of the urban system have persisted in colonial Latin America (in fact, also in most Latin American countries after their independence), despite the many concrete changes in the development of these cities. The same is true of other aspects of the structure of cities.

E
INTERNAL STRUCTURE OF CITIES

1 Bureaucratic-Patrimonial Town Planning

The predominance of forces of bureaucratic-patrimonial centrality also greatly influenced the physical layout and internal structure of most of the towns in Latin America, the majority of which were molded in the gridiron geometric pattern. To Richard Morse this pattern was an outcome of "the subordination of the streets to a central will" (1962a:320). Towns had to conform to detailed regulations and standards of design that dealt with most of the elements of town planning: the universal grid pattern (all streets running perpendicular to one another, city blocks being squares or rectangles), the width of streets, the location and size of the main *plaza*, the orientation of streets, the sites of public buildings, and, of critical importance, the subdivision into lots. The political ability to impose a master plan and standards of urban design over almost all the newly established towns in Spanish America resulted in cities as widely separated as Bogota in the north and Concepcion in the south, "and practically every city in-between have exactly the same size of block, the same width of street, the same general pattern" (Violich, 1944:28).

The first city to be built according to a regular, geometric urban plan was Santo Domingo in 1504. The model for its layout was influenced by the rectangular city plans used in Spain by the Catholic monarchs when they established new towns as military strongholds on recaptured land at the end of the Moorish campaigns (Smith, 1955:3). Precise and explicit royal instructions, dealing with the planning of towns were delivered to the new territories by Pedrarias d'Avila in 1514. It seems worthwhile to quote part of these early orders:

> In view of these things necessary for settlements and seeking the best sites in those terms for the town, then divide the plots for houses, these to be according to the status of persons and from the beginning it should be according to a definite arrangement. For the manner of setting up the *solares* will determine the pattern of the town, both in the position of the *plaza* and the church and in the pattern of streets, for towns being newly founded may be established according to plan without difficulty. If not started with form, they will never attain it [Stanislavski, 1947:96].

The phrasing and main thrust of this royal directive indicate a strong Roman influence. It emphasized the need for order in the division of land as the major factor shaping the form of the town, yet it did not include an explicit order to build the town in a grid pattern. These plans did not reflect any symbolic orientation, as the newly established cities did not represent a cosmological order on earth. Even the aesthetic orientation (indicated by the revived Italian Renaissance classicism) influenced town planning in Spanish America only later in the mid-16th century. The crown's basic intention in imposing design codices in Spanish American towns was to assume complete control of the land and to ensure its precise and measured allocation to the Spaniards who founded the towns and settled in them. It was this territorial orientation that linked the bureaucratic-patrimonial order with the unique phenomenon of strict and precise geometric design of towns, which was repeated all over the continent.

The instructions transmitted to Ovando in 1501, to d'Avila in 1513, and to Cortes in 1523 culminated in promulgation of the body of statutes governing the layout of towns in "Discovery and Settlements Ordinances" signed in 1573 by Philip II (Hanke, 1967:278-283). A significant part of these famous regulations dealt with the various elements of town design and structure, clearly summarizing the concepts and rules of the patrimonial center in regard to the physical structure of towns in Spanish America.

The royal center's regulations of town design crystallized in several steps that, basically, seemed to formalize "a situation already perfectly defined in practice" (Hardoy, 1975:30). As long as the design of the new towns fulfilled the basic patrimonial need for land control and allocation, the actual physical layout emerged gradually through the pragmatic efforts of the governors and city founders in the periphery.

2 The Grid Pattern and Physical Layout of Cities

The "Royal Ordinances for the Laying Out of New Towns" of 1573 provide an extensive source for describing the major elements of the internal structure of most Spanish American cities. On its founding, the city was planned in a rectangular grid pattern of wide streets. The city blocks were squares or rectangles of uniform size. The blocks were subdivided into lots *(solares)* that were the basic unit of land allocation. The grid pattern, having its origin in Greek and Roman town planning, possessed several important characteristics that rendered it suitable for the urban development of Spanish America.

Due to its relative simplicity, the grid pattern was best suited for carrying out the plans and designs of a bureaucratic-patrimonial order. Under the specific circumstances of Latin American colonization, many of the cities were often provisional or had to be expanded or contracted (Gakenheimer, 1967:49).

Given that the granting of land was a major policy instrument of the royal center, the grid pattern provided the most accurate tool for measured property allotment. In actual practice, the size of an individual *solar* was the basic unit determining the size of the city blocks and, hence, the size of the central *plaza.*

The grid pattern for the town layout is basically an open-ended one. A town built on a grid pattern can easily be expanded by extending the straight streets and adding more identical blocks at the periphery of the built-up area. This inherent flexibility, allowing for growth and expansion, was of special importance in many of the Spanish American towns founded by a handful of *conquistadores* and entrepreneurs that grew quite fast to become political-administrative centers. The open-endedness and physical flexibility of the grid plan could only have materialized due to the fact that most Spanish American towns were not walled. It is interesting to note that the few cases in which walls were erected occurred mainly in port towns and that, in this particular type of town, the purpose of the walls was to provide protection not so much against the local Indians as against other European countries and pirates.

The absence of walls and the open-endedness of the grid pattern are related to another important characteristic of Spanish American cities: the low level of differentiation or distinction between the town and its surroundings. This low level of differentiation is related to the system of allocation of land around each town. A good example of the operation of this system can be found in the description of the founding of Lima: The original town plan contained 117 blocks, each subdivided into 4 lots (*solares*). Lots were allocated on the day of the city's founding, and the names of recipients were marked on the city's master plan. Beyond the urban nucleus, in which city blocks were arranged in the grid pattern, tracts of land were designed as city commons (*ejido*) and pasture lands. This belt of public land was meant to provide an outlet for the future expansion of the town at the periphery of the original grid area. In this outer belt land was distributed to the persons who had received *solares* within the town's nucleus. Because of the vast empty spaces in the New World territories, the jurisdiction of a municipality could extend, in many regions, for tens or even hundreds of miles around the urban nucleus. This fact, as well as the allocation and distribution to the townsmen of land within the municipal jurisdiction (and this particular land control and allocation system), brought about minimal differentiation between the town and the land around it, between the urban center and its rural periphery.

The pattern of land use in most Latin American cities was focused on the main plaza (*plaza mayor*). This plaza, sometimes located in front of the built-up area overlooking the river or sea and sometimes within the built-up area, constituted symbolically and functionally the heart of the city. As a rule, the main plaza was flanked by public structures housing the three major powers that affected the life of the city and the province: the governor's palace, the cathedral, and the city hall (the seat of the *cabildo*). It is hard to find a similar instance of the physical expression of the basic sociopolitical power structure, as exemplified by the centrally located housing of the three major powers. The transfer of political power at the end of colonial rule merely transformed the viceroy's or governor's palace into the presidential residence. The main plaza and its surrounding centers of power were the main foci of the city's population: The viceroy's or governor's palace was the seat of all political activities of the ruling elite; the city hall provided municipal services of a limited range and catered mainly to landowners, merchants, artisans, and builders; the church was the main center of congregation for the whole population. Urban life, social organization, and, in the case of capital cities, the national and provincial life were thus centered on the main plaza.

Concentration of the political, religious, cultural, and administrative organs of power at one point by the patrimonial regime of colonial Latin America was the determining factor of the entire land-use pattern in cities. As they came into being, all government offices, archives, and record-keeping offices, and various municipal agencies were located as near as possible to the main plaza, which gradually assumed the character of a government quarter. Because of its easy accessibility from all parts of the city and the great influx of population into it (either for administrative or religious purposes), it became the city's major commercial center.

Thus, the central area of the city was shaped mainly by the forces of centrality determining the seats of government and Church; this was followed by the operation of the forces of concentration in attracting shopping and commercial activities in this area. The balance between political and religious factors and commercial factors remained in favor of the political-religious ones until the 20th century, when the balance tilted toward the forces of concentration, transforming the city center into a Central Business District.

The internal structure of Latin American cities showed remarkable continuity throughout the whole of the colonial period and, in most cities, was preserved until the present day, with some of its basic features determining the central area of the large cities.

F
SOCIAL ORGANIZATION

The continuity of the social pattern shaped by patrimonial orientations and policies in the face of constant changes and expansion can also be found in the social organization and ecological structure of Latin American cities. The structure of social hierarchies in Latin American society in general and the urban society in particular was characterized by certain basic features that can also be found, in different forms, in other patrimonial regimes.

The major attributes of status were based on the following: (1) a combination of ascriptive-primordial criteria of origin and race and those of proximity to the center; (2) access to the center, to the attributes of macrosocietal status controlled by the center through the collection of tithes and through the distribution of symbols of prestige and to the land, and by controlling access to the viceregal or provincial powers; (3) the group basis of such hierarchies, which was relatively narrow, local, or regional, although there was a vague general identification in terms of origin and race, and the degree of their social autonomy was relatively low; (4) most status sets, which were organized on narrow (local) horizontal bases with a strong tendency toward the development of clientelistic ties; and (5) the existence of a weak tendency to autonomous expression of a wider—especially political—class consciousness.

Many important aspects of these characteristics of status organization have persisted throughout the colonial period in Latin America and beyond it, despite far-reaching changes in the composition of the population and in the economic organization of society. These changes, indeed, gave rise to great differences in the concrete features of structuring the social hierarchies.

The structure of Latin American urban society was rooted in the old Iberian world, but in the new circumstances, some of the urban society's features were changed in a much more polarized direction than in the Iberian peninsula. In Latin America, a two-class "caste" social structure initially evolved. It comprised, on the one hand, an elite of landowners, mine owners, high-level bureaucrats, and clergy; on the other, the mass of workers in *haciendas,* rural communities, and tropical plantations, and the urban domestic servants. Between these two strata existed a small group of merchants, petty bureaucrats, and minor ecclesiastics. A major difference between the social structure of the colonies and the metropolis was that, in addition to income, social position in Latin America was based on power, family standing, color, and, later, the distinction between Spain and Latin America as the country of birth. At the same time, the arrival of Negro slaves, first in the Caribbean Islands and later in other parts of Latin America, added an additional ethnic factor at the lower end of the hierarchy.

About half a century following the conquest, a simple tripartite division of the colonial society evolved, based initially on racial criteria: The elite was made up of whites (Spanish-born *peninsulares*); the blacks were slaves, and the Indians were peasants who usually lived in semiautonomous communities apart from Europeans and blacks. The colonial society's economic organization followed a similar pattern of division as particular tasks were associated with the various racial groups: Indians produced food; blacks worked as unskilled laborers on sugar plantations and in the mines and domestic servants in the great houses of the major cities. The *peninsulares* made up the privileged sector, assuming all controlling and leading positions. Spaniards (*peninsulares*) and whites born in America (*criollos*) tended to concentrate in the large urban centers, fulfilling the administrative, legal, and financial functions, and they predominated in the major mining towns. The countryside was left to the Indian population, among whom small numbers of whites were scattered, living in their estates or in small towns.

A basic cleavage thus developed in the colonial society between the Indian world, with its strong preconquest continuity and its inward-looking and local orientations, and the Spanish world, carrying the Iberian tradition and interregional and international orientations. From the beginning these two categories were not, of course, monolithic (Altman and Lockhart, 1976), and this relatively simple pattern of stratification was not entirely homogeneous. In the 17th century it started to collapse because of the combined effects of mixed marriages, the internal dynamics of colonial society and economics, and its relations with the center in Spain. A distinction then emerged between the peninsular Spaniards and those born in America (*criollos*). The union of Spaniards of peninsular origin with Indian women produced a growing number of *mestizos,* who struggled to obtain a position in the social hierarchy. With the decline in the number of Spaniards arriving in the colonies and the weakening of the patrimonial regime's center, the *criollos* and, to a much lesser extent, the *mestizos* began to assume a dominant role in urban society, becoming a potential threat to the Spaniards' domination. In periods of economic recession in the Spanish metropolis and in the American colonies, competition for access to wealth and power became much tougher and the criteria of ethnic purity much more pronounced in shaping the structure of the urban society. During the 17th century, a thin stratum emerged between the white elite and the multitudes of Indians and

Negroes, consisting of the offspring of racial intermingling among whites, Indians, and blacks, in their many combinations. This stratum was subject neither to slavery (as the Negroes) nor to patronage (as the Indians) and grew continuously in number.

The color and the legally prescribed inferiority of the blacks fitted them readily into a society of *castas*. The union of white Spaniards and blacks produced the *mulattos*, who also strove to make a place for themselves in the social hierarchy. Whenever *mulattos* and free Negroes posed a threat to the elite of Spaniards and *criollos,* formal and informal barriers were raised to curb any such attempt. But it took only a few decades of interracial contacts in the New World for the *mestizos, mulattos* and *zambos* (offspring of an Indian-Negro union) to constitute a major component of the social structure.

Europeans considered themselves to be the sole representatives of the metropolitan authority, with natural rights to fulfill all the highest administrative, military, and ecclesiastic positions, while considering the American-born *criollos* as slightly inferior. The *criollos,* becoming more powerful economically, found in the *cabildos* (the town councils) an open avenue for their political aspirations. The *cabildos* gradually became the cornerstone toward autonomy and eventually toward the national independence of the colonies (Stein and Stein, 1970).

The most noble and well-established families were the veteran ones. By the time a second generation matured, the native-born Spaniards made up, in most areas, the major part of the large estate owners and occupied the seats of the town councils. They entered professions, soon constituted the majority of the lower clergy in various religious orders, and climbed up the echelons of governmental positions until they penetrated the basic organ of the Spanish American government, the royal *Audiencia.*

The role of the newcomers from Iberia constituted an important factor in the growth of Latin American society, in both large and small cities. To climb up the social ladder, the new immigrant utilized all his family and hometown connections. Applying entrepreneurial skills, numerous immigrants quickly joined ranks with the established and wealthy veteran families. By the mid-18th century, *criollos* had apparently reached a commanding position at the highest level of colonial Spanish society. They dominated the economy of most cities and regions, controlled town councils, and filled the majority of the ecclesiastic, military, and governmental positions, except for the few at the very top of the hierarchies. In order to keep the colonies under patrimonial control, the crown reduced the *criollo* majority in the *audiencias* in the second half of the 18th century by sending new waves of representatives to Latin America and ensuring that, in each province, the viceroy was Spanish-born. These were temporary measures that succeeded in slowing the *criollos* for a time but could not ultimately prevent them from becoming the major component of the elite, the dominant group in Latin America.

The exact ways in which these various groups became interwoven with these economic and political trends have been the subject of much research (Bronner, 1986). An attempt to define the socioeconomic dimension of the racial categories was made by Mörner, who claimed that the Spaniards were bureaucrats and merchants and the *criollos,* large landowners; the *mestizos* were artisans and shopkeepers; the *mulattos,* urban laborers; the Indians, peasants and unskilled laborers; and the blacks were slaves working on plantations and in mines (Mörner, 1967, 1983).

Other scholars doubt the existence of a distinct occupational pattern of these various groups, especially with regard to *criollos* (McCaa, Schwartz, and Grubessich, 1979). Most significant in this connection is the work of Seed (1982), which deals with the social dimension of race in Mexico City in the mid-18th century. Seed's work is based on a detailed analysis of the 1753 census of population, which was taken as a preliminary step in resolving some of the acute social problems created by the large influx of migrants into the capital city in previous years. Analysis of the census focuses on the precise relations between a refined racial division and the types of employment in the city. The racial framework consisted of the following five basic categories: whites, Indians, blacks, *mestizos,* and *mulattos.* Two additional groups appeared in the 17th century: *castizos* (light-skinned *mestizos*) and *moriscos* (light-skinned *mulattos*). A second level was thus added to the *castas,* giving rise to a more complex categorization of the different groups: One line was the continuum, ascending from the Indians to *mestizo-castizo*-white; the second, a parallel one, ascended from blacks to *mulatto-morisco*-white. Seed examined the socioeconomic status of each racial group through their types of employment and ownership of means of production.

The relation between race and the division of labor in Mexico City was thus based on the different economic roles played by Indians, blacks, and *peninsulares,* each of these major racial groups specializing in fulfilling the specific economic tasks already mentioned. Blacks were mainly domestic servants; Indians were employed in all forms of manual labor; and Spaniards dominated the wholesale trade and commerce and occupied the highest administrative and ecclesiastic positions. People of mixed parentage were much more likely to be artisans than the racial groups to which their parents belonged. Even though people of mixed blood were characterized as artisans, this intermediate group did not form an undifferentiated middle stratum but, rather, showed shades of differences reflecting the social origin of the parent groups. Although a few *criollos* were craftsmen, they tended to be merchants and shopowners more than members of other intermediate racial groups representing the peninsular merchants, their parent group. *Mestizos* and *mulattos* tended to be artisans more than their parent groups; the large number of *mestizos* who were laborers and the greater number of *mulattos* who were domestic servants were the representatives of their parent population. Differences in the employment structure between *mestizos* and *mulattos* were the outcome of the basic division of labor between the black and Indian communities (Seed, 1982:600).

In its details, this picture was specific to Mexico City. Great differences developed in the various parts of Latin America in the concrete composition of urban society. This stemmed from the diverse regional economic bases, the proportion of a region's Indian population, the level of accessibility of regions and towns in relation to the core areas in Latin America and the metropolis. Nevertheless, the common traits of society were much more prevalent than the specific characteristics of particular towns and regions (Altman and Lockhart, 1976:11). But in spite of these differences and variations, the basic elements determining the social structure in colonial Latin America remained unchanged throughout the entire period and even have an influence on present society. It was indeed the relations among race, social position, and economic roles that were of critical importance in shaping the urban society. In structuring these relations, the different primordial categories, as manipulated either by the imperial or local centers, were the most important ones.

G
SPATIAL ORGANIZATION OF THE URBAN POPULATION

The major characteristics of the social structure that crystallized in Latin American cities also greatly influenced the spatial organization of their urban population and, at the same time, the ecological patterns that determined the physical and symbolic access to urban resources. Control over these resources and over land in particular constituted one of the most important basic characteristics of the urban structure.

Of special importance here was the structuring of the areas surrounding the center: (a) Were they segregated or was the residential heterogeneity the prevalent pattern? (b) What were the bases of the residential segregation? and (c) Was it based on the population's racial composition, class, or occupational differentiation?

Analysis of Mexico City's population, based on the census returns and on detailed mapping, provides an illuminating example for defining the basic features of the population's spatial organization. It shows that cities, mainly those that had been established in areas densely populated by Indians, were initially designed to accommodate two distinct urban groups: the Spaniards and the Indians. Mexico City was demarcated into a central quarter, where the Spaniards lived, and a group of outlying *barrios*, intended for the Indians. This dichotomous division of the ecological structure became blurred, mainly because it failed to accommodate the rapid growth of the *mixto* groups that assumed the greatest importance with the decline of the Indian population in the 16th century. The demand for cheap labor, both domestic and unskilled, further undermined the two-tiered ecological structure of the Spanish center.

To continue with the example of Mexico City, its ecological structure evolved in the 17th and 18th centuries and comprised three different concentric zones: (1) the central core, including the political, administrative, and ecclesiastic institutions and large, well-built residential homes owned by the wealthy elite group of Spaniards and *criollos*; (2) an intermediate zone made up mainly of tenements with a high density of population; and (3) the outer zone in the outskirts of the city, where Indian communities lived, which was characterized by shacks, a low level of urban infrastructure, livestock enclosures, and vegetable patches (Moreno and Anaya, 1975).

With crystallization of a rudimentary class-based urban society at the end of the 18th century, ecological differentiation assumed the form of racial-occupational segregation. The city center was occupied by whites (both peninsular and *criollos*), their slaves and servants, as well as by people of mixed blood engaged in skilled artisanal occupations. The poor (mostly blacks or free castes), who were employed in unskilled menial tasks, lived in the peripheral zones at the cities' outskirts (Robinson, 1979).

The case study of Mexico City provides an almost "ideal type" of ecological structure of colonial Latin American cities, and the same pattern was repeated all over the continent. It can be summed up in terms of a declining gradient of the social status, highest at the city center and lowest at the outskirts. As the social status was determined by a racial-occupational differentiation, the ecological structure had a distinctive pattern of Spanish high-class center, an intermediate ring of mixed-blood groups, and an urban periphery of low-class Indians and blacks.

It is significant to note that newcomers to Latin American cities tended to reinforce the basic ecological structure; European migrants usually settled in the central parts, whereas most of the local rural migrants (mainly Indians, who began increasingly to populate the cities) occupied the continually expanding fringes. As the population of the cities grew, land became scarce and the spatial organization based on the racial occupational differentiation became blurred. There were many intrusions of low-strata quarters into the central parts of cities and a few enclaves of high-strata neighborhoods at the fringes of large cities. This basic ecological pattern, even though less conspicuous and differentiated spatially than in the early times, remained the basic feature of Latin American cities.

H
URBAN AUTONOMY

The basic features of the patrimonial order and the urban society's total structure in Latin America have also greatly influenced the nature and scope of the urban autonomy that tended to develop there.

The main characteristic of the tendencies to achieve such an autonomy was that they were focused on the attempts of local *cabildos* to usurp the powers of the central imperial authorities or to free themselves from their supervision, but they included no strong impulse to restructure the urban government or to a growing participation of the different strata in such a government. Accordingly, it was mainly the degree of the imperial government's strength that controlled the scope and intensity of the outcome of these autonomous tendencies, without changing the basic implementation; it only made it more variegated. The monolithic control of the European monarchs had cracks and gaps in it that, on the one hand, enabled the 300 years of colonial rule to endure and, on the other hand, accounted for its subsequent abolishment, leaving behind imprints that endure to the present.

All colonial cities and towns housed some representatives of the metropolitan political authority. The level and number of secular and clerical officials were directly related to the status of the city within the Spanish Empire's political hierarchy. In addition to these representatives of imperial authority, every city and town had agencies responsible to some degree for municipal self-government. It should be stressed that these local authorities were basically local extensions of the central royal power and their external authority was very limited. The degree of autonomy allowed the *cabildos* varied greatly according to their proximity to the major centers of power—the closer they were to these centers, the less autonomous they were—and to the level of their economic affluence, their regional importance, and the changes in imperial policy, as applied in different ways under the Hapsburg and Bourbon regimes (the two dynasties that ruled Spain during the colonial period).

The first two viceregal cities, Lima and Mexico, were the seats of royal tribunals (*audiencias*), having special *fueros* (rights), and their jurisdiction extended through the whole viceroyalty. Gradually, because of the tremendous difficulties in controlling the settlements in the continent's remote parts, a slow process of territorial decentralization started under the Hapsburgs and accelerated under the Bourbons, resulting in additional towns being elevated to the rank of *audiencias*. Usually the new *audiencias* resented the authority of the older ones and used every possible device to circumvent it. In general, the factor proximity/remoteness affected the level of efficiency of the control.

Cases in point were Caracas and Quito, which initially belonged to the Lima *audiencia* and then attempted to free themselves from its control. In the case of Mexico there was less conflict, for instance, between Guadalajara and Mexico, given the smaller distance between them, the better connections, and hence more efficient control (Haring, 1963:86). Such an autonomy seems to have been strongest in Venezuela, due both to its remote geographic position and to the lack of interest vested in it by the conquerors, who concentrated their early efforts on other parts of the continent.

The control mechanism of the patrimonial bureaucracy included *residencias* (public trials to which officials were periodically subjected) and *visitas* (secret or open investigation of the officials' conduct). To avoid corruption, officials were required to submit an inventory of their personal wealth and had to obtain royal consent if they intended to marry in the colonies. Furthermore, in order to prevent other allegiances and interests from taking root, officials were frequently transferred. In the few cases where *criollos* were appointed to official positions, they were not allowed to serve in the districts in which they lived (Sarfati, 1966).

Yet in spite of these measures, both in the peninsula and in the colonies, every echelon of the patrimonial bureaucracy was open to bribery and corruption either by foreign or by local trade interests (Stein and Stein, 1970). All important appointments for office were made by the king or had to be approved by him. In cases of emergency (upon an official's decease, for instance), the viceroy, often in concert with other powerful interest groups, appointed his own nominee so that by the time the king's decision arrived (between 8 and 12 months later), it was met with a fait accompli.

Through the system of *procuradores* (representatives) at the court in Madrid, the *cabildos* maintained direct contact with the king, often circumventing the viceroy's authority. The economic rights of the *cabildos* included the management of limited funds, the imposition of a few approved taxes, and the protection and regulation of food supplies. The fixing of prices and granting of building licences and lands (*mercedes*) were rights that initially had been accorded to the municipal corporation. Soon, however, lands were preempted by firstcomers, and the Church also acquired vast tracts through donations or bequests. The initial royal policy of "equal" land distribution, based on patrimonial rights and privileges, was thus gradually circumvented, creating a powerful landed aristocracy that later became the *criollo* oligarchy.

The political rights of the *cabildos* included policing, penalizing, establishing the rules governing the acceptance or rejection of people into the city, and various other minor local matters.

Encroachment on the efficient control of the *cabildos*, if it did not actually increase their autonomy, at least allowed them room to maneuver. The *cabildos'* autonomy was greatest when, at the beginning of colonization, the problems of the struggle against Indians demanded immediate, on-the-spot decisions. As the conquest became institutionalized, monarchs gradually increased their control by limiting the power of the *cabildos*. Royal directives, however, were purposely ambiguous. Although such ambiguity was intended by the crown to avoid the emergence of influential local governors, it also resulted in rather free interpretations of the directives, as expressed in a sentence that has come to be known as the epitome of colonial philosophy: "Obedesco pero no cumplo" (I listen but do not comply) (Haring, 1963:112-113).

The Hapsburgs and Bourbons exercised differing influence on municipal autonomy. The Hapsburgs were jealous of their control but were rather inefficient in its application. With the ascendance of the Bourbons, in an era of enlightenment and scientific discoveries, the *cabildos* suffered a gradual weakening of their relative autonomy. Through the introduction of the system of *intendantes,* who were often talented, vigorous, and loyal representatives of the crown, the Bourbons succeeded in centralizing political control in Madrid. Although many new *audiencias* were added (namely, increasing the territorial decentralization) throughout their rule, the authority of the *cabildos* was more strictly and efficiently controlled, a fact that was often resented by the nascent *criollo* oligarchy, especially considering that the *cabildo* was the only institution in which they were allowed to participate.

The Bourbon reforms attempted to improve and centralize control of royal authority over the entire continent. In the course of applying these reforms some moribund city corporations were abolished. To improve control over the very large areas of the colonies, the Bourbon reforms had to establish new administrative centers in the various regions. The original 16th-century structure of the two overpowering capital cities was therefore gradually transformed into one of many competing centers of power. When the imperial power split into several medium-sized cities, these could strive for greater external autonomy, each of them potentially posing a greater threat to imperial rule or calling forth the intensive use of various control mechanisms.

Thus it was the *cabildos* that constituted the major focus of any potential attempts or movements to reach autonomy. It is necessary to mention another political institution related to the *cabildos,* typical of Latin American colonial cities: the *cabildo abierto,* an open town meeting "at which resident citizenry was invited or summoned to attend, held to deal with matters and situations unforeseen by the law, to adopt policies in times of crisis, to raise troops, to receive important information and communications, to give notice of new taxes" (Pierson, 1952:590).

Controversy prevails concerning the role of the *cabildo* in general and of the *cabildo abierto* in particular, but it seems that the autonomous nature of the *cabildo abierto* as a political institution was exaggerated, it being convened to receive royal edicts, not to deliberate on policies. Nevertheless, the rebellions connected with the *cabildos* were those with the farthest-reaching repercussions.

Ironically, the colonial town, created to be a pillar of colonial power as well as an improvement on the Spanish model (which was never applied in practice), ended up in playing an essential role in ending colonial rule. "The *cabildo* was the bridge between Spain and America. . . . If Spain planted *cabildos,* she had to reap the harvest of nations" (Moore, 1954:243). And yet the nature of such attempts at autonomy had developed in a very special direction, rooted in the fact that the whole pattern of the Latin American *cabildos'* potential autonomy reflected the dominance of the core and the low level of differentiation between core and periphery in the Latin American patrimonial regime. On the one hand, minute regulations controlled almost every aspect of municipal life; on the other, we come across instances of circumvention of these regulations on almost every level.

Accordingly, such attempts at autonomy were directed mostly toward the usurpation of power from the central institutions and much less toward restructuring the mode of government, toward allowing far greater access and participation of the different groups in general, and urban ones in particular, in the reshaping of urban institutions. Basically the *cabildos* did not aim at changing the hierarchical conception of authority; at most, they would change some of the details of such hierarchy or allow a somewhat wider participation in the center without changing its structure.

The continuity of this structure could also be observed in the impact of local urban rebellions. Social and economic stress in the cities produced breakdowns, often expressed by popular rioting in large urban centers in the 17th century and in smaller towns in the 18th century. The basic roots of these disturbances greatly differed from one city to another, but most rebellions were caused by either food shortage or cruel treatment of the Indian and black populations. The deep fears of social upheavals triggered by the population of the lowest social strata and the tremendous difficulties of providing basic food products were the major issues dealt with at each *cabildo* meeting in every Spanish American city. Recurrent agricultural crises, disruption of transportation, and manipulation of the market aggravated the problems of providing food to cities. Municipal authorities were unable to cope effectively with the riots that flared up constantly and were quelled by the central or local authorities. This did not result in a permanent change in the urban society's political economic order or its standing within the general framework of the Latin American patrimonial-bureaucratic order.

CHAPTER 5

Urbanization in the Chinese Empire

We will analyze Chinese cities and urban systems as they developed within the framework of the Chinese imperial system, the most enduring in the history of mankind. This analysis will also serve as the background for a comparative study of other imperial systems that emerged in other Axial Age civilizations and the cities and urban systems that developed within them, mainly the Russian and Byzantine empires.

A
HISTORICAL AND GEOGRAPHICAL BACKGROUND

Urban development originated in Northern China during the Shang dynasty (1766-1122 B.C.) and Chou dynasty (1122-256 B.C.) and crystallized during the early imperial period beginning with the Ch'ih dynasty in 221 B.C., continuing during the Han period (202 B.C.-220 A.D.), and culminating with the emergence of fully integrated urban systems under the T'ang dynasty (A.D. 618-907). From the second half of the 8th century until imperial China reached its peak during the

Southern Sung dynasty, it underwent a medieval "urban revolution" (Shiba, 1970; Twitchett, 1966, 1968), characterized by high levels of urbanization and a large expansion in the number and economic power of small and intermediate towns. Urban development lost its momentum at the period of the Mongol invasions but revived during the late imperial Ming (1368-1644) and Ch'ing (1644-1912) dynasties. Urbanization of the late imperial period was of a lower level than previously observed, but the urban system was more mature and better integrated than in the early medieval era (Skinner, 1977:28).

Chinese urbanization processes, spanning a period of more than three millennia, were characterized by tremendous temporal and regional variations. A distinct pattern of urbanization existed, however, owing to a common geographical base—a single type of economy prevailing throughout the whole of this extensive domain—to the unifying factor of the imperial political regime, and to the continuous sociocultural system.

Developing since the 3rd century B.C. and being molded during the Han, Sui, and T'ang dynasties, the Chinese Empire possessed one of the most unique political-cultural systems of mankind. Geographically, it started along the middle course of the Yellow River (Huang-Ho) and its tributary, the Wei River, and from this core area spread out to the North China plain.

It is possible to trace the spatial expansion of the Chinese Empire by following the establishment of the *Hsien* (county) capitals. These were the garrison-administration centers established by the central government along the expanded frontier of agricultural cultivation, and their spread can provide an accurate measure of the colonization process (Chang, 1963). Through systematic colonization, Chinese civilization penetrated and occupied the banks of the Yangtze River, mainly during the Han dynasty. The southeastern province along the coast was colonized during the T'ang dynasty. The farthest provinces of Southern China—Yunan and Kwangsi—were fully colonized much later, in the 13th century. By the downfall of the Yüan Empire (1368 A.D.), the Chinese had colonized every corner of the Eighteen Provinces making up China proper, south of the Great Wall and east of Tibet.

B
THE CHINESE SOCIOPOLITICAL SYSTEM

1 Basic Premises

The political system developing in China from the Han and especially the T'ang periods was a continuous and compact imperial one and, as such, shared several characteristics with other imperial systems: the existence of a strong, autonomous center and a ruling autonomous elite—in this case, the coalition of the emperor with the Confucian *literati* and bureaucracy.

The basic features of the Chinese political system started evolving during the Han period and has crystallized since the T'ang period. The most important characteristics were the crystallization of an autonomous political center and the predominance of the emperor-*literati*-bureaucracy coalition (the military played an important role but became relatively secondary in periods of stability), and the predominance within the center of the Confucian-Legalist ideology, with a strong admixture of secondary orientations (especially Taoist and Buddhist).

The sociopolitical and cultural system that developed in conjunction with the imperial system exhibited special features that also shaped some of the specific characteristics of this, as distinct from other, imperial systems.

The Chinese tradition was probably the most this-worldly of all the Great Traditions of the post-Axial Age. Although the Confucian-Legalist framework of the imperial system allowed room for the other-worldly orientations of folk religious sectarianism or private speculation, its main thrust was to nurture the sociopolitical and cultural orders as the major focus of cosmic harmony. It emphasized this-worldly duties and activities within the existing social frameworks—the family, broader kin groups, and imperial service—and stressed the connection between the proper performance of these duties and the ultimate criteria of individual responsibility.

This tradition also emphasized a strong transcendental aspect of individual responsibility, but this was expressed largely in terms of the importance of political and familial dimensions of human existence. Moreover, as enunciated in the center's official ideology, Chinese tradition stressed the basic affinity between the societal order represented by the center and various types of peripheral collectivities. Orientation to the center and participation in it constituted an essential component of the collective identity of many local and occupational groups. All these orientations greatly influenced the structure of the Chinese center and the major elites and strata in Chinese society. The center was absolutist, though with unifying ties to the periphery, but it attempted to control the channels of this solidarity and of access to itself.

2 *Literati* and Bureaucracy

The *literati,* the major group or stratum linking the imperial center to the broader society, were very important in the crystallization of the main institutional features of the Chinese imperial system and in the processes of structuring center-periphery relations (Balazs, 1964; Ho, 1962; Kracke, 1953, 1957; van der Sprenkel, 1958). This elite was constituted by the people who passed the Confucian examinations or studied for them and comprised a relatively cohesive collection of groups and quasi-groups. These groups shared a common cultural background characterized by the examination system and by their adherence to classical Confucian teachings and rituals. This relatively widespread elite was recruited, in principle, from all strata, even from the peasantry; in fact, the literate class was mostly (though not entirely) recruited from the gentry. Its single and most organizational framework was almost identical to that of the state bureaucracy (which enlisted 10 to 20 percent of all the *literati*) and, except for some schools and academies, had no organization of its own. Moreover, political activity within the imperial and bureaucratic frameworks was a basic referent of Confucian ethical orientations (Fairbank, 1968).

The *literati* and the bureaucrats evinced specific features that are of special importance to our analysis. They were the major carriers of the Confucian (or Confucian-Legalist) world order and orientations, the articulators of the models of cultural and social order, and as such were relatively autonomous, especially symbolically, vis-à-vis both the broader strata and the political center, even if they were rather closely related to them. The *literati* were recruited and organized according to criteria directly linked to, or derived from, the basic precepts of Confucian-Legalist canon and were neither interfered with nor controlled by the broader strata of society or by the emperor himself.

They were not, however, merely intellectuals performing some academic functions; they constituted a central power elite exercising control over two crucial aspects of the flow of resources in the society: information related to constructing the definition of social and cultural worlds, and the reference orientation of the major social groups. The *literati* thus exercised a virtual monopoly over access to the center; that control, however, was not (as in Russia) based on coercion alone but also on solidarity ties that the *literati* regulated (Yang, 1959).

3 Control Mechanisms and System of Stratification

The *literati*—or rather the bureaucracy in coalition with the emperor and, to a smaller degree, with the heads of the military—constituted the group or social category exercising control over the production and flow of resources in China. The most crucial aspect of this control was the dissociation between ownership of resources—vested in the peasantry, merchant groups, and so forth—and the control of their use and conversion in macrosocietal frameworks, vested in the center.

This mode of control was also evident in the system of stratification that developed in China, whose major characteristics were (a) the development of this center as the central focus of the system of stratification; (b) the relative predominance of political-literary criteria in the definition of status (that is, the official prominence of the *literati* and officials and the growing importance of Confucian ideology in defining the criteria of stratification); (c) the relative weakening of the aristocracy and growing social and economic predominance of the gentry; and (d) the evolution of several secondary patterns of structuring social hierarchies.

4 Structure of Social Hierarchies

The Confucian ideology regarded society as a hierarchical order of four groups: scholars, farmers, artisans, and merchants. Each group was evaluated on the basis of its productive contribution to the ideal society. The highest prestige was bestowed on scholars, as they were the only ones who knew and perpetuated correct living. The farmer's duty was to provide the scholars with their livelihood. Merchants were not endowed with high prestige because commerce was regarded as "nonessential" and "frivolous" (Ch'ü, 1957; Fairbank, 1957; Pulleyblank, 1960). The actual structure of the society was, of course, much more complex and the hierarchy less clear than presented by the Confucian image.

Neither the upper nor the lower class was homogeneous. In the upper class subgroups were differentiated according to birth, level of scholarship, and political power and in the lower class according to occupation and wealth. In the upper group the criteria of subdivision created three partially overlapping categories: aristocracy, *(literati)* gentry, and officials. The prestige of the aristocracy, in relation to the *literati*, fluctuated from time to time until the end of the T'ang dynasty, when the aristocracy became less important and almost totally disappeared (Balazs, 1964; Eberhard, 1952, 1967). Within this social structure, special importance was accorded the gentry class, which comprised a stratum of landowning families fitting between the earthbound masses of the peasantry and the officials and merchants who ran, respectively, the administration and commerce (Fei, 1953). The gentry did not constitute a "feudal" class in the exact sense of the term. The Chinese peasant was free to sell and, when possible, to purchase land, both according to the law and in practice. He was earthbound as a result of many circumstances, but not through any institution similar to those that existed in feudal Europe.

Each occupational group in the lower classes possessed its own prestige hierarchy. Wealth accentuated the internal differentiation of the occupational group and influenced the status of each occupation (Balazs, 1960, 1964). "Classical" (Confucian) education was the legitimate channel of mobility from the lower to the upper class (Chang, 1955; Ho, 1959; Nivison and Wright, 1959). This path, however, was legally closed to merchants, but wealthy merchants managed to direct their most gifted sons toward scholarship. Merchants had two other means of moving into the gentry class: purchase of land and marriage. Given that farmers were free to sell their land, merchants had the opportunity to invest in land and thereby gain both a steady income and social prestige.

Legitimate entrance into bureaucracy was gained only through examinations, but office could also be acquired either by purchase or through the direct bestowal of a title (mostly to a military leader) by the emperor. This expedient varied in importance at different periods but mostly constituted an important channel of mobility (Eberhard, 1962; Menzel, 1963).

In addition to the two classes defined as being integral and necessary parts of society, a third group, castelike, consisted of "mean" outcast people (slaves, prostitutes, entertainers, government runners, servants, beggars) whose occupations were branded as useless to society (Ch'ü, 1957; Yang, 1956). These people were barred from contact or intermarriage with the lower class and were excluded from competing in the examination system, and it was therefore impossible to change their menial status.

All these characteristics of the Chinese social structure were, of course, closely interwoven in its economic system.

C
THE AGRICULTURAL ECONOMIC BASIS

From its early history, Chinese economic structure was based on agriculture, but at the same time, a well-developed commercial system evolved. The cultivated areas were the large river valleys and the surrounding hillsides, and agriculture was very intensive (Tawney, 1962). The control of water resources constituted a central problem, and agriculture was heavily afflicted by frequent floods and droughts. The connection between agriculture and irrigation created specific social needs as well as patterns of power relationships: The control of canals was an important issue in the struggle between the emperor and feudal lords. The maintenance and repair of the canals and dikes were the central administration's main duties and became symbolic of a well-organized and efficient administration (Wittfogel, 1957).

The peasant village was the organizational unit for agriculture, and it evinced great stability in its organization and characteristics, irrespective of the changes that occurred in the laws of tenure and ownership. The traditional property-holding unit was the family that either owned the land or occupied it as tenant (for the state). In the first T'ang period, the Sui's agricultural system was adopted, allotting equal-sized fields to each married couple. This tenure was, in principle, strictly bound to the demographic structure. Changes in the household—such as death, birth, marriage, and disability—entailed redistribution, which occurred annually, according to the data provided by the Households' Registers (Balazs, 1953, 1954; Friedman, 1970; Maspero, 1950; Wang, 1956).

The gentry's and officials' property rights were less restricted than the peasantry's. The former acquired their property, either as a gift from the emperor (in which case it became hereditary and tax-free), or as part of a benefice (also tax-free) bestowed accompanied by a grant of office (Schurmann, 1956).

Despite all attempts to maintain a relatively equitable type of agricultural organization, the major systems of holdings—equal allotments and so forth—usually broke down because of errors made by the administration and the rapacity of officials who wanted to extend their properties and privileges and were thus not eager to enforce laws limiting their prerogatives. Reforms of the property system accordingly formed a recurrent pattern in Chinese agrarian policy, and each new dynasty tried to tackle the land problem anew (Zen and De Francis, 1956).

From about the 3rd century B.C., land property was, in principle, marketable and the whole system of agriculture became very intensive, producing not only for subsistence but also for the regional (rural and urban) markets. These markets were also a product of commerce, which was highly developed in China, especially from the time of the Sung dynasty.

The Chinese imperial system thus evinced two basic characteristics that were very important for purposes of our analysis. The economic system, based on intensive marketable agriculture and a highly developed commerce, generated a network of wide-ranging and multiple economic markets. The relatively open political and cultural system also generated the emergence of political and cultural outlets that gave rise to relatively wide and cross-cutting institutional markets. The center, however, exercised a complex system of control over these markets (Willmott, 1972).

It was within the framework of the interaction between these two major characteristics of the social structure that various aspects of the urban structure emerged: different types of cities and special urban systems.

D
THE URBANIZATION PROCESS

1 Basic Components of the Urbanization Process

The urbanization processes that evolved in China were strongly influenced by its geopolitical setting and the different constellations of its forces of concentration and centrality. The early stages of urbanization can be traced to the Shang dynasty period (ca. 1766-1122 B.C.). Archaeological evidence, mainly from the Anyang area on the banks of the Huang Ho (Yellow River) in the Northern Provinces, indicates the existence of a well-developed civilization by the middle of the second millennium B.C.

The first component of the urban system was the establishment of one capital city dominating and controlling the entire kingdom. The second component was the founding of military-administrative centers (following the kingdom's territorial expansion), as new areas, spreading out from the core region of the Northern Provinces toward the Yangtze and the Southern Provinces, were cultivated, guarded, and serviced by the central government authorities. On all regional levels (including even the county), administrative-military urban nuclei were thus established and, on the county level, were designated as *Hsien* capitals. After the Chou period these urban centers became the backbone of the entire urban system, and their spread throughout agrarian China can be regarded as a process of spatial diffusion, of proliferation of new rural counties. Each of these constituted an almost self-contained economic as well as political unit, and virtually all central functions were concentrated in the *Hsien* capital (Skinner, 1977:18). The establishment of this administrative urban hierarchy is a supreme example of the operation of forces of centrality as a stimulus to urbanization processes.

2 The *Hsien* Capitals

The *Hsien* capital was the administrative center of the county—the spatial representative, as it were, of the central government authority that reached out to every province and county of the vast Chinese territory. With expansion of the frontier regions to the south and west, new *Hsien* capitals were established until a complete urban system of about 2000 administrative centers covered the entire territory of China (Chang, 1963:124). These small- to medium-sized towns were usually square-shaped, with double walls. They fulfilled a defensive role—guarding the area against waves of invaders during the feudal period—as well as an administrative one of tax collection, military garrison, and dispensing public services during the imperial periods (Balazs, 1964:67).

The colonization process, however, was not based only on military-political conquest. It was also one of continuous extension of agriculture that, in turn, shaped the processes of concentration emerging within it. In colonizing their subcontinent over a period of 3000 years, the Chinese did not regard the city as the prime instrument of their penetration into new territory. The expansion of agriculture, centered on peasant villages, constituted their main vehicle. At a later stage, when the land was tilled and the rural population became larger and more stable, one of the stockaded villages was reconstructed as the center of imperial authority and provided the nucleus of an urban center (Wright, 1977:33).

Thus the development of urban systems in China was also strongly linked to the establishment and expansion of agriculture and to the emergence and continuity of a free peasantry, constituting the basis of a producers' and consumers' market. The transformation from peasant marketing system to urban central place hierarchy that developed in China was effected through the increasing population density and intensive cultivation, which led to the establishment of new markets and to the growth of the existing ones (Skinner, 1964-1965). The markets continued to grow, and within the framework of a given set of centers, the larger ones added more days to their periodic schedules. Finally, in a process extending over a few centuries, the markets stabilized and their periodicity ended. Further growth in population density was accompanied by an increased number of permanent market centers and by reduction of the distance separating them. The total volume of trade augmented, as did the number of marketing hours per week. Permanent establishments replaced the mobile stalls, and the degree and scope of economic specialization also increased.

The linkage of peasant households to growing urban markets for food and handicrafts as well as to urban sources of exotic goods also changed the nature of the central place hierarchy. Successively higher levels of centers in the hierarchy of markets were characterized by a larger volume of trade, more marketing hours per week, a higher proportion of permanent as compared to mobile units, and an increased degree and scope of economic specialization. This process of transformation involved the extension of these characteristics to the lower-level market centers. In the late 19th century, influenced by the extension of urban demands over the countryside and by the reduction of household self-sufficiency, the peasant economy likewise became gradually commercialized, increasing the volume of marketing done by the household. The outcome of this slow process within the pattern of spatial organization was that the marketing system operating in trading areas of the major city and in provincial capitals became commercialized. The system of periodic markets was then transformed into a stable, central place hierarchy. Needless to say, there were continual changes in the patterns of such hierarchies and in the rank of different places within them. Some of these changes were related to transfers of capital cities, others to processes of political superiority, and still others to shifts in the centers of economic activities (Rozman, 1973).

E
POLITICAL-ADMINISTRATIVE AND ECONOMIC URBAN SYSTEMS

1 Introduction

From the combination of these forces of centrality and concentration and the processes they generated developed the major characteristics of the two urban hierarchies (political-administrative and economic) and their interrelations and interconnections. The main features of this interconnection were the strong, continuous interweaving of the political and administrative urban hierarchies and the lack of any distinct religious or cultural hierarchy.

As early as the Shang (1766-1122 B.C.) and Chou (1122-256 B.C.) periods, the beginning of a commercial hierarchical system, based upon the administrative hierarchical structure, can be identified. The administrative structure provided the spatial framework for the urban system stimulated by commercial development, and it ensured safety and security for the buyers and sellers moving among the various centers. With the full development of this system, imperial rule supplied security for the strong spatial interactions among the centers and allowed merchants in the towns at the lower level of the urban hierarchy to utilize the benefits of economic concentration. The marketing system's prosperity was ensured by the imperial government's political stability, whereas the relative autonomy of the rural society encouraged the peasants' full participation in the economic process of marketing products and buying goods, thus providing a continuous stimulus for the activities of the urban commercial centers.

The combination of forces of centrality and concentration was evident in the fact that the *Hsien* capitals and, of course, all the higher-ranking cities in the administrative hierarchy by decree were market towns, fulfilling a major role in the economic development of each region. Beyond this common characteristic, however, at different periods and in different regions of the Chinese Empire the urban system was shaped by the relative strength of the forces of centrality and concentration. Accordingly, the precise shape of this interweaving, the continuous relative predominance of forces of centrality within it, and their impact on the structure of cities and urban systems can be understood only by analyzing the special characteristics of the forces of centrality and concentration developing in China.

2 The Capital City

The effect of forces of centrality on constructing cities and urban hierarchies in China was most clearly manifest in the establishment of main capital cities. The combined forces of centrality and concentration were revealed mainly in the emergence of different markets and the setting of capital cities at all levels: country, province, prefecture, and district. These capitals—walled towns, symbols of power and means of defense ("Ch'eng" means both city and wall in Chinese)—represented the might of the empire at every level (Chang, 1977:76).

Atop the administrative hierarchy stood the capital city, the emperor's seat, the supreme political and spiritual authority of the empire. The following characteristics of the capital city are of special importance to our discussion: the continuous shift in their location; the importance of the geopolitical and economic considerations causing these shifts; the predominance of cosmological-political considerations in the construction of those cities; the fact that their location was determined by the empire's political center and by the emperor's personality and not by the geographical or economic importance of any particular site; and, finally, the fact that the shift of the capitals did not affect the relative stability of the urban system as a whole (Wright, 1965).

The transfers of capitals from one location to another can be traced from the beginning of Chinese history and are closely related to strategic and economic reasons, to the ascent to power of new dynasties, to changes in the external political situation and in the spatial economic base, and to the progress of new technologies and production methods. It was not until late Shang times that a capital city began to assume regional importance. In 1401 B.C., Anyang was established as the capital of the Shang Kingdom and located in the fertile loess soil of Honan on the Huang Ho, the major artery of North China. Anyang was situated in a central position within the Shang Kingdom and reflected the central political and economic role of the country's capital city. The growth of Anyang and of later capitals demonstrates the fact that the larger the territorial extent of the kingdom, the stronger was its central authority and the more prominent its capital city.

3 "Gate" and "Heart" Functions

The transfer of the capital from Anyang to Ch'ang-an and later to Loyang, back to Ch'ang-an, and thereafter to several other sites, previous to Peking's emergence as the imperial capital in the 13th century, reflects the vacillating role of the capital between the "gate" function (Ch'ang-an, Hangchow, Peking) and the "heart" function (Anyang, Kaifeng, Nanking), between an inward-looking and outward-looking orientation of the political regime. This emphasized the role of the capital in blocking incursions—especially of nomads—from beyond the borders, or stressed the links of the existing rulers with their former homeland, as in the case of the Mongols (Yüan dynasty).

The shift of the capital to Ch'ang-an and Loyang also reveals the tension that existed between rival geopolitical and economic roles of the capital city. China expanded and prospered, but threats from nomads in the north and northwest increased. At the same time, the country's economic center moved steadily toward the south and east. Location of the capital at Ch'ang-an reflected a geopolitical orientation toward the northern border, whereas the Loyang location indicated an administrative and economic orientation toward the major concentration of population and economic activities within China. Ch'ang-an remained capital of the empire for 530 years, in the period between the end of the Chou dynasty (256 B.C.) and the fall of the T'ang dynasty (A.D. 907), emphasizing the major importance of the western gate to China at that time. It was not until the economic heart of the country shifted to the Yangtze Valley that Ch'ang-an lost its primacy as capital (Tregear, 1965:94).

Transfers of capitals also reflected changes in the territorial base of the political power. This was evident in periods of political strife and disunity, when China was divided into several kingdoms of regional nature. This fragmented structure prevailed in the period between the fall of the Han dynasty (A.D. 220) and the emergence of the Sui dynasty (A.D. 589), which reunited the entire country under a single imperial regime. The T'ang dynasty ruled over all of China between the beginning of the 7th century and the beginning of the 10th century and reestablished an imperial regime centered in the capital city of Ch'ang-an on the Wei River (Wright, 1967).

Between the T'ang and the Yüan dynasties (A.D. 907-1279), China underwent another long period of disunity and political fragmentation. This was expressed by the absence of even a single primate city; its place was filled by a number of regional capitals of similar magnitude and importance. The Yüan dynasty resumed the imperial regime and placed its capital in Peking, which retained this role throughout most of the Ming and Ch'ing dynasties until the present.

With the exception of Hangchow, capital of the Southern Sung dynasty (1127-1279), it is significant that none of the capital cities was ever located on the coast. This is due to the basic continental orientation of the Chinese Empire, which ruled over a huge continental mass, was economically and culturally self-contained, and in many respects constituted a closed market of vast proportions. Construction of the capital cities was based on a religious-political orientation according to which the city was regarded as the symbol of the cosmic order (Wheatley, 1971:477-482). In this respect capitals differed greatly from other types of cities that developed through the impetus of forces of concentration.

F
CAPITAL CITIES AS CENTERS OF ADMINISTRATIVE AND ECONOMIC CONTROL

Capital cities were not only symbolic religious centers but also foci of political power and control and centers of administration. In the capital the imperial bureaucracy—which served as the chief agent of control and as the focus of the status orientations of the entire population—was concentrated and controlled. The major spatial locus of such control, beyond the capital city itself, was the administrative town that represented imperial authority on all regional levels, from the county capital (*Hsien*) to the provincial capital. The population of the *Hsien* was well aware of belonging to it, the urban capital of the *Hsien* and its surrounding rural hinterland constituting the basic spatial unit of government and economic activities.

The basic subdivision of the empire into *Hsien* was recorded in the enormous compilation of *Hsien Chih*—local *Hsien* histories. From these detailed records it is possible to reconstruct the evolution of the administrative urban system and the function of towns at each level of the administrative urban hierarchy (Chang, 1961:39).

Although there was some disagreement among scholars concerning the degree of overlap between the commercial and administrative urban hierarchies, it is possible to analyze their interactions systematically, especially on the basis of Skinner's most comprehensive research on this subject (1964-1965, 1977). We can conclude therefore that throughout the urban hierarchy, the administrative and economic networks of cities in traditional China were not entirely independent of each other. From the bottom of the scale to the *Hsien* capital (the lowest echelon of the administrative network), the two networks merged into a single one and formed a "natural" network of central places, growing in number as it ascended the urban hierarchy. From the level of the *Hsien* capital, however, an economic network with different structural features was imposed from above and expanded "artificially" from the center toward the periphery (Skinner, 1977:275-346). In spite of the structural differences in the two networks above the level of the *Hsien* capital, there appears to have been a high correlation on these levels between the economic and administrative status of the cities.

G
THE ADMINISTRATIVE URBAN SYSTEM

The administrative system was a complete urban hierarchy composed of capital cities that were three levels of administrative units below the imperial capital—that is, the provincial capital, the prefecture capital (*Fu*), and the county capital (*Hsien*). The hierarchical order of the provinces, prefectures, and counties formed the standard administrative structure throughout the Chinese Empire. This hierarchical structure was nested, meaning that the territory of a province was completely covered by the prefecture-level units and the territory of a prefecture was covered by county-level units. Every spatial unit, every village and town was thus unequivocally controlled by a capital city within the urban field administration. The main tasks of the field administration at all levels, from county to province, were to ensure the regular flow of revenue into the imperial capital, to defend the various parts of the realm against internal and external enemies, and to promote social order. In the densely populated regional core areas, the imperial administration was involved mainly with taxation; along the regional and national frontiers, the main preoccupation was with defense and security. In both cases, however, the imperial administration had to obviate the concentration and consolidation of local power that might pose a threat to the central control (Skinner, 1977:307). This crucial point greatly influenced the relations between the administrative and economic urban systems.

Evolution of the administrative urban system resulted from the spread of central authority throughout agrarian China and can be regarded as probably the purest example of an urban system shaped by forces of centrality. As the creation of administrative towns followed agrarian colonization, the regions that were colonized earlier developed an administrative system before those colonized later. It is therefore possible to identify a clear north-to-south sequence of evolution of administrative centers (Chang, 1963). This geographical sequence has great historical significance, as it reflects the developmental process of an urban system that originated at the core and spread to the periphery, applied on both a national and a regional scale. The establishment of new walled towns (*Ch'eng*) to serve as political-administrative centers identified the frontier, the new periphery at a particular period of time, and was carried out at the initiative of the center, thus legitimizing and enforcing the central authority's control over the periphery.

Due to the ongoing colonization process of the Chinese subcontinent and China's continuous population growth, it might be assumed that the size of the administrative urban system would increase to accommodate territorial expansion and population growth. It is therefore rather surprising to discover that the total number of county-level capitals (*Hsien*), which constituted the cornerstone of the administrative system, remained almost constant throughout imperial history.[1] As the frontier was pushed forward and new counties were instituted, it can be concluded that the number of counties in the already settled areas constantly diminished (Skinner, 1977:19). We witness the continuous consolidation of county-level units in the national and regional cores. The amazing fact that the number of administrative cities remained more or less constant leads to the far-reaching conclusion that the effectiveness of the administrative urban system declined continuously from the T'ang period to the end of the imperial period, through steady decline in the functioning of the basic-level administration (Skinner, 1977:19). This "freezing" of the administrative urban system can be explained on the grounds that any significant enlargement of the system not only would have meant proliferation of the urban bureaucracy (*yamen*) but would also have loaded communication facilities to the breaking point and have posed control and consolidation problems whose solution would have been beyond the capabilities of any premodern agrarian society.

With the level of administrative cities being kept almost constant, on the one hand, and the continuous growth in population and economic activities, on the other, increasingly more nonadministrative urban centers emerged. This was particularly the case in the lower levels of the economic urban system—that is, the standard market towns, the intermediate market towns, and the central market towns, the vast majority of which were not included in the administrative urban system. This brings into focus the differences and relations between the two urban systems, which we will now examine.

H
RELATION BETWEEN ADMINISTRATIVE AND ECONOMIC URBAN SYSTEMS

The relation between the administrative and the economic urban systems underwent significant changes throughout the imperial periods. In the early period, until the early T'ang dynasty, the county (*Hsien*) was an economic and administrative unit, and central functions, both economic and administrative, were monopolized by the county capital (Twitchett, 1966:226). Government regulation ordered that an officially controlled market had to be located in the county capital and none was permitted outside it. The economic transactions thus were regulated by the political-administrative structure, representing the dominance of the centrality principle over the forces of concentration.

The official market system of the early T'ang period, based on the county capital and the higher levels of the administrative system, was an elaborate apparatus for regulating the empire's commercial activities (Twitchett, 1966:207). In capital cities, at all levels, the official markets were walled, forming an integral part of the walled-ward system, which enabled officials to control and regulate all facets of the commercial transactions carried out within the clearly demarcated market area. The market director, a bureaucrat of rank in field administration, assumed responsibility for all aspects of commerce.

The county capital—the backbone of the administrative urban system—usually functioned also as the basic unit of the urban economic system. Due to the concentration of unproductive consumers in it, to its nodal position in the local transportation network, and to its strong administrative control of commerce, the county capital became the center of collection and distribution of local produce and constituted the connecting link with the trade flows emanating from large distances. These conditions facilitated the evolution of an economic central place system, emerging at the lower levels and developing to its full extent at the higher levels of the economic urban system.

The gradual social and political processes that occurred during periods of change, beginning with the late T'ang and culminating in the Southern Sung, significantly altered the relation between administrative and economic urban systems (Shiba, 1970; Twitchett, 1966). These changes were described by Elvin as a "medieval revolution in market structure and urbanization" (1973:chap. 12). Their basic features were the following: (a) relaxation of the rule that each county could maintain only one market, which had to be located in the capital city; (b) breakdown of the official marketing organization; and (c) gradual disappearance of the enclosed marketplace and its replacement by a less rigid and confined pattern of commerce that could operate almost anywhere (either within the city or in its outlying suburbs). These changes were accompanied by an increased volume of intra- and interregional trade, by increased monetization of taxation and commerce, by the growth of financial institutions, by a marked increase in the number and wealth of the merchants, and by a softening of the official attitudes that disparaged trade and the merchant class (Twitchett, 1968:74-95). The new urban system that evolved in the medieval period provided the spatial and organizational framework for economic and commercial expansion during that period. With the number of administrative centers remaining constant, urban growth, based on economic development, took the form of numerous new marketplaces, towns, and cities of a nonadministrative nature. The urban systems that emerged during the medieval period included about 6000 central places as opposed to 1500 during the early T'ang. The innovation was that the vast majority of these new urban places were not capital cities of any kind. A new system thus matured, based mainly on the forces of economic and demographic concentration.

Since the medieval "urban revolution" in the T'ang-Sung period, it was possible to distinguish between the two hierarchies of central places and their associated territorial systems—that is, between a formal urban system created and regulated by the imperial bureaucracy for purposes of political control and field administration, and an informal system created through economic transactions. The first reflected the bureaucratic structure of imperial China; the second, the economic and social processes of Chinese society.

All capital cities fulfilled significant economic functions for their hinterland, and almost all those at the upper levels of the economic hierarchy were capital cities. In the late imperial period, cities of the first (central metropolis) and second levels (regional metropolis) were capital cities within the administrative urban system. Almost 90 percent of the cities of the third level in the economic hierarchy (regional city) and fourth level (greater city) were capital cities, whereas those of the fifth level (local city) corresponded quite well to the county capitals, more than 75 percent of which were *Hsien* capitals. Below this level of the economic hierarchy occurred the break with the administrative system, as the vast majority of towns of the three lowest levels—the central, intermediate, and standard market towns—were nonadministrative centers (Skinner, 1977:390).

We can conclude, therefore, that the further we study the imperial system, we find that the forces of concentration were quite strong and effective in shaping the urban system until the late imperial period, when the administrative capitals became a subset of the economic central places. The administrative system fit quite well within the upper levels of the economic urban hierarchy, but it was absent in the lower levels of the economic urban system. The two urban systems merged at the county capital level, which in most cases fulfilled the economic role of local city. This feature enables us to regard the *Hsien* capital as the locus of articulation between the official structure of imperial control and the unofficial one of societal management and economic activities.

Despite this merging of the administrative and economic systems at their upper levels, the boundaries of the administrative units seldom coincided with the marketing areas of the same cities. Each city thus fit into two distinct zones of influence. This spatial discrepancy between the two systems was instrumental in preventing the local gentry and merchants from becoming more independent of the bureaucratic government. In areas where the concentrated local power constituted the greatest potential danger, administrative boundaries were deliberately fixed in such a way as to divide the local economic power. Provincial boundaries were so drawn that the regional core was divided among two or three provinces in order to split the metropolitan trading area between several administrative units (Skinner, 1977:343). The nonoverlapping of the administrative and economic territorial division resulted in fractionizing a region's powerful gentry into competing elites and sharply differentiating the hierarchy of administrative systems—to which each group of provincial gentry was related—from the economic territorial hierarchy that furnished power to the merchants.

China clearly provides an illuminating example of a high degree of fusion of the two urban systems—the administrative and the commercial—but not of total identity between them. The administrative urban system was the relatively dominant one, serving as the spatial locus of the function of control determined by the political-administrative-cultural elites. But this control could not eradicate the vitality and relative autonomy of the forces of concentration and of the economic urban system created by it. It should also be noted that given the complete identity of the cultural and political elite in China, a distinct religious or cultural urban system did not have to evolve there.

Both China's administrative and economic urban systems were usually characterized by a transition from a primate system—which developed in the early periods of the Ch'ih dynasty—to relatively full-fledged rank-size systems that flourished especially during unification of the empire, while in times of disintegration they tended to become truncated.

I
INTERNAL STRUCTURE OF CITIES

The two types of cities that developed in China—the political-administrative cities headed by the capital city, and the economic cities—differed greatly in their internal structure. The large capital cities of imperial China, which were generally imperial or provincial capitals, were the seats of governors, officials, subofficials, and their service suppliers. Commerce flowed from the imperial capital toward the imperial court, its servants, and the existing structure of officials, soldiers, and the like (Balazs, 1964:79-101). The administrative-political function of the capital cities at all levels, as the representatives of imperial power and authority, caused most of them to be built according to cosmological principles derived from Han Confucianism (Wheatley, 1971). They were always walled cities constructed according to a preconceived plan, in a regular and formalized pattern.

The cosmological principles applied to the Chinese cities were made up of elements drawn from the core ideology of the "Great Tradition," sometimes enriched by borrowing from the peasants' and artisans' "Little Tradition," but scarcely by the addition of elements from alien civilizations. This core ideology was perpetuated by the *literati* elite, but it would seem that the cosmological ideas of the elite in regard to cities were shared as well by other groups in the society. Artisans, anonymous perpetuators of a continuous architectural tradition, were responsible for applying the cosmological ideas to city building and undoubtedly influenced them in many ways (Wright, 1977:34).

The following short review will trace the evolution of the cosmological principles applied to the Chinese cities with respect to their location and structure. The spatial structure of the Chinese capital city was conceived as an ecological symbolization of the cosmic order. The tendency developed to reproduce a minor model of the cosmic order on the city site. The ancient Chinese city was thus founded as an *axis mundi*, an *omphalos* that symbolized the powerful centripetal forces active in the universe. Most of the important buildings and structures of the Chinese city were erected along the central axis known as the "Celestial Meridian Writ Small" (Wheatley, 1971:456). The ceremonial center was the primary and most ancient focus of power and authority in Chinese society. It was the material expression of the religious concept of the "general order of the universe," of the "sacred." The cosmic order, as materialized in capital cities, provided the framework for social action and had a definitive integrative function. By submitting to the authority of the deity whose dwelling was the ceremonial center, peasants and artisans were subjected by that very fact to the economic control of the priesthood. That group was concerned with the material aspects of religion and represented the population's interests vis-à-vis the deity.

In building new cities, the ritual-symbolic practices were already beginning to emerge in the Shang and Chou periods, and it is possible to discern several elements that were of ritual-symbolic significance in the cities of these early times. The first practice was to place the city in precise alignment with the direction of the four winds; the second was to build the city walls in the form of a square or rectangle, in accordance with the four main directions. South being ritually favored, cities faced south, and their main gate was located in the south wall. The third practice was to choose the site of a new city by means of divination, chiefly through the oracle of bone inscription (Wright, 1977:38).

Among the innovations emerging during the Eastern Chou period (mainly since the 8th century B.C.), was the articulation of city plans into functional areas. In the Eastern Chou cities, a central area, usually walled, contained the places and main buildings used by the nobility. Surrounding this central core was a second walled area accommodating industrial and artisan quarters, popular residential quarters, commercial streets, markets, and some farmland. This double enclosure, which became a cornerstone of the normative theories of city building that developed in later periods, exemplifies the effect that the forces of centrality had on the internal structure of Chinese cities.

The Chou cult of Heaven worship was practiced with solemn rites on a round altar located in the favored southern side of the capital city. Later an altar for Heaven worship became a common feature in all capital plans. Religious legitimation was bestowed on the ruler—the emperor—not on a specific site, and a place derived its sacredness from the fact that the emperor resided in it.

During the Han dynasty a systematic urban cosmology emerged, which prevailed throughout the imperial period and was a synthesis of the early Shang and Chou traditions in city building and the newly established Han Confucianism. The new cosmology was based on four principles (Wright, 1967, 1977):

(1) *Archaism*—an elaborate scholastic system, encompassing ancient venerated symbols, ideas, and institutions in order to gain popular acceptance
(2) *Organicism*—the interaction of all human and natural worlds and elements
(3) *Centralism*—the centrality of the emperor in the world of men and the centrality of China, the Central Kingdom, in the universe
(4) *Moral justification*—the belief in the emperor's moral right to rule and the moral rightness of the social hierarchy

These four principles affected the cosmological significance of town building and the internal structure of capital cities from the Han period throughout the duration of the Chinese Empire. According to imperial conception, the site of the capital was the epicenter of an orderly spatial grid extending to the boundaries of civilization. This was the perfect expression of the principle of centralism. The precise orientation of the city walls according to the main directions (indicating the forces of nature controlled in the interests of the whole realm) reflected the principle of organicism. The square shape of the city was in accord with the principle that the emperor, ruler of all under Heaven, should live in a structure that was a replica and a symbol of earth itself.

The city was divided into quarters by nine streets directed north to south and nine streets directed east to west, with a corresponding number of gates in its walls. This element reflected the archaic theory of emblematic number, developed before the imperial period but incorporated into Han Confucianism. The disposition of the principal structures within the city reflected the principle of centralism. The palace of the emperor or prince occupied the very center of the city; the audience hall was south of the palace, and the market was north of it. This disposition—the assignment of a location of least honor and minimum positive influence to mercantile activity—reflected the value system of Han Confucianism.

After the Han dynasty the capital cities of imperial China incorporated many of these cosmological attributes in their location and internal structure. The last example—and perhaps the most complete application of the symbolic principles in building a capital city—was Peking, reconstructed in 1421 as capital of the Ming dynasty in closer accord with imperial cosmology than the capitals of the Sui, T'ang, and Sung dynasties. Both in its details and broad outlines, Peking conformed to the most ancient precedents of cosmological significance. In general, the internal structure of the major as well as the smaller capital cities was based primarily on such cosmological considerations and only secondarily on economic ones.

At the same time, the religious-cultural and political role of the city was not linked to any particular place; it was the seat of the emperor, who endowed a particular site with its religious role and meaning, symbolized in the city's structure. Frequent transfers of the capital city did not hamper its fundamental religious-cultural role.

In contrast to the administrative capital city, shaped mainly by political-religious forces of centrality, the economic city was designed by the commercial and economic forces of concentration. In most cases the economic city retained the characteristic street pattern of the unwalled market town. Its shape was irregular and its layout informal. The streets curved along waterways and old roadways, around mounds and hilltops, and skirted public buildings such as temples, which might have existed before the town expanded. In the economic cities that were not strongly influenced by the forces of centrality, the cosmological-ritual component was of little significance in shaping their urban structure, but the following factors were of importance in this respect: convenience of transportation of people and goods; accessibility between the city and its trading areas and cost of land—determined by the urban accessibility pattern—and the level of amenities available in different parts of the city.

Thus we can identify the two "ideal" types of Chinese cities: the capital city, planned, regular, and formal, and the "economic" market town, unplanned, sprawling, irregular, and unconfined by walls. It is sometimes possible to trace a hybrid type in which some degree of planning has been superimposed on an economic city but too late to have any decisive influence (Mote, 1977:107).

It is significant that North China's cities tended to be of the planned, regular type, reflecting the fact that most of them were established in the preimperial and early imperial periods, mainly as administrative centers. By contrast, a much higher proportion of South China's cities were of the unplanned, irregular type, as most of them were established during the medieval and late imperial periods when the urban economic system became predominant and most of these towns were of nonadministrative origin.

J
URBAN AUTONOMY

1 Urban Administration

Some of these basic characteristics of the Chinese political system and social structure that defined some crucial aspects of the structure of urban hierarchies and of the internal structure of cities also explained the nature of the internal social structure of Chinese cities and the degree and type of their autonomy. Compared to most other societies or civilizations analyzed in this book, an almost total absence of any kind of urban autonomy was observed in China, together with a strong development of different types of merchant associations or guilds that were tightly controlled by their own members and the government. This most striking characteristic of urban China—its almost total lack of urban autonomy—was thus described by Mote: "The Chinese city had no citizens, it possessed no corporate identity and never had a government distinct from that of the surrounding countryside" (1977:114). China never developed the concept of a city as an independent political unit or corporate social identity. The city was always regarded as the administrative center of a region. There was no concept of the city as a municipality with its self-government, and no concerted political action was taken or city dwellers organized.

The internal urban administration was subordinated to the central authorities that appointed the officials in charge of each quarter. As it was not self-administered, the city never had its own laws or charter and could not fight for its rights against the central government (Eberhard, 1956:267). Chinese cities did not have a town hall, there being no need for a place in which citizens could assemble and exercise their political rights (Mote, 1977:115). The city, however, did not have to defend itself; its defenses were built as part of the nationwide defense installations by the central government, to which all citizens were equally subservient.

There were no specific or autonomous municipal institutions in the traditional Chinese city, but several voluntary organizations were connected with different quarters of the city. These organizations had no administrative function, and the zonal organizations within the city did not operate on the overall urban level. From the administrative point of view, the city did not constitute an independent unit but formed an integral, inseparable part of the region. In this respect, however, the imperial capital was a conspicuous exception in having its own independent administration that served it exclusively; it was not subordinate to the regional authorities but was controlled by the central bureaucracy (Eberhard, 1956:265).

The strength of these factors was also manifest in the specific nature of the central administrative control of urban economic activities. Administrative control of the city was very encompassing, including taxation of markets, arbitrary fixing of prices and rates of interest on loans, and taxation on loans and contracts. The severity of state control was well illustrated by the limitations set on the production of certain kinds of textiles and by strict supervision of the type of garments that various urban strata were permitted to wear (Eberhard, 1956:267).

The strict control of the markets during the T'ang period included evening curfews, officials stationed in the markets, prohibition of transactions outside the market area, and the imposition of a system of fixed prices. Shopkeepers and middlemen were required to keep exact and detailed records of all transactions and to submit them to the local official at month end. Special rules covered the transaction of commercial activities. The vast majority of these rules were applied through middlemen, and the law prescribed that important deals be conducted only through licensed middlemen. Every application to the local government for a middleman license had to be accompanied by a guarantee of liability from another established middleman. Moreover, this license was revoked if its holder was found guilty of making unfair appraisals, embezzling, or exacting unreasonable commissions.

2 Merchant Guilds

The lack of a public law system, apart from the imperial law that protected only landed property and bureaucratic status, left the merchants with neither legal protection for their property nor security when undertaking hazardous economic transactions. To the extent that merchants managed to obtain concessions from the central administration, even these achievements were not accorded legal status; it was possible, at the stroke of a pen and by means of an arbitrary governmental decision, to abolish the merchants' modest accomplishments. Paradoxically, the necessity of having to function under such circumstances brought into being new types of merchants' and artisans' associations or guilds; because one of the main reasons for their foundation was the need to resist administrative pressures, these organizations were forced to impose new forms of social control on their members (Van der Sprenkel, 1977).

The establishment of guilds, however, served the interests of both the government and merchants and artisans. Merchants believed they could best defend their professional interests by organizing into corporations, and the political authorities benefited by presenting their overall demands directly to representatives of the guilds. By regulating these organizations the authorities could exercise tight control over the merchant strata. Peter Golas (1977) points out three basic differences between the Chinese and the European guilds: (a) the much more subordinate position of the Chinese guilds vis-à-vis the political authorities, (b) the greater emphasis placed by Chinese guilds on a common geographical origin as a cohesive force, and (c) the far lower incidence of clashes and battles among the guilds in China.

Strong solidarity was demonstrated in these guilds, often based on common geographical origin, traditions, and customs. In addition to the guilds' stress on consensus, the high value placed in China on respect for authority was also a strong cohesive factor. Most of the guilds were small associations, seldom numbering more than 30 members, each of whom, under such conditions, knew all the other associates.

Strong restrictions on group memberships existed, however, and during the late Ming and early Ch'ing periods these were limited to people having a common geographical origin. Most Chinese guilds of the time grouped people who were native of an area other than the city in which the guild was located. This phenomenon was related to an ancient Confucian political-cultural orientation that, in order to prevent favoritism, gave preference to bureaucratic service in an area removed from the one of permanent residence (Golas, 1977).

The tendencies described above were reinforced by the general characteristics of political control and supervision, which tended to explain why merchants' wealth was so often transformed into land or exchanged for official titles. In contrast to Medieval Europe, most real estate—urban land, industrial and commercial buildings, workshops, and shops—was not privately owned but was state property (Balazs, 1954). By these means, the merchants' dependency on the central administration was reinforced. Guilds and corporations in traditional Chinese cities were forced to provide the government with the demanded quantity of products at lower than market prices, including payment for transportation.

Moreover, it was most significant that many merchants (even if dissociated from the gentry class by their origin) aspired, after accumulating some wealth, to acquire an official title for themselves or their sons and thereby enter the ranks of the bureaucracy (Eberhard, 1956:267). Merchants' inability to transform their economic power into political influence was a result of the vicious circle created by the systems of control, in general, and stratification, in particular. These systems prevented the emergence of a politically powerful autonomous class of merchants, which, it was reasoned, could destroy the traditional stratification system upon which the imperial structure rested.

This tendency was strengthened by the fact that, contrary to Medieval Europe, the bulk of government revenue was derived from land taxes. Merchants therefore possessed no clear leverage for obtaining concessions. In rare cases in which the central administration borrowed money from merchants (usually in periods of crisis), this was regarded as a "compulsory loan imposed from above" rather than as one granted by the private sector under favorable terms (Balazs, 1964:77).

Our analysis indicates that the Chinese city—as splendid, majestic, and rich as it could be—never became the focus of a social or cultural creativity distinct from the major forces of centrality—or, to express it in another way, an autonomous force of centrality having its own specific characteristics and dynamics. In contrast not only to Europe but also to Japan, a specific bourgeois class consciousness or ideology never emerged, and this can be observed in two major negative aspects of city life. The first was the lack of any particular civic consciousness; the other, contrary to other civilizations, was that even in periods of the central government's decentralization and decline, the city never usurped power to such an extent that it could compete with new dynasties, warlords, and the like.

NOTE

1. That is, 1180 *Hsien* in the Han period; 1255 in the Ch'ih period; 1235 in the T'ang period; 1230 in the Sung period; 1115 in the Yüan period, 1385 in the Ming period; and 1360 in the Ch'ing period (Whitney, 1970:75).

CHAPTER 6

Urbanization in the Russian Empire

A
INTRODUCTION

In Chapter 5 we analyzed the characteristics of the Chinese Empire's urban systems and cities. Our analysis was of interest not only in its own right but also as a comprehensive study of cities and urban systems in imperial regimes.

We will now analyze the Russian and Byzantine imperial systems and, subsequently, will touch upon some imperial systems that developed within the realm of Islamic civilization. These shared some characteristics with other imperial systems: an autonomous center that attempted to penetrate the periphery according to its own basic premises, relatively wide institutional markets, and autonomous political and cultural elites that carried and articulated some strong transcendental orientations and exercised relatively flexible control over the production and flow of resources in these markets. They differed from the Chinese Empire, however, in the concrete manifestation of their common characteristics, the predominant cultural orientations articulated by their respective elites, and their economic foundations and political ecological settings.

These differences have greatly influenced some of the most important aspects of the structure of their cities and urban systems. In the following chapters we will analyze some of these aspects—especially those that are most pertinent to our analysis and that have adequate documentation.

B
HISTORICAL BACKGROUND

1 Pre-Kievan and Kievan Periods

The political development of Russia, characterized by transition from a relatively decentralized, semifeudal system to a highly centralized imperial one—albeit based on a relative heterogeneity of elites under the ultimate absolute predominance of the political system—has influenced several crucial aspects of its cities and urban systems.

Urbanization processes emerged in the 8th to 9th century, during Russia's pre-Kievan period. The first towns then established were located on a north-south line from Novgorod to Kiev. These towns were the outcome of the consolidation of eastern Slavic tribes and the emergence of multitribal political centers; they were of very small-scale with a limited economic base. In the 9th and 10th centuries a particular urban form emerged: the fortress town. The fortress (*kreml*) towns were established along the Novgorod-Kiev axis and fulfilled the function of collecting tribute from the rural population of the area for the local ruler. Most of these towns lacked a stratum of merchants and artisans and were sustained by locally produced agricultural foodstuffs. A few of the fortress towns began to include a *posad*—a nucleus of commercial and artisan activities. Cities possessing a *kreml* and a *posad* were mainly those located along the navigable rivers, and they benefited from better access to the products of larger areas and, eventually, from foreign trade. Throughout the pre-Kievan period, fortress towns were highly independent of one another, functioning as centers of almost entirely self-sufficient local areas; they were scattered throughout Russia without unified control and did not form an urban system.

The ascent of the Kiev principality to a dominant position among the many princedoms marked a new stage in Russia's urban history. Kievan Russia had not yet assumed the characteristics of an imperial regime. Despite the country's unification under the Kievan prince, no extensive state bureaucracy was established; rather, the state exhibited many characteristics similar to those of European feudalism. Local princedoms were not organized in a hierarchical system. The local princes strengthened their rule over the local population and tried to secure as vast a territory as possible, controlling it from their fortress towns. Trade was limited to a local scale, and interurban or international trade was insignificant. The techniques of governing large rural areas were not well developed, and the princes had to parcel out their direct control over much of their holdings (Rozman, 1976:48).

The establishment of a large number of churches and monasteries during the 11th century, however, gave great impetus to urban growth, as these not only fulfilled religious functions but gradually evolved into major centers of production, consumption, and large-scale trade. The urban structure of the Kievan period in the 11th and 12th centuries can be characterized as a two-level hierarchical system, the lower one comprising local towns and the upper consisting of Kiev, the national center, and Novgorod, the secondary capital. Commercial interaction between upper- and lower-level centers, as well as among the lower-level towns themselves, was not of significant magnitude; each prince tightly regulated his city market. Cities and towns on both levels of the hierarchy evolved according to a similar pattern, made up of two centers: the political center, or *kreml*, which, in Kiev, consisted of the Great Prince's residence and, in the other towns, the residences of the local princes subordinate to Kiev's Great Prince. The second, the economic urban center, was the market, the focal point of the *posad*. The market was usually situated just outside the fortress walls, and beyond it, extended the *posad* area, which included the work areas and residences of the merchants and artisans. In some larger towns an outer wall was built to enclose the market and *posad* area. With its dense population and increased economic activities, Kiev had to develop two markets to accommodate the large volume of transactions conducted there.

Kiev began to lose its dominance toward the end of the 12th century. The city had been sacked by invading tribes (1169), was bereft of much of its military and political clout, and by the eve of the Mongol invasions the upper level of the urban system had weakened considerably.

2 Mongol Invasions

Throughout the 13th and 14th centuries, Russia was repeatedly invaded by Mongol tribes who sometimes attacked the northeastern part of the country and sometimes the southern regions. The Mongol invasions influenced the Russian settlement system in varied ways, hastening some processes and slowing others. Basically the invasions broke up the political hierarchical organization, thus augmenting decentralization of the political power and control, and Russia's political structure became similar to its very early one, which was made up of small, isolated cells. The Mongols' constant demands of tribute and submission to their rule, however, may have been instrumental in helping the princes to consolidate their hold on their local areas, thus strengthening the individual cells (Rozman, 1976:49).

The devastation brought about by the Mongols disrupted the economy. In areas under continuous attack, urban trade and handicrafts dwindled to insignificance. Towns lost much of their economic base and became mainly military outposts; the trading activities moved to rural areas. Most of the handicrafts shifted to the rural estates; commerce was either conducted in periodic markets or moved to small towns, thus descending the scale of the urban system.

During this period, aside from its spiritual leadership, the Church began to assume a dominant role in the economic realm. It could dominate much of Russia's internal trade because merchants of the Church, exempted from payment of taxes to the princes, could use free labor and easily transport large supplies of food from one periodic market to another. By comparison, urban merchants and craftsmen lacked an effective form of organization (Hamm, 1976:25). They conducted their activities under the direct control of the princes and were helpless when facing competition from tax-exempt monastic centers. The churches and monasteries thus accumulated great wealth due to their commercial activities and their control of large tracts of land. This was clearly attested to by the magnificence of the religious buildings found in the *kreml* as well as in the *posad.*

The two and a half centuries of subordination to the Mongol invaders were characterized by the frequent establishment and disappearance of cities. Local princes, eager to control their territories in periods of strife and upheaval, built satellite towns and converted existing settlements into satellites of their own. The preferred locations for these new towns were the boundaries of the princedoms, which constituted bases for defense and extension. Sometimes an in-between town located on or near the border of a princedom became the capital when its territory expanded. The new towns of Moscow and Jizhnii-Novgorod are examples of those settlements that became the centers of princedoms (Rozman, 1976:50).

The rivalry among the princes under the Mongols endured until the early stages of the succeeding Muscovite period and frequently caused towns to change hands from one prince to another. When the towns were not located strategically, they were readily abandoned. The destruction of villages by waves of invaders led to the disappearance of some of the small towns that constituted the service centers of their rural areas. These rapid changes augmented even more the decentralized character of the settlement system, which was composed of numerous small towns at the lower level of the hierarchy and did not possess a unifying structure controlled by a major center.

3 Ascent of Moscow and Establishment of the Russian Empire

The roots of Moscow's rise are deeply ingrained in the Mongol period. Moscow started as a minor fortress town, later becoming the center of a small princedom and then expanding until it rivaled Tvar in preeminence among the northeastern cities. Finally, in the early 15th century, it became Russia's undisputed primate city.

Moscow's ascent to its dominant position within the Russian urban system reflects the strengthening of imperial power by the tsar in Moscow. From the end of the Mongol period, in the middle of the 15th century and for the next two centuries, an autocratic government took shape in Moscow. This process of political and economic concentration culminated in the *Ulozhenie* (code) of 1649, which made the cities in general and Moscow in particular the expression of direct imperial control. Until the end of the 19th century, they remained the central form of imperial dominance.

The first two centuries of the Muscovite period, which superseded the Mongol era, were characterized by extensive territorial expansion and the growth of the Russian Empire's population. By the middle of the 17th century, the empire had achieved most of its outward expansion (Rozman, 1976:57). The extension of political boundaries under the imperial regime was substantiated by opening virgin lands for cultivation and by the establishment of military administrative centers. This is evidenced by the increase in the number of the *uezd* centers—the basic administrative territorial unit. Their numbers rose from roughly 100 in 1500 to approximately 165 in mid-century and to about 230 at the beginning of the 17th century (Rozman, 1976:58). By the middle of the 17th century, many of the original fortress towns had gradually been transformed into typical *uezd* towns, serving as centers for the agricultural hinterlands.

With the growing authority of the tsar in Moscow, forces of centrality began to shape the urban hierarchy and the composition of the cities. *Uezd* cities were molded to accommodate the military, administrative, and fiscal needs of the imperial regime, and the basic step was the transformation of cities from their previous control by private individuals into state property. In the late 15th century, Ivan the Third was especially active in implementing this transformation. Private suburbs on the outskirts of *uezd* cities were gradually placed under state control. By the middle of the 16th century, many of them were already under the tsar's control, and the *Ulozhenie* of 1649 mostly abolished private control of the suburbs. This process had a strong impact on the commercial and artisan population of the *posad* that gained new legal status under the tsar's direct rule. Toward the end of this period, the cities that previously had been fragmented into sections and whose residents had been granted privileges according to the standing of their lords, gradually formed a single unit under central control.

Throughout those two centuries, Moscow grew rapidly and made up a significant proportion of the total Russian urban population. Moscow's evolution to Russia's primate city reflects the ever-growing concentration of political power in the hands of the tsarist regime. The tsarist government made special efforts to establish Moscow's economic dominance over Russia. Rich merchants from Novgorod, Pskov, and Smolensk were forced to settle in Moscow in order to improve its commercial ties with all parts of Russia and with foreign countries. These superior trade connections succeeded in rapidly turning Moscow into the country's major economic center. Political centralization, augmented by economic integration, attracted large numbers of inhabitants to the capital.

Moscow's fast growth led to a tremendous expansion of its built-up area. Its *kreml* became the political and religious center of the Russian Empire, and high imperial officials and noble Boyars found residence within its walls. The *posad* area expanded considerably to include a large number of markets as merchants and artisans flocked to the city. The great population pressure on Moscow resulted in the accelerated construction of walls to surround the expanding city, and this occurred three times in succession in the 16th century. The city's fast growth as the political and economic center of the Russian Empire strongly established the primacy structure of the Russian urban system that endured until the 20th century.

C
RUSSIAN IMPERIAL SYSTEM

It was around this period that some of the distinct characteristics of the Russian imperial system crystallized (Pipes, 1974; Seton-Watson, 1952)—characteristics that distinguished it from many other imperial systems, especially that of China. In common with other imperial systems, however, the Russian system was characterized by an autonomous center and relatively autonomous elites. But the structure of these elites, in connection with the cultural orientations they carried, presented some very special traits. Being part of the Christian civilization, the Russian imperial system was distinguished by the prevalence of strongly interwoven this- and other-worldly orientations, as well as by the concomitant heterogeneity of the political, religious, and economic elites.

After the Mongol defeat, the structure of these centers and elites in the Russian Empire revealed some specific features. The center was based on strong coercive orientations that had a far-reaching influence on the structure of control. The combination of such heterogeneity, together with the impetus to economic development, gave rise to a somewhat differentiated social structure, characterized by relatively wide and potentially cross-cutting markets.

The center succeeded in attaining a relatively high degree of subordination of the cultural to the political order and a relatively low degree of autonomous access of the major strata to the principal attributes of the social and political orders. The political sphere became the monopoly of the rulers, whereas the economic one became less central. The economic activities were left, to some degree, to their own autonomous development, as long as they did not impinge directly on the center; insofar as they did, however, they were largely controlled.

In this context the center tried to monopolize those activities that were of central importance from the point of view of maintaining the cosmic and social order and, above all, political activities. The broader strata, however, were granted autonomy in other mundane activities, primarily economic, without being permitted to imbue them to any large extent with wider meanings, in terms of the basic parameters of cultural religious spheres. Furthermore, they were not allowed any direct, autonomous political access to the center.

To this end the center vigorously segregated access to those attributes of the cosmic order (to salvation) that was allowed to all social groups through the comparatively weak mediation of the Church, from access to attributes of the political and social orders that, after the post-Mongol period, were almost totally monopolized by the political center. Furthermore, the center kept a firm segregation of the loci of this-worldly resolution of the tension between transcendental and mundane orders.

Religious heterodoxies became either other-worldly oriented or dissociated from the political sphere. Sometimes, however, as in the case of the true believers, they impinged to a certain degree on the economic sphere (Gerschenkorn, 1970).

In order to maintain its monopoly, the center had to wage war against those strata (above all, the aristocracy and the free city-states) that, especially in the earlier Kievan period, were the carriers of more autonomous access to the attributes of the social and cosmic orders. This struggle gave rise to strong power orientation and coercive policies on the center's part. Specifically, access of these elites to each other and especially to the center was limited by the central political elite, although not altogether successfully; within each of these groups there remained some latent orientation toward each other and toward the center.

As a result, a sharp distinction emerged in Russia between ownership of resources vested in the major strata and the control over their use and conversion in the macrosocietal setting vested in the center (Eisenstadt, 1971; Raeff, 1966).

Although the center was unable, given the strong emphasis placed on the tension between the transcendental and mundane orders, to prevent the emergence of certain autonomous orientations to the essential attributes of the cosmic and social orders among different potential institutional entrepreneurs and collectivities, it attempted to curb the autonomous political expression of such orientations and, above all, the possibility of linkage among the various types of social protest and between them and the potential elites. For a long period the Russian center was successful in these attempts and accordingly minimized the transformative potentialities of Russian society.

From the interaction between the control tendencies of the center and the social structure, the specific character of the social stratification evolved. This was also important for understanding the numerous aspects of the Russian urban structure.

Some tendencies emerged in Tsarist Russia (Pipes, 1974) that were seemingly similar to those in Western Europe, in terms of the evaluation of position in connection with the structuring of social hierarchies. That is, relatively strong emphasis was placed on the autonomous access of different strata to the major attributes of economic and social orders. These tendencies were counteracted, however, by strong attempts of the ruling elites to emphasize functional contribution to society's welfare as perceived by the center and, by proximity to it, to minimize the autonomous expression of class consciousness and to segregate the different occupational and social groups that could coalesce into broader horizontal strata.

The elites of Tsarist Russia tended to encourage the segregation of lifestyles and patterns of participation among the different local, occupational, and territorial kinship groups (Eisenstadt, 1971; Raeff, 1966). These elites attempted to minimize the "status" or "class" components of family or kinship groups' identity as well as the family's autonomous standing in the status system. They tried to establish a uniform hierarchy of evaluation of major positions, especially with regard to access to the center. They also aimed to make this hierarchy a relatively steep one within the center, between it and the periphery, and, to some degree, among the peripheral groups as well, by discouraging the development of any countrywide class consciousness among most groups and strata.

The monopolistic bias of the central elite tended to exert severe control over the legitimation of the life-styles of each subgroup. The center continually attempted to limit and break up the efforts of any stratum to transcend its own life-style beyond restricted parochial scope, or to claim legitimation for its life-style in terms of wider central values independent of the elite. Different occupational groups in Tsarist Russia have also been inclined to social segregation and to emphasize their own distinct occupational or professional goals. Such groups often tended to coalesce into relatively closed semistrata, each stressing its separateness from other groups.

As a result of all these tendencies, a relatively low level of autonomous, countrywide class consciousness evolved in Tsarist Russia, as well as the political expression of such consciousness and organization, although tendencies in that direction can certainly be found.

D
PROCESS OF URBANIZATION

1 Major Features of Russian Urbanization

The processes of urbanization—the concrete shape of the cities and urban systems that evolved in Russia during the long period of imperial rule—were formed by the continuous interaction of several forces. On one side stood the various forces of concentration, the social groups that participated in the different economic and social aspects of city life—mainly the urban groups proper (merchants and artisans, the aristocracy and landowners who often dwelt in the cities, the Church, and the peasants moving into the cities)—according to their relative strength and to the cooperation or conflict between them. On the other side were attempts by the rulers to regulate these different forces.

With respect to city life, the aims of the rulers focused mainly on their attempts to maximize their own revenues and, because of the lack of bureaucratic personnel, to minimize the cost of raising that revenue. The tsars also aimed to control the political, social, and, to some degree, even economic activities of the various groups, especially their social and political autonomy, while again minimizing the cost of such control.

From the attempts to implement the rulers' aims, which were closely related to the tsars' general political orientations and policies, developed the central characteristics of Russian cities: those of the "service city," which was defined as serving the state and primarily constituted a major source of tax revenue for the tsar's government (Hittle, 1979; Miller, 1976:34).

The rulers attempted to maintain these characteristics of a service city by exercising control over the access to trade and manufacture. They also regulated the internal life and structure of the cities in general and the various social groups that composed them in particular, specifically urban groups.

In principle the major characteristics of the urban structure and urban systems in Russia were shaped by the rulers' broad mechanisms of control. However, their shape at any particular period was greatly influenced by the concrete relations existing at any time between the social groups characterizing the city on the one hand and the rulers on the other.

Continuous struggles developed between these different social forces over access to trade and manufacture and the accrued benefits. There were also difficult relations with the rulers due to the latter's efforts to regulate all aspects of urban life.

The strength of each of these social forces varied during this long period, according to the relative strength of demographic changes and population flow, the development of internal and external trade, the strength of these groups vis-à-vis the rulers, and the latter's success in controlling the former's activities.

The rulers' major weak point, from the viewpoint of maintaining their ideal service city, was rooted mostly in the fact that they were often too feeble to exercise adequate control over all groups. Many of the nonspecifically urban groups, therefore—the aristocracy, the Church, the peasants—either moved into the cities or appropriated many of the trading and manufacturing activities. The rulers were thus deprived of revenue and of their ability to control if not the autonomous political expression, at least the economic activities of these groups. By so doing, many of the proper urban groups were dispossessed of the bases for their economic activities.

Such situations occurred repeatedly in the Russian Empire's history, giving rise to one of the most important aspects of the rulers' relations with the cities: their continuous efforts to legislate what may often be seen as reforms, the most important of which were the *Ulozhenie* of 1649, Peter the Great's Urban Reform of 1699, and Catherine the Second's urban reforms of the second half of the 18th century. These reforms were characterized by several basic, often contradictory, features: (1) by attempts to grant the right to engage in trade and manufacture only to specifically urban groups; (2) by attempts to involve these urban groups in some degree of self-control and, especially, responsibility for payment of taxes, and for proper social and political behavior; and (3) by limiting such self-control in order to maintain the upper levels of such control in the hands of the rulers (and later through their bureaucracy) and not to allow these groups any real political self-expression.

It was due to no small degree to these contradictions that the reforms were never fully successful and that a shifting balance existed between the unregulated development of cities and the goals of the reforms.

The interrelations between these various forces and orientations also affected the internal ecological structure of Russian cities, their autonomy, and the structure of the urban system.

2 The Service City: Limits of Urban Autonomy

The basic characteristics of the interrelation between the centrality and concentration forces, as they evolved in the Russian Empire, were the ones that shaped the internal structure of the cities—their autonomy, internal composition, institutions, and conflicts.

The heavy-handed policy of the state, implemented through the imposition of its aims on the cities, constituted a decisive factor in preventing the townsmen from developing into legal corporate units with some urban autonomy.

The starting point for understanding the very limited and highly restricted autonomy that evolved within Russian cities is analysis of the *posad*, which constituted one of the major features of the traditional Russian cities. The *posad* was the area of settlement situated between the original fortress and the outer fortifications. It was the source of the urban tax paid by the *posad*'s inhabitants, who were mainly engaged in small trade and handicrafts. During the Kievan, Mongol, and early Muscovite periods the *posad* had been a territorial entity, a separate city quarter adjoining the fortified town; it became more of a legal entity in the late Muscovite and early imperial periods. A significant number of towns in Muscovy, however, did not possess a *posad*; conversely, *posad* settlements were found in the countryside, especially near monasteries.

The *posad* formed a legal entity because its members were forced to bear collective responsibility to fulfill their tax and labor obligations (Pipes, 1974:201). The few affluent members also had to assume the extra duty of collecting taxes and tariffs. The *posad* member's status was hereditary and his descendents were forbidden to leave the *posad*. The land on which the residences and workshops of the *posad* members were built belonged, in the early periods, to the prince and later to the tsar and could not be sold or disposed of by the *posad* people; thus a main avenue to accumulate capital was shut off. In many respects, the status of the *posad* members—who constituted the backbone of the Russian Empire's trade and handicrafts—was inferior to that of the rural serfs. Given their low status and the heavy taxes imposed, it is understandable that despite all the prohibitions, the *posad* members attempted to escape from their communities and find a landlord or monastery that would permit them to trade and produce goods without having to bear the heavy tax and work burdens. The government took drastic measures to prevent the desertion of *posad* people and imposed heavy penalties, but with minimal effect (Pipes, 1974:202). It should be emphasized that the taxes paid by members of the *posad* constituted the largest contribution of cash revenue to the Russian state, and most of the armed forces were financed by them (Hittle, 1979:27).

The critical feature in the situation of the *posad* people was the fact that they had a service relationship to the tsar, similar to that of the rural society (Hittle, 1979). At the same time, because the Muscovite state did not possess the resources to maintain a large fiscal bureaucracy, the urban servitors—the *posad* people—were forced to fulfill public functions, such as tax collection, town administration, and the building and maintenance of public works. In exchange, the state was expected to support its servitors' claim to the monopoly of urban trade and production of goods. Yet the urban economic base of the *posad* was constantly threatened by the economic activities of peasants, nobles, and clergy, who did not reside in the towns and were therefore exempt from paying the stiff urban taxation.

E
URBAN SOCIAL STRUCTURE

The social composition of the cities and the social organization of their population clearly reflected the basic orientations and policies of the state toward the urban population. The major social groups living in the cities derived their identity and position in relation to the other strata of society from their particular service relationship to the state (Hittle, 1976:54). This stimulated, as was true of all groups in Russian society, a high degree of status segregation as well as narrow status consciousness focused, to a large extent, on the state.

The social organization of Russian cities reflected the basic segmentation of Russian society into two main estates, both defined by their service relation to the imperial authority: the nobles, who served in a military or administrative capacity, that formed the "service estate," and the other group made up of peasants, artisans, petty traders, and various manual laborers. This group of commoners formed the estate of *tiaglo* bearers, the *tiaglo* representing the load of taxes and labor that commoners owed the tsar. The clergy was a separate social order that was exempt both from taxes and from rendering services.

The distinction between servitors and commoners was fundamental in the social organization of Imperial Russia. Even though the servitors could not be regarded as nobility in the Western sense (because they lacked corporate privileges), they still enjoyed real material benefits collectively, the most important being a monopoly on land and serfs. The commoners' estate consisted of people who enjoyed neither personal rights nor economic benefits and whose main obligation was to produce goods and provide labor needed to sustain the monarchy. The gap between these two estates was unbridgeable; the imperial regime was interested only in services and income and was determined to keep the people within their particular estate, thus ensuring maximum social rigidity. This social immobility was further strengthened in the 16th and 17th centuries by the geographic immobility decreed by the laws prohibiting peasants from leaving the farms and tradesmen from leaving the towns. The cumulative effect of these measures was to make the social status in Imperial Russia fixed and hereditary (Pipes, 1974:87).

Among the various urban social strata, the townsmen (*posadqie liudi*, the *posad* people, the merchants, and the artisans residing on state-taxable land) constituted the most important segment of the urban population. Their main service function was, as already mentioned, to carry the tax burden (*tiaglo*) in the form of heavy direct taxes and to collect the two largest indirect taxes, the customs and liquor duties.

The combination of the occupational orientation to production and exchange of goods within an urban setting, with the common bond of taxpaying and services obligation, made members of the *posad* (initially at least) a coherent, identifiable social group, potentially the most cohesive among the various forces of concentration in Russian cities. Their essential role in the country's economic development led, especially in later periods, to their aspiration to greater participation in the political process and to greater urban autonomy. But the service relationship between them and the state prevented the development of autonomous intermediate institutions and any autonomous corporate organization and consciousness of the city as such. Under the Russian imperial regime, these aspirations could only be realized through imperial municipal reforms (to be analyzed shortly) that, however, would never allow the development of full autonomy.

The social fragmentation of the Russian city found its counterpart in the institutional structure of city administration. Residence in the city did not imply urban citizenship. Except for the military governor—the tsar's appointed agent—no institutions were responsible for the entire population of the city, and each social group was administered separately by the appropriate government bureau.

Within Russian cities, guild-type frameworks emerged that did not lead to any significant autonomy of the urban society. Craftsmen living in the cities were organized in loose-knit guilds, differing from those found in Medieval Western Europe. They possessed the traditional division of masters, journeymen, and apprentices, but craft regulations were scarce and the quality of the merchandise was controlled neither by artisans nor by city officials. Even more important was the fact that the guilds, motivated by the need of the urban taxpaying community to count as many participants as possible in order to render the tax burden easier to bear, did not maintain an exclusive character but accepted new members. As the technical level, quality, and cost of the goods produced by the urban craftsmen were similar to those of their rural counterparts, guild members had to face fierce competition from the countryside.

Peter the Great's Magistracy Reform introduced some innovative features into the social organization of urban society, which was divided into two major groups: "regular" and "irregular" citizens. This new division was applied only to those urban inhabitants subject to the Magistracy authority—basically the *posad* population—as gentry, clergy, Church dependents, and foreigners were specifically excluded from the Magistracy's jurisdiction. The first group of regular citizenry included two categories called *gildii* (guilds): one consisting of wealthy merchants, bankers, and professionals such as doctors, apothecaries, and merchant ship captains and one made up of people engaged in petty trade and urban craftsmen. The inhabitants who did not qualify for membership in either of the guilds and were mainly hired laborers performing unskilled work fell into the second category of irregular citizens.

The establishment of the first guild citizenry under the Magistracy regulations, granting them privileges and securing for them a dominant role in city life, reflected the changing base of the urban social organization. This change indicated the social structure of Western European cities, as perceived by Peter the Great. During Peter's reign, the two privileged merchants' corporations (the *gosti* and the hundreders) rapidly lost their importance and fell almost into oblivion. Peter's policy thus caused the privileged corporations to vanish gradually and moved the first guild, comprising talented professional people with taxpaying capabilities, to the forefront of urban society.

Peter's reforms provided a guild-type social organization, not only for the more successful and ambitious members of the *posad* but also for the rest of the *posad* people, ordering that each art and craft establish its own guild. In order that the guilds be similar to their prototypes in Western Europe, they were instructed to elect aldermen, keep records specifying their rules and privileges, and have their elected officials supervise their members' activities.

Yet in spite of their formal resemblance, the Russian guilds differed in some basic features from those of Western Europe. They were not compulsory for all artisans; their membership was not an exclusive right passed on by inheritance; they did not enjoy special privileges in the marketplace; and, above all, they had not come into being through the initiative of the craftsmen themselves but had been created by the central government. Despite Peter's efforts to reorganize urban society and allow each group some corporate life of its own, the guild legislation proved to be mainly a means of deepening governmental control over the *posad* population, the most important component of the urban society (Hittle, 1979: 87).

F
INTERNAL STRUCTURE OF CITIES

1 Basic Components of Russian Cities

These basic relations between the state and the urban groups also shaped the internal ecological structure of the cities, which evinced a high degree of segregation, ultimately uncontrollable.

The basic ecological structure of Russian cities was made up of three major elements: the *kreml*, the *posad*, and the suburbs (*slobody*), each representing a particular aspect of urban life. Their importance varied according to the relative strength, at any given time, of the forces shaping them.

At the center of nearly every Russian city stood the *kreml*, the fortress. Surrounded by walls, well defended, and located in a strategic site, the *kreml* contained the centers of administrative, military, and ecclesiastic powers. The existence of the fortress as the nucleus around which the city developed was not limited to the border towns where the military situation was precarious, but characterized almost all Russian towns. The military presence was ensured by the arsenal, the powder magazine, the stables, and barracks, accommodating the leading military servitors. The ever-present cathedral and the few dwellings of the leading clerics (representing the role of the Church in urban development) were also situated within the fortress boundaries. The *kreml*, comprising the major institutions of the central government and the Church, was the supreme physical expression of the critical importance of the forces of centrality in shaping the internal structure of Russian cities.

The second component in the ideal-type structure of Russian cities was the *posad*, which we have already analyzed. We have seen that the original meaning of the *posad* related to the area of a city that lay beyond the walls of the *kreml* and was often surrounded by the outer land walls that served as the city's first line of defense.

Gradually, however, the *posad* assumed a second meaning, differing from the original geographic one. It became a fiscal administrative definition including the city's entire population having a state-service obligation (*tiaglo*). Despite this new meaning, the people living under the yoke of the *tiaglo* and engaging in trade and manufacture tended to concentrate in the same areas of the city, and the original territorial sense of the *posad* did not, therefore, disappear entirely.

The suburbs were the third component in the ecological structure of Russian cities; they developed mainly in Moscow but existed to a lesser extent in other Russian cities of any considerable size. They emerged in the early periods of settling the northeast as the physical expression of the competition for a limited supply of manpower (Hittle, 1979:29). The nobility and Church officials offered protection and privileges to those who would move to settlements on their lands and engage mainly in agricultural activities. Later, with the rise of Moscow and the growing importance of towns in administering the Russian state, new suburbs were established in the immediate vicinity of cities, again through the initiative of princes, Boyars, Church officials, and monasteries. Unlike the earlier suburbs, the new ones, mostly private quarters, were also the sites of trade and handicrafts production and were subject to different regulations and jurisdiction than were the *posad.*

In the early 16th century, another type of suburb emerged, this time under the auspices of the government: The central government authorities settled groups of contract servitors on parcels of land adjacent to the towns' walled limits but not within them. As the number of contract servitors grew, these state-sponsored quarters expanded considerably to become an important part of the built-up urban area.

This ideal type of structure underwent significant changes resulting from the continuously changing and expanding urban population and its relations with the imperial government. From the second half of the 16th century onward, the social composition of the suburbs, mainly of those sponsored by the state, was transformed by the influx of *posad* members who came to reside in them, side by side with the contract servitors for whom these suburbs had been established. Parallel to changes in the social composition of the state suburbs, the central government transformed the land of those suburbs into taxable property, and their residents thus became taxpayers along with the original *posad* members. The ultimate stage in this process, occurring simultaneously with the expansion of cities, was absorption of the suburbs within the general urban framework of the *posad*, not as a territorial but as a legal-administrative category (that is, the state suburbs and the *posad* were incorporated into a single body).

In contrast to the state suburbs, which gradually merged with the *posad*, the private suburbs remained a constant feature of Russian cities throughout the tsarist period. The nobility and clergy's spacious urban estates, with their luxurious buildings and well-kept grounds, were scattered among the residential areas of the *posad* people and contrasted sharply with their poor surroundings. A large number of dependents lived in these private suburbs, serving their masters and working partly in handicrafts and trade. Benefiting from the privileged position of the estates' owners, these dependents became serious competitors of the *posad* population with whom they shared common trade activities and handicraft production.

Foreign citizens constituted an important element of the urban society. Even though they were not numerous, their role in the provision of international contacts in commerce, new technologies, and cultural norms accorded them major importance in the Russian economy and society, although their presence was often a source of concern and irritation to the central authorities, the Church, and the native *posad* townsmen. Foreigners lived in few Russian cities, but their small number in most of these cities did not justify the establishment of a separate quarter. In Moscow, however, where several thousand foreigners tended to congregate, a new element—suburbs of the foreigners—was added to the internal ecological structure of the city. Since the 15th century, the famous Nemetskaia quarter had been a distinct feature of Moscow's urban structure, and more quarters for foreigners were established at later periods. The inhabitants of the foreigners' suburbs were subject to Russian law and jurisdiction but exercised a certain degree of autonomy over their internal affairs and enjoyed numerous privileges and exemption from state obligations (Baron, 1970).

Thus it can be observed that two of the basic components of the ecological structure of the Russian city—the *kreml* and the private suburbs—remained strong and unaltered and kept their dominant position. The *posad*, which at the early stages of urban development had been a territorial unit, gradually changed its nature and became the administrative definition for taxpaying citizens, including the population of the state suburbs. Except for the walled *kreml*, no regular physical pattern can be discerned in Russian cities, especially in the second half of the 16th century when all the above-mentioned processes of change started to evolve. The *posad* and the suburbs merged into a continuous built-up area of poor quarters with a disordered street pattern, the major transportation access radiating from the *kreml*. Areas of commerce and handicrafts were intermingled with the residential quarters.

2 Ecological Differentiation

Within Russian cities some division of labor emerged in commerce and had an effect on the cities' ecological structure. Major commercial transactions were carried out by the central government, monasteries, guild merchants, and foreigners; most of the *posad* dwellers were confined to being artisans and small shopkeepers with a very limited scale of operation.

The merchant courts (*gostinyi dvor*) were the physical expression of the large-scale commercial operations being transacted in most cities, and their number grew with the size of the city and level of its economic activities. The merchant court was built in the form of a wood or stone arcade containing both shops and storage areas for use by exporters, importers, and wholesalers. The concentration in the court of the numerous commercial activities facilitated the collection of customs duty and tariffs and improved bargaining and goods selection by retailers and clients. The merchants' courts, which were the focus of the urban commercial activities, were not concentrated in a single area but were distributed all over the *posad*.

The *posad* people carried out their small-scale retail and handicraft activities in small shops arranged in rows according to their specific product, profession, and trade. With the increasing number of shops, the trade rows grew to occupy an extensive area, creating a commercial sector in the urban internal structure. The best example of such a quarter was the Kitai Gorod, located in Moscow east of the *kreml* and north of the Moskva River. It originated as an open-air market within the *kreml* under state control. Evicted from the *kreml* by tsar Ivan the Terrible, the market was relocated on the other side of the Red Square, in Kitai Gorod, where it rapidly assumed a permanent and durable form, to become the city's major commercial quarter. Trading was first conducted in stalls that gradually evolved into enclosed shops arranged in elongated rows, which were later linked by passages. This complex structure encompassing hundreds of stalls, shops, and workshops somewhat resembled oriental markets and bazaars.

Similar to the bazaars, the Kitai Gorod market displayed a high level of specialization, with the rows differentiated according to the products and services offered. It was in the commercial quarter of the Kitai Gorod, in the shadow of the *kremlin* walls, at the junction of the major transportation routes, that the main impact of the *posad* on the internal structure of the Russian city can be observed (Gohstand, 1976:160-181).

This amorphous structure of the Russian city, with its low level of urban facilities and markedly limited infrastructure, occasionally interrupted by conspicuously affluent private suburbs, reflected the weakness of most of the urban strata, especially those related to the *posad*, the lack of almost any common social bonds among the different strata inhabiting the city, the feeble urban government, administration, and resources, and the ultimate inability of the state to control this continuous expansion.

G
ATTEMPTS AT URBAN REFORM

1 The Law Code (Ulozhenie) of 1649

The continuous weakening of the control exercised by the rulers, the concomitant dissatisfaction, and even the uprisings that were generated led to attempts at reform. This was evident in the new types of legislation that could be called urban reforms. In earlier periods these aimed at controlling the diverse elements of the urban population and later even at increasing their participation in the urban administration. Yet all were caught up in the internal contradictions of the imperial system.

The first such legislation was initiated in 1649, following the 1648 riots that began in Moscow and spread rapidly through Muscovy. The riots forced the government to reexamine the relationship between the state and the cities and to issue a new law code, the Ulozhenie of 1649. This code constituted a landmark in structuring Moscow society in that it established Moscow's social hierarchy and, until the early 19th century, provided the legal framework for administrative and economic activities. The two short chapters of the law code dealing with townsmen formed the legal basis for urban and rural life in Russia until Peter the Great introduced the 1699 Urban Reforms.

The 1649 Ulozhenie decreed that the state granted townsmen monopoly over trade and the production of goods, favoring them over lay and clerical establishments. Several measures were taken to increase the number of townsmen in order to enlarge the urban tax base. A major step introduced by the law code was the confiscation of land, households, and servants from the Church and monastic establishments. It was at that time that the greatest privilege granted the *posad* people seemingly occurred: The new law code, which forged Russian society, attempted to protect the economic interests of the state's urban servitors by granting them monopoly over trade and manufacturing. The townsmen then became a closed hereditary estate and were restricted to their 1649 place of residence. The central government's heavy hand thus ensured their economic security but imposed limitations on their mobility, legal rights, and acquisition of property (Koenker, 1981).

The post-Ulozhenie period did not, therefore, contribute to the flourishing of the cities. Plagues and fires took a heavy toll on the urban population; moreover, the effects of these disasters were aggravated by the recurrence of those problems that the 1649 law code had attempted to resolve. The promised monopoly could not, in fact, provide *posad* communities with the economic security they sought. Even though from 1649 they enjoyed the exclusive right to produce articles for sale and to maintain their shops, this represented a meager economic advantage, as other groups, also engaged in trade and the production of goods, did not have to pay heavy taxes. The *posad* continued to face stiff competition from both the peasantry and the clergy: Peasants living on the property of lay and clerical landlords set up and traded in regular markets (*slobody*) in most cities but did not bear a share of the tax burden. Competition from the peasantry and clergy caused great bitterness and inflamed conflicts in Muscovite cities. The central government occasionally took steps to placate the *posad* population but without much success, and the *posad* was never able to shake off the deadly rivalry of the tax-exempt groups. The townsmen's commercial monopoly was successfully challenged by peasants, and the churches and monasteries fought to regain their privileges and participate actively in the commercial markets. The lucrative foreign trade was also strongly encouraged by the imperial authorities, further cutting into the townsmen's share of the urban economy.

Despite the 1649 reform prohibiting the purchase of the townsmen's property by privileged members of Muscovite society, such as the aristocracy and the Church, the latter found numerous ways to circumvent the law and continued to acquire and control large tracts of urban land. By the end of the century, the government came to terms with the noble groups and permitted them to possess urban land held even before the 1649 reform.

A similar development took place in the control of foreign trade. The Trade Statute of 1667 abolished duty-free trade and increased taxes on imported goods. Trading by foreign merchants inside Russia—a constant threat to town merchants—was banned outright and foreign traders were restricted to port trading cities. The central government hoped to increase its tax revenue from the cities through these reforms but the acute need for additional income drove it to grant special privileges to foreign merchants, hence invalidating most of the reforms.

Major parts of the Ulozhenie were thus nullified, reaffirming the continuous power of the privileged sector of society over the townsmen and also indirectly increasing the rulers' control and revenue. The dominant force shaping the urban scene continued to be the imperial government, which involved itself in all significant aspects of urban life, such as property relations, administrative institutions, tax matters, and social structure. From this developed the service city (Hittle, 1979), which displayed all the problems and weaknesses we have indicated.

2 Urban Reforms of Peter the Great

These developments occurring toward the end of the 17th century led to a second cycle of urban reforms, those of Peter the Great and Catherine the Second.

These reforms were indeed the outcome of the central government's policies and did not reflect a local urban initiative. They gave rise to a steady flow of decrees and regulations, aiming at reshaping the towns and the townsmen's lives, which changed to no small degree the coherence of urban society, the economy, and the relationship between the Russian city and the central government.

Peter's attempts at urban reform were shaped largely by his vision of the proper type of city, which had been greatly influenced by his encounter with the West. Peter's ideas and goals, which motivated the various urban reforms, found symbolic expression in the establishment of the new capital city of St.-Petersburg. The structure of the new city, the composition of its population, and its basic economic activities—all these innovative features were implemented forcefully by the tsar and were designed to act as a showcase, a physical example to be followed by all cities in the Russian Empire.

Peter based the city on large-scale industry, mainly shipbuilding, and made it an important center for international commerce by compelling prominent merchants to transfer their activities to St.-Petersburg. In this way, he demonstrated vividly his urge to transform the Russian economy by borrowing Western European technology and by encouraging the townsmen to lead the new economic development. By moving the seat of his government and the central institutions of the Church to St.-Petersburg, Peter indicated the supreme role of the central government in the political, economic, and social realms. By building the city according to a master plan, by giving great attention to the design and physical appearance of the streets, canals, public squares, and residential quarters, Peter expressed his wish to apply Western European norms to Russian cities. Through his conscious effort to populate the city with people of varied skills and background, Peter demonstrated his concept of a balanced and efficient population composition that would be the most conducive to economic growth and development. In all these respects, the building of St.-Petersburg probably had a greater ultimate impact on the development of urban institutions and society than whatever could have been achieved by Peter's various urban reforms.

The reforms he introduced constituted an attempt at radically changing the traditional role and institutional structure of the Russian city. However, the new legislation and regulations reflected the continuity of past trends in Russian society and the traditional ambiguous and ambivalent attitude of the government toward the city and urban groups.

Peter's central aim thus continued to be the maximization of city revenue and the maintenance of his control over the urban groups. Unlike in former periods, however, these reforms took into account the demand for some urban autonomy without weakening the attempt to assert the ruler's control.

The *posad* struggled to maintain its position between the landed aristocracy and the increasing flow of peasants into the city. By introducing reforms, the imperial power attempted to strengthen the *posad* and thus maintain balance between the competing forces and ensure more efficient control over the various sectors of the urban population. Needing the economic benefits contributed through the *posad*'s activities, the central government succumbed to pressure and initiated municipal reforms that, granted grudgingly and often withdrawn, increased some aspects of the urban autonomy but never allowed it to materialize fully.

Peter's urban policy was implemented in three stages by the three major administrative reforms: (1) the Burmeister Reform of 1699, (2) the Provincial Reform of 1708, and (3) the Magistracy Reform of 1721-1722. All the reforms were motivated by the acute fiscal problems existing during Peter's reign and attempted to make the cities more responsive to the treasury's needs. The frequency of the reforms, representing the ongoing search for innovative administrative structure and institutions, testifies to the difficulties of keeping up with increasing demands for funds of the militarily expanding imperial state (Hittle, 1979:78).

The first of these reforms, the 1699 Burmeister Reform, inspired by Peter's impressions of the city government administration of Amsterdam, called for the election of local administrators (*Burmistry*) for each city and empowered them to manage financial and judicial matters. The jurisdiction of the *Burmistry* extended over the *posad* people but not over the clergy, nobles, and other city residents. This reform can be characterized as creating an elected leadership for the taxpaying townsmen. The decreed establishment of the Burmeister institution did not aim to increase urban autonomy for its own sake but rather to allow the *posad* people to evade the avaricious yoke of the military governors. By applying this administrative reform, the central government hoped to improve the level of cooperation of the *posad* people and thus raise their taxpaying potential. The transfer of control and tax collection from the military governor to a body of officials elected from among the local population proved to be a successful innovation as it considerably increased the taxes levied and forwarded to Moscow. All revenues collected by the local *Burmistry* were transferred to the Moscow Burmeister Chamber, which allocated them to the government agencies responsible for maintaining and equipping the army.

The Burmeister Reform can thus be regarded as a major reorganization of the imperial regime's tax collection apparatus, encouraging local control at the collection stage while centralizing it at the stage of accumulation and allocation. The initial introduction of the Burmeister Reform in January 1699 exacted a high price from the townsmen in doubling their previous tax levy. Because of the cities' reluctance to accept it, this reform was temporarily shelved, but it was reintroduced at the end of the same year without doubling the former taxation. The predominating fiscal considerations of this reform materialized, as evidenced by the institution of the Moscow Burmeister Chamber in the wake of implementation of the urban reform. This chamber acted as chief administrative organ for Moscow's townsmen, collected the revenue from the local *Burmistry,* and distributed it to the appropriate state agencies. Within a short period the state's income from the cities was boosted by a significant amount.

Peter the Great's urban reforms veered in a new direction with the 1721 Magistracy Reform. Basically, this new reform aimed at the reinstitution of central control over the urban population engaged in trade and manufacture—the *posad* people—and made the magistracies (town councils) responsible for its implementation. It was focused on the town council, whose size varied according to the number of households in the city. The magistracies were supervised and their activities were watched and controlled by the St.-Petersburg Main Magistracy, thus limiting the possibility of their becoming the autonomous agency of local self-government.

As already mentioned, the Magistracy Reform recognized two categories of citizens: the "regular" and the "irregular." The regular citizenry comprised merchants, artisans, and professionals such as doctors and apothecaries. The poor people made up the irregular citizenry. The magistracy was elected by and among the regular citizens but held jurisdiction over both groups.

Because Peter the Great's death occurred soon after the initiation of the Magistracy Reform, the reform was implemented only to a very limited degree. But its spirit clearly indicated the tsar's realization that the best way to ensure the flow of increased tax revenue was to allow the city to develop in an orderly manner and make it physically attractive and economically prosperous—the same themes that had motivated the tremendous effort expended in building St.-Petersburg as a modern and well-designed capital.

But here also the basic limitations of Peter's vision became apparent. Despite the apparent increase in urban autonomy resulting from the Magistracy Reform, this did not constitute a sharp break in the institutional structure and role of Imperial Russia's cities, just as it did not change the basic structure of urban society. The reasons for this were twofold: The reform did not change the social organization of urban society, which remained composed of several segregated social groups, the identity of each being defined by its relationship with the state. The reform was unable to bring about the establishment of a category of urban citizenship that would encompass all inhabitants of the city. The other reason for the very limited changes that occurred in the wake of the Magistracy Reform was that it applied only to the urban population dealing in trade and manufacture, the *posad* dwellers, and not to the other groups: the aristocracy, Church, and foreign inhabitants.

The Magistracy Reform was basically the instrument with which the central government attempted to streamline its administrative apparatus by directly linking cities to it and eliminating intermediate agencies that decreased the flow of tax revenue into the state coffers (Hittle, 1979:84). The fact that the concept of the Magistracy did not emerge from the *posad* but was created by the central government again emphasized the role of the forces of centrality in shaping urban institutions and society.

3 Reforms of Catherine the Second

The years following the death of Peter the Great witnessed the faltering of the Magistracy Reform. The rulers were not successful in generating the development of the cities' various forces of concentration in the direction they desired, nor were they able to maintain the privileged position of the *posad* people. The townsmen, however, continued to perform their service duties, and the "service city" still constituted a dominant feature of Russian society throughout the first half of the 18th century (Hittle, 1976:61).

A number of societal changes occurred at that time, however, that gradually undermined the bases of the service city, the first of which related to changes in population size. In spite of severe obstacles, large numbers of migrants flocked into the cities, altering radically the proportion of the *posad* people in the total urban population and rendering obsolete the former identification of the city with the *posad*, which thenceforth ceased to be even a remote approximation of the city's social composition.

In a parallel development occurring in the economic realm throughout the 18th century, townsmen lost ground in their struggle to curtail the competition of peasants and privileged manufacturers. Although the peasants had never been granted the legal right to engage in trade, this practice still became universal. Abolition of the internal customs duty in midcentury removed a major obstacle to small-scale commercial transactions in the cities and especially in the countryside. The 1775 government decree, permitting anyone to establish a manufactory without obtaining an official license, put an end to the merchants' aspirations to secure the exclusive right to own manufactories. Numerous such enterprises were started in the countryside, hampering the economic monopoly of the urban *posad*.

As was fully apparent in the early years of Catherine's reign, these various processes gradually resulted in the obsolescence of the conception of the service city and gave rise to yet other attempts at reform. Catherine's urban reforms, initiated in the second half of the 18th century, were motivated by the failure of the traditional service system to meet the imperial regime's growing needs. The attention paid by Catherine's government to the cities' problems was prompted by the declining effectiveness of the city as a source of revenue and, in principle, the basic attitude toward the city as the outpost of royal authority in the countryside (Pipes, 1974). An important example of this basic approach was Catherine's massive effort to increase the number of cities in the empire and her success in doubling their number in a single decade (1775-1785). But what is significant is the method used to accomplish this feat. It was the simple procedure of reclassifying villages as towns by bestowing urban status upon them without changing the structure of their economic base. The royal authority was also expressed in an opposite direction as, for instance, when Catherine, wishing to discipline several dozens of cities, deprived them of their urban status.

Within this broad framework of orientation to the city, closely connected with the segregation of different occupational and social groups in it, a major shift (mostly organizational or administrative) took place when identification of the city with the *posad* was abandoned. For the first time the "general survey" was established and the city was regarded as a spatial entity, entirely separate from the surrounding countryside. Equally significant was the permitted election of a mayor—the representative of all the inhabitants of a city. The provincial reform of 1775 also created a single authority for law enforcement in each city and town. Catherine's government redefined the functional role of the city: Besides constituting a prime source of revenue, the city was to become a distinct unit in a somewhat decentralized administrative apparatus whose main goal was to preserve law and order in it and its surroundings.

The cities' new duties required that changes be introduced in several of the existing urban institutions. The magistracy emerged from the reform chiefly as a court of law, its jurisdiction extending only over the townsmen. Many responsibilities, formerly shared by the magistracy and the military governors, were transferred to the newly created municipal agencies, which were fully integrated within the bureaucratic structure of the central government, thus heralding the end of the "service city."

The 1775 reforms did away with the main justification for the existence of the *posad*. Extension of the state bureaucracy to the municipal realm left the *posad* with little to regulate in relation to services. The principles of mutual responsibility constituting the core of the *posad*'s community were abolished, thus shattering its rationale.

At the same time, the central government sought to involve all social groups in the city administration. Given the main orientation of the center, however, and the nature of Russian society—deeply stratified into segregated social estates oriented to the center—these intentions of the legislation could not be realized, but this did not give rise to anything like the autonomous Western city, despite some external resemblance. Indeed, the implementation of Catherine's reforms was impeded by inherent conflicts between legislative intentions and administrative and social realities. Whereas the legislation aimed to draw a larger number of urban citizens to participate in urban affairs, a fast-growing state bureaucracy further extended its authority over the cities and stifled local initiative.

The various attempts of 18th-century monarchs to reform the institutional structure of the Russian cities and, seemingly, to foster in Russia Western-type cities to be inhabited by the urban bourgeoisie were not very successful. The various elaborate urban reforms bore little relation to reality, and only seldom did they produce results. Their most conspicuous failure was their inability to achieve their aim of consolidating the city inhabitants into a cohesive and legally recognized group enjoying some form of self-government. At the end of the 18th century the cities were better organized institutionally than they had ever been, but they remained under the heavy hand of the imperial regime (Hittle, 1976:67). The government failed, however, to control the continuous development of forces of concentration, of economic initiative beyond the cities—again undermining some of the basic aims of the reforms.

While Catherine thus transformed villages into cities, she allowed numerous large commercial and manufacturing enterprises to retain their rural status as a concession to the landed aristocracy. This resulted in exempting the serfs employed in the latter's trading and producing activities from paying taxes—except, of course, for the soul tax—and many industrial estates employing thousands of industrial workers thus legally retained their village status. Economic development did not occur in urban centers, mainly because trade and industry were steadily moving to the countryside. Lacking the necessary economic and autonomous political base, an urban bourgeoisie could not be induced to emerge either by imperial decree or by well-intentioned urban reforms. Urban economic development therefore lagged behind, and this was clearly evident in the structure of the urban population which comprised more peasant serfs than merchants and artisans. Even in Moscow, the Primate City, at the end of the 18th century, there were three times as many peasant serfs as merchants and artisans, and this ratio was even higher in the cities at the lower levels of the urban hierarchy.

4 19th-Century Urban Reforms

The cycle of failure and new attempts at reform, which repeated the old contradictions, recurred in the 19th century with a growing emphasis on municipal self-government and the possibility of reconstructing Russian cities on a Western model.

An important step toward greater involvement of city dwellers in the administration of their city was taken when the 1846 Municipal Statute was first applied in St.-Petersburg and, later, in 1862 in Moscow. This statute was based on the 1785 Charter of the Cities issued by Catherine the Second but proved to be slightly more viable. Various segments of urban society were more ready to participate in city government; in addition, the central government had a pressing need to discard some of its fiscal obligations to the cities. The 1846 decree, similar to Catherine's 1785 charter that preceded it, was aimed at establishing a municipal administration based on the class organization of the urban society. The large municipal council, the General Duma, was organized along legally defined class lines representing five groups: (1) hereditary nobles, (2) property owners (who were honored citizens), (3) guild merchants, (4) small entrepreneurs and clerks (having the status of permanent city residents), and (5) persons listed as members of the Society of Artisans. The large number of peasants who lived in St.-Petersburg but who owned no property and were not registered as members of an urban class were not represented in the municipal council. These severe limitations upon franchise brought, in 1846, the number of the qualified electorate of St.-Petersburg to just over 6,000—less than 2 percent of the actual total population of that city.

The 1846 statute allowed for the establishment of a small executive Duma that would implement the guidelines set by the General Duma and supervise the city administration's daily work. Presiding over both municipal bodies was the mayor—chosen by the tsar between the two candidates proposed by the General

Duma. The rights and realm of authority granted to the St.-Petersburg city government in 1846 differed only slightly from those laid down in the 1785 charter, which, in view of the growing complexity of the tasks facing the municipal administration in mid-19th century, rendered them insufficient. The 1846 statute for St.-Petersburg and the ensuing version applied in Moscow in 1862 placed the city government's various activities under the direct and close supervision of the central government's various local and imperial representatives.

Despite the very limited self-government granted, in the wake of the 1846 urban reform, to the few large cities of the Russian Empire, a much greater level of involvement in municipal affairs of certain groups of the urban society ensued. This lessened to some extent the overwhelming control and supervision of urban affairs by the imperial government.

New social forces emerged, however, making an impact on the structure of cities and on the outcome of these reforms. Even though the emancipation of the serfs (decreed in 1861) took many years to materialize, it had a direct influence on urban growth in Russia in that it increased considerably the flow of peasant workers into urban areas. This rapid urbanization process increased the complex administrative, social, and economic problems of the cities with which the feeble and ineffectual city government was unable to cope. Moreover, the complete dependence of the local government upon the central authorities meant that increasing demands were made on the imperial budget, proving a growing burden. It could be argued that one of the major motives behind enactment of the urban reforms of the 1860s and 1870s was the desire to reduce the pressure brought to bear upon the state budget by minimizing the outflow of state funds to take care of urban needs (Hanchett, 1976).

These reforms, and especially the Municipal Statute of 1870, allowed the maximum statutory freedom to municipal self-rule in the Russian Empire's history. Whereas all the former urban reforms were concerned mainly with regulation of the structure and functions of the various urban classes, the 1870 law was the first to deal with the city as a whole. The urban class organization continued to subsist and to shape and influence urban society in all its aspects, but the new system of municipal government was much less related to this class organization. The 1870 statute defined the management of the municipal economy as the major sphere of activity of the city government and determined the latter's responsibility to maintain the urban infrastructure and buildings and to provide public services for commerce, education, and culture. In order to fulfill all these responsibilities, the city government had to impose taxes on city dwellers and thus reduce the central government's financial burden.

The 1870 statute was outstanding in that it clearly stated, for the first time in Russian urban history, that the municipal government had some right to function without interference from the central government; it also established the criterion of legality as the dominant principle for use by the central authorities in evaluating municipal actions.

In some crucial areas, the central authorities maintained control even though they forced the city government to cover the expenses. The most important example was the so-called city police, which was financed by the city but was under the imperial regime's direct control. Despite these limitations, however, the 1870 reform constituted the culmination of all the urban reforms initiated in the 18th and 19th centuries, granting the Russian cities some measure of urban autonomy and self-government.

Yet even these reforms could not attain their aims and were caught between the contradictory orientations that guided them. Indeed, the contradictions between the various social forces and the government's basic policy orientation, as well as those that plagued the government itself, became most visible in this period of great economic, social, and political change.

Toward the end of the 19th century there was growing dissatisfaction with the recently established municipal government. Town officials complained of the burden of compulsory expenditure imposed upon the city by the state government and the numerous legal limitations set upon powers granted to the municipality. On the side of the central bureaucracy, strong concern was expressed about the lack of total state control over municipal affairs. Indeed, toward the end of the 19th century attempts at reform halted with the 1892 Municipal Statute, which turned back the balance of power between the central and local governments. This new statute treated the municipal government as an integral part of a unified system of state administration, completely subordinate to the central authorities, thus entirely abolishing the meager benefits derived from the previous reforms. The 1892 Municipal Statute greatly limited the franchise allowed to cities and expanded the basis that gave state authorities the power to interfere in and cancel the enactments of the municipal governments.

The electorate designated by the 1892 statute was less representative of the total urban population than under any previous legislation, and the freedom of decision and action was more restricted than under any of the other 19th-century urban reforms. It can be concluded that the "counterreform" of the end of the 19th century fully reinstated the historical role of the Russian cities.

The failure of these—as of the former—reforms was due to internal contradictions in the government's attitude toward the urban strata: the contradictions between, on the one hand, its orientation at control and its denial of any autonomous political access and, on the other, its attempts to promote some urban corporate identity and organization. It was because of these contradictions and the rulers' attitude toward the forces of concentration that the tsars were unable to regulate the development of economic forces and their impact on the cities. Thus both the bases of any autonomous urban organization and the central government's ability to control the cities were undermined.

CHAPTER 7

Urbanization in the Byzantine Empire

A HISTORICAL BACKGROUND AND MAJOR PREMISES OF THE BYZANTINE EMPIRE

The processes of urbanization in the Byzantine Empire provide yet another comparative case of urbanization in imperial systems. We will be analyzing these processes—especially in the earlier period of the Byzantine Empire—from its inception at the beginning of the 4th century to the beginning of the 8th century. Special emphasis will be laid, first, on the development of its different urban hierarchies and, second, on the internal structure of its major, imperial city Constantinople. Together with Constantinople's own "successor" Istanbul, as well as with Peking, these can serve as prime illustrations of full-fledged imperial cities.

Historically, the Byzantine Empire dates from the year 330. The Roman emperor Constantine the Great inaugurated the city of Constantinople as the capital of the empire and as his residence, making this eastern capital the seat of the senior emperor. The Byzantine Empire lasted until 1453, when Constantinople surrendered to Mehmet II, the Ottoman sultan.

This long span of 11 centuries of Byzantine history can be divided into four major periods, following Hussey's (1957) periodization. During the first, from

330 to 717, the empire was shaped. Its foundations were laid by Constantine the Great (324-337), and the full crystallization of its basic premises was achieved two centuries later, under the reign of Justinian the Great (527-565). The emperors of the Heraclian dynasty (610-711) forcefully established the international role of Byzantium against the Persians, the Arabs, and the Slavs in the Balkans.

The second period is the medieval stage of the Byzantine Empire and extended from 717 to 1056. The years 717-842 are noted for the iconoclastic controversy that finally led to the crystallization of the relationship between state and Church. The second part of this period (842-1025), which was mainly under the rule of the Macedonian Dynasty, can be described as the apex of Byzantine expansion, culminating in the reign of Basil II (976-1025).

The third period, beginning with the death of Basil II and lasting until 1204, witnessed the fundamental changes that occurred in the power of the central authority as opposed to that of the landed aristocracy. Under the Comnenian Dynasty (1081-1185), imperial power was temporarily reaffirmed, only to be crushed by the Latins who captured Constantinople and dismembered the empire during the fourth crusade in 1204.

The fourth and last period, extending from 1204 to 1453, witnessed the final agonizing throes of the Byzantine Empire, resulting from civil wars and diminishing material resources, and its final collapse under continuous Ottoman onslaughts.

The initial crystallization of the Byzantine Empire arose out of what has been called the decline and disintegration of the late Roman Empire. Indeed, some of the major characteristics of its urban development can best be understood against the background of the Roman Empire.

Two aspects of this background are of special importance. One is the tradition of the city-state, which had survived in several ways in the Roman and, to some extent, in the Byzantine Empire. The second aspect is the major way in which the processes of urbanization in the Byzantine Empire were related to or influenced by developments in the Roman Empire, in which economic, political, and military processes led to its decline and to the emergence of the Byzantine Empire in the eastern part of the Roman territory.

Yet, notwithstanding the great influence of the Roman background, the Byzantine Empire evinced some unique characteristics as an imperial system. The most important of these features was the tendency toward greater centralization, to maintain—or at least to attempt to maintain—relatively more compact boundaries than the relatively fluid ones of the Roman Empire (Luttwak, 1976).

A second tendency, to reconstruct the periphery, much stronger than prevailed in Rome, was closely related to a third characteristic. This was the development of Christianity as state religion, with the concomitant evolution of a new institutional nexus, that of the Byzantine Church and monasteries, as well as the accompanying change in the basic cultural orientations and bases of legitimation.

As in all Christian civilizations, the major premises that became prevalent in the Byzantine Empire were distinguished by (a) the emphasis placed on the tension between the transcendental and the mundane orders, (b) a relatively close interweaving of this- and other-worldly orientations of salvation, and (c) a relatively high level of commitment to the maintenance of the sociocultural and social roles related to such a conception of salvation. These premises were also characterized by the tension between, on the one hand, the stress laid on the relative autonomous access of all sectors of society to these attributes of salvation and to the political order and, on the other hand, the Church's and emperor's mediation of such an access. An additional factor was the conjunction of the relative standing of the Church and the emperor on such a mediation and the relatively close relation between the Christian and Hellenic components of the society's collective identity. In that respect, the political tradition of the city-state was also of great importance—the tradition of the citizens' participation in city governance and the seeming accountability of the ruler to the population.

Although the concrete manifestations of this tradition weakened quite early in the Roman Empire—especially in its later phases—some of its impact survived there in several aspects of life and, later, in the Byzantine Empire. It endured, above all, in the importance of the urban population, or plebs, in Rome (and, to some degree, also in Constantinople and in the Byzantine Empire's other cities) and in the prominence of such various urban factions as the Blues and Greens that were very active around the hippodromes of the city. It survived as well in the continuous political importance of the army due to its origin and image as a citizens' army.

In the Byzantine Empire, these cultural orientations were connected to the development of rather special constellations of social and political forces in general and of those of centrality and concentration in particular.

B
FORCES OF CENTRALITY IN THE BYZANTINE EMPIRE

The major pivotal forces of centrality were the emperor, the bureaucracy, the army, and the Church.

Being the representative of God, the emperor naturally had a special relationship with the Church. It became customary, after the mid-5th century, for each emperor to be crowned by the Patriarch of Constantinople, the Byzantine Church's highest ecclesiastic authority. Everywhere and on all occasions, the emperor's unique—and, indeed, his sacred—position was emphasized (Barker, 1957; Baynes, 1955; Brehier, 1949; Hussey, 1957:100-114).

The very conception of a Christian Empire implied that the Church was part of the polity and was, in all respects, under the general care of the emperor, even though certain specific functions could not be performed by him. The dichotomy between what was Caesar's and what was God's was not so obvious in the Imperium Christianum of East Rome as it was in Western Christendom and surfaced only if the emperor was a heretic. The emperor's special responsibility for maintaining law and order among his ecclesiastic and lay subjects was recognized; the imperial novels contain many references to regulations concerning this responsibility.

This did not mean, however, that the Church was not in many ways an autonomous institution. The Byzantine Church had very strong autonomous and universalistic orientations and a strong organization of its own. Its autonomy was greatly restricted by its strong ties with the state and by its acceptance of the state's and the emperor's powerful position in the religious sphere. Yet the Church participated actively in the central political institutions in the senate, the court, and the bureaucracy. Because it contained strong other-worldly elements, however, and because of the rise of politically passive monasticism and the Church's acceptance of the state's cultural position, its political activity was largely confined within frameworks and organizations established by the existing political institutions where the Church could play an important role. Moreover, it initiated basic political issues (e.g., during the iconoclastic wars), fulfilled many vital political and administrative functions, and was often concerned with wider issues of policy. In general, the relation of the emperor to the patriarch, the secular to the ecclesiastic, was best expressed in Byzantium by "interdependence" and not by the misleading "ceasaropapism" (Barker, 1957:26-54, 86-89; Baynes, 1955; Brehier, 1949, 1950; Hussey, 1937, 1957:85-100).

The same could be said to some degree of the monasteries that constituted a basic part of Byzantine culture and society and a basic aspect of Byzantine religion. Throughout the Middle Ages, monasticism remained an integral part of Byzantine life; from its inception, all Byzantines, great and humble, were fervently attached to the monastic institutions. Monasticism served the Byzantine polity in many ways. It did not have to provide education to the extent that Latin monasticism did in the early Middle Ages, but it was a useful ally in social services. Monastic life was not specifically organized toward the pastoral ministry in the Byzantine Empire (for the most part, monks were laymen) or with a view to charitable work or "social service."

At the same time, the monasteries were very important politically, as was demonstrated during and after the iconoclastic wars. The fact that monastic intervention in political-religious disputes was so often successful was due not only to the influence of a few outstanding personalities but also to the wide popularity of the monastic body as a whole.

The Church and the monasteries, moreover, played a crucial educational role, both generally and in inculcating broader orientations through which the Christian and Hellenistic identity of the empire was implanted. In this process they were, to some degree, independent of the emperor. Indeed, the Church and, to a lesser extent, the monastic institutions shared one significant trait: Despite their close relations with the state and the secular world, they survived the downfall of the Byzantine State and continued to thrive thereafter.

Side by side with the emperor and the Church, the bureaucracy and the army also evolved as very strong continuous social formations in the Byzantine Empire and as part of its forces of centrality. Both were highly organized. In principle, they were under the emperor's control but in fact often exhibited strong autonomous tendencies, especially in the political area (Eisenstadt, 1963; Ostrogorsky, 1956).

C
MAJOR SOCIAL GROUPS IN THE BYZANTINE EMPIRE

The major social forces consisted of the aristocracy, to some degree the bureaucracy (which was part of the center as well as a distinct social group), the peasantry, and various urban groups.

The structure of Byzantine society was relatively highly differentiated (Brehier, 1949; Charanis, 1944, 1951a; Diehl, 1929; Hussey, 1957; Runciman, 1933). The aristocracy—comprising mostly big landowners, many of whom came from established families—and the bureaucracy stood at the top of Byzantine society. Many members of these top groups, however, were relative newcomers who owed their membership to frequent political upheavals. Because there was no legal definition of the status of aristocracy, it comprised a group of landed property owners. Other members were often recruited from the higher echelons of the civil bureaucracy; military leaders and great magnates of the provinces could also join the aristocratic ranks. The aristocracy was characterized by its members' similar interests and ways of life and was fairly accessible to newcomers (Beck, 1956, Charanis, 1951b; Guilland, 1947; Lemerle, 1958; Ostrogorsky, 1954).

The bureaucracy was one of the most powerful groups in Byzantine society. Until about the 12th century, possession of talent was one of the means for entering its ranks. The bureaucracy was a main channel of social mobility; many of its members were recruited from the different rural and urban strata that entered imperial service (Andreades, 1926; Brehier, 1949:39-166; Stein, 1954).

The peasantry was a crucial part of the Byzantine social structure, and the emperors always carefully maintained the peasantry's freedom and independence from the aristocracy. Between the 8th and 11th centuries the free peasantry, concentrated in village communities, flourished and provided an important source of recruits for the army, the bureaucracy, and even the professions.

The primary feature of post-7th-century Byzantine society was the free village community. Each peasant owned his land and cultivated it himself. He could, however, let it to another member of the community, either on a share basis or against cash payment. For taxation purposes, each such community formed a fiscal unit; if one farmer failed to meet his obligations toward the treasury, his neighbors were held responsible unless measures were taken to relieve them of this responsibility. On the whole, these villages were prosperous, at least in the

8th and 9th centuries. Of course, many of the villagers were poor, but a considerable number were well-to-do and a few even became very wealthy.

During this period, for the nonservile elements of the population freedom of movement was another feature of the empire's rural society, but this mobility was probably not very extensive. The vast majority of people grew up and died in the community where they were born unless they were moved by the authorities or driven out of their homes by a foreign incursion.

A constant struggle was waged by the free peasants against the aristocratic landlords who wished to encroach on the peasants' property. Throughout the periods of the empire's greatest strength, the free peasants were able to preserve their freedom with the help of the emperors, but after the 11th century, the power of the aristocracy grew and the free villages began to disappear (Andreades, 1935; Charanis, 1951a, 1951b; Dölger, 1953; Guilland, 1947; Lemerle, 1958; Ostrogorsky, 1954, 1956; Stein, 1954).

By the 9th and 10th centuries, the free village communities were seriously endangered by the tremendous expansion of large lay, ecclesiastic, and monastic estates. This was clearly perceived as a threat to the interests of the state by the enlightened emperors of the 10th century. The free peasantry began to decline, after the 11th century, as a result of the aristocracy's continuous encroachments and the weakening of the central power. The tremendous expansion of large estates during the 9th and 10th centuries, which gradually absorbed most of the small peasant holdings, had similar effects on the military estates. Emperors of the 10th century who tried to save the small peasant landholders also sought to protect the soldiers, but their measures proved as ineffective in the latter as in the former case.

In the 13th century, the vast majority of the rural population of the empire was constituted by dependent peasants known as *paroikoi,* whose number had greatly increased on state lands—a manifestation of the state's diminishing interest in the free peasants.

In medieval eyes, Constantinople appeared as a great commercial and industrial city. This impression was enhanced by the great importance of the capital as a center of international trade, its geographic position being inestimably beneficial in this respect. Despite occasional interruptions caused by wars, Byzantine foreign trade grew progressively until the international convulsions of the 11th century. The imperial authorities' interest in it was confined mainly to their concern for revenue and for acquiring the raw materials needed by imperial factories. The emperors always aimed at controlling trade, as is evidenced by the very elaborate system of guilds, which we will describe in detail when analyzing the structure of Constantinople.

Internal trade primarily consisted in supplying Constantinople with the necessities of life and a growing number of luxuries. Both trade and industry thus constituted important elements of Byzantine economy and society but formed a basic and dynamic part of the Byzantine social structure.

Byzantine trade began to undergo drastic changes at the end of the 11th century from which it never recovered. The Seljuk conquest of Asia Minor (in A.D. 1071) altered the whole of Byzantine economy by robbing Constantinople of the source of its corn supply and of a great part of the peasant population that constituted one of the chief foundations of its strength. The constant wars in Asia Minor upset Byzantine land and coastal trade. Moreover, the Normans, who at that time had invaded Southern Italy, were conducting piratical raids on the prosperous Greek peninsula and posed a very real threat to all the European provinces of the empire.

D
CENTER AND SOCIAL GROUPS IN THE BYZANTINE EMPIRE: CONTRADICTIONS OF THE IMPERIAL SYSTEM

As with all other imperial systems, the Byzantine Empire was caught between its dependency upon the relatively free resources generated by the various forces of concentration of the center and its attempts to regulate those resources through complicated economic, social, and political policies (Eisenstadt, 1963).

Compared with other imperial systems in general—particularly with those of China and Russia, analyzed earlier—as well as with the Ottoman Empire, which directly succeeded the Byzantine Empire in its own territory, one fact appears to be of crucial importance: A relatively high degree of corporate organization and even consciousness tended to develop in principle (such tendencies also existed in China) within the various groups, including the central elites; but more important, certain (though limited, especially when compared with Medieval Europe) tendencies developed toward autonomy. These groups tended to define their status in their own terms, according to autonomous criteria not entirely dependent on the center. They also tended to obtain some autonomous access to the centers of power. As already indicated, these tendencies toward autonomy were rooted in the traditions of the (Roman) city-state as well as in Christian principles.

These tendencies affected the legitimation of the rulers. When the structural conditions appeared relatively propitious, various groups (especially the aristocracy and, to a smaller degree, the Church) were able to impinge upon the center and obtain some participation in it, even if such participation was indirect.

Between the 6th and 11th centuries, the aristocracy thus participated in party politics at court. The senate and the bureaucracy generated certain articulated political organizations and activities in the form of cliques and were active in their own groups and parties. During periods of aristocratic preponderance and decline of the empire, the aristocracy tended to concern itself primarily with the direct usurpation of power and asserted its ascriptive territorial rights against the central government's demands.

Although the participation of social groups other than the Church, the bureaucracy, and the army was not as active and open, they also took part in the political process. Compared with other imperial systems—especially those of

China and Russia—these urban groups' political participation was more extensive and articulated, fostered by their orientations to wider issues and by their attempts to acquire legal political rights. The more active urban groups tended to promote some articulated political activities (Bratianu, 1938; Brehier, 1949; Diehl, 1929; Ostrogorsky, 1956; Vasiliev, 1952; Vryonis, 1963). The economically less developed urban groups often participated in sporadic, politically unarticulated rebellions and uprisings. In some cases, especially among the Byzantine "Zealots" of the 14th century, diverse religious, social, utopian elements and orientations were included (Barker, 1957; Charanis, 1941; Ostrogorsky, 1956).

E
BASIC PARAMETERS OF BYZANTINE POLITICAL AND SOCIAL HISTORY: IMPACT ON THE DEVELOPMENT OF URBAN SYSTEMS

1 Introduction

The very existence of these various groups intensified the process of political struggle in Byzantium and, hence, also the fortunes of the center and of the different groups.

The major stages of the Byzantine Empire's history were characterized by different constellations of the concentration and centrality forces. The first period, from the 4th through the 7th and 8th centuries, witnessed the disappearance of the free peasantry and the preponderance of great landed estates, either private or belonging to the emperor or the Church. In the second period, the free peasantry and international trade flourished. From the 7th to the 11th century, the emperors tried to restrain the aristocracy from becoming too strong and from encroaching on the free peasants; they were also able, on the whole, to pursue the extension of trade and even, to some degree, of internal trade. In the third period, however, from the 11th century, these imperial efforts were less successful. Different aristocratic groups gained the upper hand, both economically (through the extension of their landholdings) and politically. The last dynasties—the Comnenians (1057-1059 and 1081-1185) and the Paleologis (1261-1453)—were greatly dominated by different aristocratic groups. It was also during that period that trade declined steadily.

These various forces of centrality attempted, through diverse means, to regulate the forces of concentration upon which they still widely depended. The different fortunes of the centrality forces greatly affected the development of cities and urban hierarchies.

We will start our analysis by studying the impact of such developments on the structure of the urban hierarchies in the first period of Byzantine history that lasted from the 3rd to the 8th century.

The first basic constellation pattern of these various forces emerged during the first stage of transition, from the Roman to the Byzantine empires, and has been relatively well researched. As we already mentioned, the major changes that took place then were the establishment of the political center and the tendency to growing centralization and relative compactness of its borders, which contrasted with the basic pattern of the Roman Empire (Luttwak, 1976). There was also a weakening of the autonomous economic forces, especially the free peasantry and urban groups benefiting large landowners and the imperial and ecclesiastic (or monastic) estates, a situation that was to occur again in the last period of the Byzantine Empire. This specific constellation of social forces had far-reaching repercussions on the structure of cities and urban hierarchies. The single most visible event was, of course, the establishment of Constantinople, one of the most illustrious imperial centers in the world.

At the same time, the standing of many other cities—especially middle-level cities—experienced a relative decline due to two interconnected factors. The first was the loss of their political autonomy and, particularly, the political tradition of the city-state (Jones, 1940). The second factor was the far-reaching changes in the structure of the urban system, related to variations in the empire's economy.

The most important process in the transition from the Roman Empire to the Byzantine Empire was the relative disintegration of the urban hierarchy, with its strong tendencies toward quasi-primacy, toward a situation in which the middle and lower levels had weakened, their functions being dispersed among the economic, administrative, and ecclesiastic hierarchies. This general trend took place through the following transformation of the urban system: shifts of the system from one dominant function to another; ecologic separation of the various urban functions within and among cities; changes in the differential capabilities of some cities to fulfill various urban functions; and the interactions between towns and hinterland (Charanis, 1961).

In the later Roman and early Byzantine empires of the 3rd century, most of the cities fulfilled, in varying degrees, four major functions: (1) economic (distribution and production), (2) administrative (central and provincial authorities), (3) ecclesiastic (according to the Church hierarchy), and (4) military (as determined mainly by strategic conditions of the frontiers and the main directions of the Barbarian invasions). We will deal separately with each of these functions of the urban system as well as with the interactions among the various functions that brought about transformations in this system.

2 The Urban Network of Economic Centers

Prior to the political crises of the Roman Empire in the 3rd century, a well-developed urban network existed in the eastern part of the empire, with a well-structured hierarchical order fulfilling mainly economic functions. The following features made the existence and performance of the economic centers possible: (a) the political stability reigning throughout the Roman Empire; (b) the absence of strong military pressures on its frontiers; (c) the maintenance of long-distance routes enabling the development of commerce and specialization among the various parts of the empire and beyond it; (d) a relatively lenient fiscal policy allowing capital accumulation in the various centers of the provinces; (e) a large-scale open market encompassing the extent of the empire; and (f) the existence of a free peasantry providing the base for the empire's agricultural resources (Charanis, 1944).

This network of economic centers was made up of all the levels of city size: small, medium, and large. The density of economic centers differed in various parts of the empire; it was much higher in the eastern provinces, depending on the relative level of economic affluence and production and the extent of integration of a specific province within empirewide markets.

The cities fulfilled the following main economic functions: They were centers of regional distribution and collection; they were points of transfer for international or interregional commerce; and they produced goods for consumption in the city and its hinterland. These three major functions underwent severe transformations with the decline of the Roman Empire and the emergence of the Byzantine Empire. The most important of these changes was the weakening of the economic urban system's specific functions. The main reasons for the decline in the urban economic functions in the Roman Empire are attributable to the following processes: the political crises of the 3rd century due to the large turnover of emperors, civil strife, and mounting pressure on the borders, all of which had a devastating impact on the empire's economy, which, in turn, had a direct influence on the variability of the cities fulfilling a main economic function. The deterioration of transportation and communication networks within the empire resulted in the gradual severance of economic ties among the provinces and culminated in the emergence of almost self-contained local markets. The decline in interregional and interprovincial commerce divested most of the cities of one of their main economic bases.

Another major agent of change in the economic functions of cities was related to the gradual disappearance of the free peasantry and the concomitant emergence of large-scale domains based mainly on serf labor, the peasantry having lost its freedom of movement to a great extent. The large estates, owned by individual gentry, the Church, or the state itself, were almost self-contained and self-supported; they grew almost all their foodstuffs and were provided with products and goods by the artisans and craftsmen who had left the cities and were living on the estates. The large estates' tendency to economic self-containment was concomitant with the gradual agrarization of the cities. (City dwellers cultivated small plots of land outside the city limits, which is referred to as the *"proastia"* phenomenon.) The almost total disappearance of the free peasantry significantly curtailed the economic function of the small- and middle-level towns that had previously constituted the lower level of a commercial central-place hierarchy (Jones, 1964:844-855; Kurbatov, 1971:99-102).

The weakened economic function of the urban system mainly affected the lower and middle levels of the urban economic hierarchy, causing some of the small towns to dwindle to village size or be abandoned entirely by their inhabitants, who migrated to large estates or large cities. In the 6th century, the number of small towns in Asia Minor—in Macedonia, Greece, Dacia, and other parts of the empire—diminished significantly. A similar process occurred in North Africa, where the number of small towns—some 600 at the beginning of the 3rd century—decreased to about 400 during the 6th century (Jones, 1964:716-718; Kurbatov, 1971:58, 76).

In addition to the growth of the estates, whose serf labor made them self-sufficient in the production of goods, two additional major structural changes enhanced the progressive decline of the cities as centers of economic activity. The first was the strong military orientation of the imperial regime requiring the mobilization of heavy resources. This ever-growing need was filled by oppressive taxation levied in the cities by the council members who were compelled to serve in office. As the affluence of the cities' inhabitants and merchants gradually declined, the imperial regime launched its own sources of supply. The main outcome of these steps, aiming at achieving a measure of economic self-sufficiency, was to bypass the regular market channels going through the urban economic hierarchy and weakening the cities, especially those in the middle and lower levels of the urban hierarchy.

The second structural change was related to the far-reaching transformations in the role of the Church and monasteries in the economic realm. From the 3rd century, monasteries were built on the outskirts of cities and in remote rural areas, and they began to accumulate land and attract peasants and craftsmen. They thus emerged, in the 5th century, as major centers of economic activity. Gradually the monasteries penetrated the cities proper and fulfilled a major role in the urban production and distribution processes. In the 6th century, numerous monasteries began to replace small towns as the main economic centers of the regions. Direct exchange of commodities and goods among the monasteries themselves reduced the cities' economic functions even more (Jones, 1964:926-932; Koledarov, 1966:44; Kurbatov, 1971:21-36).

The common denominator for all the processes analyzed above was the dramatic reduction in the volume of resources flowing through the channels of the markets, thus canceling the need for the cities—mainly lower and middle size—to act as economic centers (Jones, 1974:129; Oertel, 1939:256-259; Teall, 1967). During the long economic transition from the 3rd to the 6th century, the urban system had indeed changed from a well-developed hierarchical economic system to one with a strong primacy distribution, as only the largest cities survived the contraction of regional commercial interaction due to the imperial government and Church bypassing the markets.

3 Changes in the Administrative Structure of the Empire and Their Impact on the Urban System

These changes affecting the empire's economic structure were parallel to those occurring in the imperial administrative structure. All culminated in the government administration reforms initiated by Diocletian and Constantine at the end of the 3rd century and beginning of the 4th century. Several facets of these administrative reforms are important for our analysis. One of these was the establishment of new levels in the administration, creating a new vertical administrative structure.

Diocletian divided the Roman Empire into 12 dioceses, each of which was subdivided into provinces. Constantine added another level by dividing the empire into four prefectures, each of which comprised several dioceses. The result of this reform was the creation of a more articulated hierarchical administrative system through the central authorities' initiative, bringing about the establishment of an all-encompassing administrative hierarchy.

Another facet of these reforms was that they intensified the density of the network of administrative centers. At the end of the 3rd century there were about 100 provinces; in the 5th century their number increased to 120. The number of dioceses grew from 12 in the 3rd century to 14 in the 4th century. This elaboration of the administrative structure resulted in an increase in the number of capital cities in each administrative level, although the zone of influence of each of them was reduced.

These reforms in the administrative structure were followed by a significant change in the character of the imperial city administration, moving further away from the tradition of the city-state. In this context, the most important changes were the transfer of authority and administrative functions from the urban councils to the central government officials and the functional division of the imperial administration into separate civilian and military authorities (Liebeschuetz, 1972).

With the transfer of authority to imperial officials, the city councilmen (*curiales*) gradually lost their political power, becoming forced hereditary tax collectors for the imperial government and having only a slim chance of moving into the imperial bureaucracy's ranks. The governor of a province, whose seat was the provincial capital, became fully responsible for all administrative but not military functions. The administrative functions performed in these capital cities were mostly oriented toward the transfer of local resources to the central government and toward province-wide implementation of instructions emanating from the center. The allocation of resources in the city and its surroundings was thus determined by the central government's interests and, to a much smaller extent, by local requirements and initiatives, thus further undermining the economic situation of the cities.

The concurrence of these administrative changes with the process of economic decline generated a special type of relationship between the urban system of economic centers and the administrative hierarchy. Initially those cities that were the most developed economically were selected as capitals of the numerous new provinces. Because of the large number of capitals and the economic decline of many of them, the administrative functions were more predominant than the economic ones.

Governors of the capital cities attempted to consolidate their positions in their respective capitals in several ways—among others, by sponsoring public works and building. The financing of these projects was made possible by the imposition of heavy taxes on the small towns, negatively affecting their economic base. Thus these activities also contributed to the continued deterioration of the small towns at the lowest level of the economic hierarchy until their almost complete extinction, when their population eventually migrated to the provincial capitals.

The second result of these administrative changes was stabilization—but only temporary—of the economic functions of cities at the middle level of the urban system by the infusion of a large number of administrative functions, given that their fulfillment required the employment of numerous officials supported by imperial resources. The administrative functions thus provided a channel for augmenting the resource base of the cities. Yet, unaccompanied by production and commercial activities of a significant level and connected to the weakening of small towns and even their disappearance, these functions alone could not provide a sufficiently strong base for an urban economy. The combination of administrative and economic functions in these cities, however, was only a temporary phenomenon (Calude, 1969:chap. II).

Toward the beginning of the 6th century, the economic contribution of administrative activities to these cities dwindled significantly. This process was also reinforced by the new administrative reforms initiated at that period: consolidation of provinces into larger units, which brought about the abolition of many provincial capitals. At the same time, the allocation of resources by the imperial government to the capital cities was reduced, impoverishing the administrative centers' economic base. From the 7th century onward, many of the capital cities came to fulfill the single role of taxation centers constituting a channel for the transfer of resources from the periphery to the empire's capital, and were devoid of any autonomous function or impact on the center (Claude, 1969:chaps. II, IV; Jones, 1964; Kurbatov, 1971:83-84, 104-105; Liebeschuetz, 1972).

4 The Ecclesiastic Network

The ecclesiastic urban system that evolved in the Byzantine Empire was a perfect example of a hierarchical system with all the characteristics of such a system. Its lowest level was the seat of a bishop that had already been instituted in the 2nd century. Almost every town of any significant size had a bishop, and there were bishops even in other population centers, such as large villages, army camps, and neighborhoods of large cities. The lowest level of the ecclesiastic network was thus very dense and the bishoprics were spread throughout the empire.

A second level, approved formally at the Council of Nicea, that emerged due to the need to coordinate and support the individual bishoprics comprised the Metropolitans (bishops of provincial capitals). In the period when Christianity became the official religion of the empire (the first half of the 4th century), the ecclesiastic provinces coincided with the administrative provinces of the empire.

The next level of the Church hierarchy—that of the Patriarchate, which was a combination of provinces—was established toward the 6th century and comprised four Patriarchates. Its formalization was followed by fierce struggles within the Church. Great differences existed among the Patriarchates in their size and in the Patriarchs' authority. On this level the Patriarch of Constantinople held the top position in the ecclesiastic hierarchy (Jones, 1964:871-894; Kirsten, 1958; Loewenstein, 1973).

The functions that the urban ecclesiastic network came to perform were closely related to the gradual decrease of urban autonomy as the Church encroached on and finally took over many of the functions fulfilled until then by municipal organizations—religious services, education, welfare, and philanthropy—thus weakening both the internal cohesion of the cities and the standing of many of them in the hierarchy. The transfer of welfare functions from the municipalities to the Church was followed by a basic change in the fulfillment of those functions. In the hands of the municipality, welfare had been limited to the city's inhabitants, whereas Church welfare was extended to the city's inhabitants and to those of the surroundings alike, thus further weakening the cities' internal cohesion. Entertainment and sports, which in the Roman Empire had been provided by the municipalities, similarly lost their importance because of the Church's influence. The organization of spectacles was increasingly taken over by the imperial administration, and the performances lost their local ritual character.

At the end of this process, representation of the local population vis-à-vis the central authority was moved from the *Curia* (City Council) to the Church. In the early stages the Church had represented only the Christian population vis-à-vis the city council, but gradually the bishop came to represent the total local population to the central authority. It can be concluded that in the course of this process, the Church preempted the functions and administrative role of the municipality (Claude, 1969; Kirsten, 1958; Kurbatov, 1971:134, 183-188; Liebeschuetz, 1972; Loewenstein, 1973).

The emergence of the ecclesiastic network also affected the economic standing of many middle- and low-level cities. In its early stages of evolution, the only meaningful contact between the ecclesiastic and the economic networks was through the influence of the economic hierarchy on the respective level of a religious center within the ecclesiastic hierarchy. Very soon, however, the Church became a major economic power in its own right, deeply influencing the economic urban hierarchy and weakening its middle and lower levels. The Church's independence was ensured by its vast landholding; with a ready supply of foodstuffs and agricultural produce, the need for market transactions was thus overcome to some degree. With the gradual migration of craftsmen from small towns to churches and monasteries, the Church was almost entirely self-sufficient and altogether bypassed the markets. This process seriously affected the small towns of the economic hierarchy, causing their almost total disappearance.

In the large cities, the Church had a lesser impact on the economy, as it primarily constituted an important source of demand for various commodities, thus encouraging the cities' economic development. The building of numerous churches and monasteries required the employment of numerous builders, masons, and other craftsmen. In the upper level of the economic hierarchy, the Church was therefore a major employer of professionals, forcibly contributing to the economic bases of the large cities while simultaneously depleting the resources of the urban hierarchy's middle and low levels (Charanis, 1948; Jones, 1964:896, 904; Kirsten, 1958; Kurbatov, 1971:35-37, 65-66; Liebeschuetz, 1972; Ostrogorsky, 1959; Teall, 1967).

5 The Military Network

The network of military cities, in contrast with the others, did not cover the entire empire but was limited to the regions along the frontiers, which were under constant threat of attack. This feature of the military network, which had initially been regional in nature, changed during Justinian's reign (527-565), when the imperial center assumed a strong military orientation with a powerful and aggressive ideological tendency. This led to the articulation of the military strategy in central planning, which produced, for a short period, a dense network of military strongholds possessing a distinct hierarchical structure.

The urban settlements could fulfill a variety of military functions. They could be garrison towns, shelter the civilian regional population in case of attack, or be a production center to provide the needs of the army, as well as constituting strategic points within the regional or even imperial overall defense plan. Large cities fulfilled the above functions; small cities were mainly garrison towns or played a role within a regional defense plan. With the gradual decline in the political authority of the municipalities and their diminishing role as economic centers, many small towns contracted until only the fortress (*kastron*) was left. This process was reinforced by the fact that military considerations were of a national nature in the imperial regime, whereas local defense needs were considered insignificant. The large cities, the backbone of the imperial regime, were well protected, while, in periods of Barbarian invasions, the small towns were abandoned.

The military administration was distinct from the civilian one, thus creating a separate network not entirely overlapping that of the "civil" administration. The separation of these two networks until the end of the 6th century reflects the central authority's concern to prevent the provincial governor from gaining too much power. The urban centers of the lower level thus degenerated into mere army strongholds (Claude, 1969:74; Kirsten, 1958:19; Kurbatov, 1971:61, 68).

In periods in which the imperial regime had a strong military orientation—as under Justinian and Heraclius—the military network became more hierarchical, penetrated into the periphery, and was strongly controlled by the imperial center. In periods of weak military orientation, the military network lost its hierarchical structure and became less dependent upon the central government, reaching the point at which military centers in the periphery were almost self-supporting, producing their own foodstuffs and equipment (Claude, 1969; Jones, 1964:773-781; Loewenstein, 1973:479-483; Ostrogorsky, 1956:86-88; Teall, 1967).

6 Some Comparative Indications

As indicated earlier, a rather special type of urban system evolved in the Byzantine Empire, especially when compared with such other imperial systems as those of China and Russia.

In comparison with the Chinese system, the most distinctive characteristics of the Byzantine imperial system were the weakness (especially economic) of the hierarchy's low and middle levels, the division of the middle and lower levels between different networks (administrative, ecclesiastic, military, and economic), and, hence, the strong tendency toward semiprimacy. This difference was closely related to the structure of the concentration and centrality forces of the respective empires. In China, the strong unified elites, combined with the continuous development of the agricultural and commercial activities regulated by the center, contrasted, in the Byzantine Empire, with the multiple central elites, the relative autonomy, and constantly changing fortunes of the major strata as well as their less effective regulation by the center.

In comparison with Russia, the Byzantine urban system was characterized by a much stronger tendency to shape its contours and foster closer interrelations between the different hierarchies affected by the regulation of the center, at least in each of the systems. In Russia a de facto stronger caesaropapism developed than in Byzantium, the Church as well as such other components of the center as the bureaucracy were much weaker than in the Byzantine Empire, and the political elite had no competitors with respect to its central regulative functions, but at the same time, the center was not always very effective in controlling the more spontaneous social forces.

The unique role of the capital city, Constantinople, in which all the functions of the urban system were concentrated, unchallenged by any other city in the empire, contrasted in some degree with China and Russia. A much greater difference existed, on the whole, among the various cities in the Byzantine Empire than in China, because few of these cities combined different urban factors, and each tended to be related to or to emphasize a role in one network (at most two). The concrete structure of the different networks appeared to change at different periods of the Byzantine Empire, according to the varying constellations of the social forces outlined above. It seems, however, that a central feature that greatly influenced the changes in the development of the Byzantine Empire's urban systems was the relative strength of the free peasantry, which also affected the scope of agricultural production and marketing. The stronger that peasantry was, the closer were the relations that tended to evolve among the different urban networks.

F
CONSTANTINOPLE—THE IMPERIAL CITY

1 The Creation of an Imperial City

One of the most interesting aspects of Byzantine urbanization is the great difference that existed between Constantinople's greatness and structure and those of the empire's other cities. Constantinople constituted the apex of all urban life in Byzantium and was "the" imperial center, the center of the empire's political, administrative, religious, and economic life.

Constantinople was created in the 4th century A.D. as an imperial Christian city, as the one locus where these two powers merged. Through this imperial stroke, Constantinople became the prime political and spiritual center of the world, replacing both Rome and Jerusalem.

Based on Miller's (1969) periodization, several stages are discernible in the evolution of Constantinople's shape and functions. The first stage, extending from the founding of the city to the end of the 6th century, is characterized by the establishment of its major outlines. Constantine defined the site and its first limits and designed the basic network to conform to his vision of the city as a replica of a Roman imperial capital. The definitive boundaries of the city area were determined by Theodosius the Second, who constructed its prime defenses in the form of the huge walls and fortification system at the western end of the triangular peninsula. The monumental architecture, defining the major spaces and masses, indicated the culmination of this stage, which added the final touch to the Roman image of a capital city.

The following analysis focuses on the internal ecological structure and social organization of the city at the end of the 6th century, the time of crystallization of the first stage of Constantinople's development. During this period the process of establishing its major outlines and architectural features took place.

The founding of Constantinople was connected to the deteriorating political and military situation of the Roman Empire in the 3rd and 4th centuries, when it became clear that Rome was no longer suitable to fulfill the role of imperial capital. At the end of the 3rd century, therefore, Diocletian decided to move the capital of the empire (the seat of government) to the east, nearer the Asiatic frontier, where the Persian power was posing a perpetual threat, and the Danube frontier, along which the danger of Barbarian invasion was imminent. Diocletian's initial plan to establish the new capital of the empire at Nicomedia was not implemented but served for Constantine, his successor, to point clearly to the east for the establishment of the new capital. The fast decline of Rome and its replacement by Ravena, in the west, as the capital of the empire, as well as the geopolitical need to move the center of the empire eastward, led Constantine to make the major decision to build the new single capital of the Roman Empire on the site of the small Greek colony of Byzantian.

This site was carefully selected by Constantine. Byzantian was located at the end of a small peninsula of the eastern coast of Europe, where the Bosphorus flowed into the Sea of Marmara. The peninsula was of triangular shape and protected by water bodies on the north, east, and south, which left only its western side to be defended on land. Its location also offered superb advantages. It commanded the easiest land route between Europe and Asia, and the sea passage between the Black Sea and the Mediterranean afforded it, on an intercontinental basis, the highest accessibility. A better location than the site of Byzantian could not have been found for ruling and controlling an empire extending on both continents. From a regional perspective, the neighboring provinces of Thrace in Europe and Bithynya in Asia possessed rich and fertile agricultural land and would be able to provide food for the urban population. The wooded hills and valleys bordering the site were endowed with springs and could be relied upon for a permanent food supply. The very large landlocked harbor of the Golden Horn could accommodate the vast imperial fleet and shelter the innumerable ships needed by international commerce to be centered in the capital city.

In addition to all these advantages, there was an additional consideration of a symbolic nature: the site was almost devoid of former habitation, and the town of Byzantian was small enough to be absorbed completely and vanish in the splendor of the new city. By establishing his capital there, Constantine signaled a complete break with the past, heralding the end of the pagan Roman Empire and the beginning of the Christian Roman Empire. The founding of Constantinople thus embodied the imperial will and vision, in both the political-military and the spiritual realms, making the new capital a perfect illustration of the operation of centrality forces in shaping urbanization processes.

2 Conception of Cosmos and Empire and the Internal Structure of Constantinople

Constantinople (later Istanbul) thus provides an excellent example of urban development under an imperial regime in constituting the ideal type of imperial city. Its imperial features were decisive in all phases and facets of its urbanization process: its origin, the forces shaping its growth, its plan and ecologic structure, its daily life, and its relationship with the Church.

Its creation, physical pattern, and symbolic metaphors all stemmed from a relatively unified view of the cosmos, which was perceived as a unitary political framework in which the all-encompassing empire was headed, embodied, and represented iconically and symbolically by its center, the capital city, and by its creator and ruler (Miller, 1969:5). The imperial character of Constantinople was reflected mainly by its size and the mode of control of urban space. Its unparalleled size, both in area and population, was due to the city's mastery over the empire's material resources. The effective control was realized physically through the planning of the city, its major outlines and spaces being perceived and executed according to the imperial will.

Despite the fact that from its inception, Constantinople was a truly Christian city, filled with churches, monasteries, convents, orphanages, and old-age homes maintained by the Church, the basic features of its ecologic structure were unequivocally classical, its plan, buildings, monuments, and the structure of its public places testifying to its Greco-Roman origin. Constantinople retained many characteristics of the urban civilization of the Roman Empire, with its great and illustrious cities at a time (in the early Middle Ages) when Western Europe was barely advancing as a self-sufficient rural economy, with its cities becoming small provincial market centers.

Constantinople's internal structure was defined by the various functions fulfilled by the city as the imperial capital, the most important being protection of the polity and the population; provision of food and services as an expression of the emperor's wealth and his responsibility for the welfare of his subjects; control of movement and activities within the urban framework; and the symbolic-ritual role of structuring the environment, in which the emperor figured as God's representative on earth.

The protective role of Constantinople as an imperial city was manifest both in the physical features of the triangular site—the Sea of Marmara and the Golden Horn surrounding the peninsula on two sides—and by the complex system of walls, towers, and fortifications on its western (land) side. To strengthen the defensive system even more, walls were built along the sea side of the city to offset any maritime attack. The walls built by Theodosius the Second, from 413 to 447, defined and delimited the urban space for almost a thousand years, constituting the single major stable and enduring feature of the urban structure. They not only fulfilled a defensive role against external threat but also provided a major element of control of the urban population. Movement into and out of the city through the gates could be channeled and supervised, allowing the imperial authorities to check and regulate the flow of travelers and merchants into the capital.

Those living in the imperial capital were privileged with receiving some of their basic needs through the emperor's generosity. Many facilities and structures existed in the city from which the distribution of food and services was carried out, and their bestowal by the emperor not only stemmed from charitable motives but was also aimed at abating any anarchic tendencies of the mob. The emperor acted as if he were the recreator, the divine representative, whose duty was to continue life on earth and support it. *Philanthropia* was one of the major traits of the emperor, and the daily provision of food and water in several parts of the city was its main manifestation (Barker, 1957:84). In this respect, the older tradition of the city-state in its prime, the popular one of *panem and circensen*, merged with special Christian-Oriental features.

The most visible expression of the imperial city's role of provider was the great aqueduct of Valens, which cut through the city on a longitudinal axis, from the Andrinople gate in the west through to the palace area at the eastern apex of the urban space, transporting water from Thrace to Constantinople. The city was therefore capable of sustaining long sieges without having to ration water significantly. The imperial responsibility of ensuring both the supply of water (the spring of life) and the distribution of bread gave the city an aura of regal wealth and munificence, an impression that people's most basic and vital needs were accessible to all inhabitants.

The street pattern of Constantinople had two distinct levels. The upper level, planned and directed by the central authorities, was well demarcated and well maintained and provided the main arteries for processions and traffic in the city. This upper-level array of main streets wedged in from the various city gates, focusing on the *Augusteum* (the apex of the triangle) at the meeting point between the palace and the church of Haghia Sophia. This street pattern, with the *Mese* forming its backbone, comprised the ceremonial sacred route, the governmental administrative axes, the specialized commercial avenues and major squares, as well as the main points of interaction in the city.

Branching out from the broad avenues of the upper level were alleys serving the various neighborhoods and constituting the lower-level street system. Due to scarcity of space and to the structures that provided protection from the severe climate and intrusion by foreign elements, these alleys were narrow and winding, well fitted to the topography and to the needs of the local population. These lower-level streets did not form a complete urban network; they were used only by local inhabitants and did not provide main links between the different neighborhoods.

A most significant feature of the imperial capital was the existence of a monumental axis in the upper-level street system: the *Mese,* which was a major street, the spinal cord of the entire road network. This wide avenue ran all the way from the main royal gate on the western wall—the Golden Gate—through the city, reaching the apex of the triangle at the *Augusteum,* the most important forum area, located between the royal palace and Haghia Sophia. This major artery differed from the rest of the road network; it had been planned and designed on the establishment of the city, was very broad, and its two sides formed a straight and organized front. It was well paved with heavy stone blocks and flanked on both sides by tall colonnades supporting a roof that provided shelter from sun and rain. These roofed sidewalks were lined by shops, making the *Mese* the city's most attractive thoroughfare.

The major functional and symbolic significance of this central avenue was further enhanced by several forums placed at intervals along this central axis. Built as large squares surrounded by colonnades, the forums gave Constantinople a definite Greco-Roman character. These large, well-defined spaces were the main public meeting points of the urban population, preserving the old Roman traditions and providing the physical setting for exhibiting the most prominent symbols of imperial power and authority: triumphal arches and tall columns supporting statues of the emperors. The forums were thus central points of interchange, interaction, trade, and entertainment, as well as the foci of major political symbolism and participation. All activities were concentrated within a controlled space and under the symbolic supervision of the emperor, a special type of public space combining the functions of centrality with a relative openness of access to them and a place of open public participation.

Along the *Mese* there were several forums with similar architectural features that served various public functions. This could be observed in the structure of the *Mese*, the central avenue emanating from the *Augusteum,* the route used by the emperor for ritual processions and triumphal marches when returning from the battlefield. The *Augusteum,* with its surrounding buildings, symbolized the center of the capital and therefore the heart of the empire. It was flanked on one side by the gate of the Great Palace, which was not only the residence of the emperor but also the seat of government, administration, and treasury of the empire. The Senate House, rebuilt many times in a classical architectural style, stood on another side of the *Augusteum*. The Constantinople Senate, although it had lost most of its political power, still maintained the ceremonies of its ancient traditions, thus linking the empire to its Roman heritage. Toward the third side of the *Augusteum* was an entrance to the great hippodrome and its adjacent public baths. This large hippodrome held a major role in the life of Constantinople's population as it was the main center for public gatherings, ritual convocations, public protests, and all types of entertainment. The hippodrome could be considered as the symbolic center of the public life and power of the people. It was here that the major factions *(demes)* fought their fights and organized their demonstrations. On the north side of the *Augusteum* soared the majestic church of Haghia Sophia, the main religious center of the capital and the empire. The church, built by Justinian to replace earlier churches on the same spot, was constructed in a bold and innovative design. It dominated the urban landscape by its size and massive shape and became the most prominent building in Constantinople, symbolizing the spiritual roots of the "God-guarded city."

In the square itself were the two symbols of measurement: the horloge (great clock), which controlled time, and the *Milion* (navel stone), which controlled distance, thus marking the point from which all distances were measured. Owing to these two instrumental emblems, the *Augusteum* assumed the role of omphalos of the world, the meeting point of the cosmos and the earthly reality. The *Augusteum* was thus the prime example of the operation of the forces of centrality, bringing together in a single defined urban space the three mainsprings of life in the imperial capital: the emperor, the Church, and the people, all under the emperor's control.

Issuing from the *Augusteum* and leading all the way through the city, the *Mese* evinced some characteristics of a Sacred Way. This was most evident in its first section, called the *Regia* (Royal Way), between the *Augusteum* and the forum of Constantine. Being closest to the palace and connecting the *Augusteum* with the forum of the city founder, this section had the highest symbolic significance: Along it was located the *Praetorium,* the headquarters of the prefect, governor of the city. The forum of Constantine differed from the other forums by its oval shape, a sort of physical representation of the sun shield, the source of power and energy (Janin, 1964:62).

From the Constantine Forum, the *Mese* led to a series of public squares where the numerous activities of daily public life of the city were held. Each of the squares assumed different functions (Downey, 1960:19). The various forums and the hippodrome of Constantinople thus continued the tradition of open public life of the city-state, but they were controlled here by the imperial center and organized, to some degree, in a semihierarchical mold.

The lower urban groups were organized into *demes*, or "circus parties," such as the famous factions of Blues and Greens, which had political aims and were not merely sport organizations. These popular parties, whose leaders were appointed by the government, fulfilled important public duties, serving as city guard and participating in the repair of the city walls. The Blues and Greens played a very important role in all the big cities, providing a means through which people could voice their political opinions (Bratianu, 1937, 1938; Bury, 1910; Dvornik, 1946; Guilland, 1966; Maricq, 1949). The *demes* were apparently effective only from the 5th through the 7th century. Their direct political impact subsequently diminished and they gradually became merely recreational groups.

But the power of the urban populace in the political life of the empire had still not waned in the 10th century. Suppressed for a while during the glorious days of the Macedonian dynasty, the masses reemerged in the 11th century, and the last representative of the Macedonian dynasty owed them the recovery of his throne from Michael V. Moreover, with the weakening of direct Roman influence and the establishment of the imperial regime, the urban groups' center of strength and political power shifted to the guilds. While these guilds were regulated and controlled by the emperor and the bureaucracy to no small degree, they were not yet entirely politically passive or uninfluential (Charanis, 1944; Mendl, 1961; Vryonis, 1963).

3 Social Differentiation and Guilds

The population of Constantinople, which rarely fell below 500,000 (Charanis, 1966:4; Jacoby, 1961), was of variegated origin and reflected the national and ethnic mixture of the Byzantine Empire itself. Among the numerous ethnic groups living in Constantinople were Anatolians, descending from Cappadocians and Phrygians, Greeks from mainland Greece and the islands, Illyrians from the Adriatic coast, descendents of Goth and Celt tribes, Copts from Egypt, Syrians, Armenians, and Jews. All of these groups mingled with the descendents of the old Roman Empire and formed an urban population that called itself Roman and spoke Greek. The loose criteria for citizenship were belonging to the Christian Church and using the Greek language, which predominated in the Byzantine Empire.

The social ecology of Constantinople was spatially relatively undifferentiated. The population of the various neighborhoods was quite heterogeneous and included a full range of its occupational and socioeconomic sectors. This implied the absence of both aristocratic residential quarters and impoverished ones. The only significant differentiation affecting the urban ecology was of a national religious character and caused the concentration of the Western Latin elements of the population in a quarter north of the Golden Horn, well separated from the main body of Constantinople.

Manufacturing, trade, and provision of services in Constantinople were highly organized within the framework of centrally controlled guilds. A most valuable source for analysis of the guild system is provided by the Book of the Prefect, a compilation of government regulations dealing principally with the most sensitive and socially important guilds (Boak, 1929): It depicts the situation of the tightly regulated and stratified trading society of Constantinople, with a very precise and meticulously preserved sphere of action for each guild.

It is possible to clarify the rank order of the stratified guild system from the small but distinctive symbolic measures differentially applied by the central authorities to the various guilds (Miller, 1969:63). The most honored was the guild of the notaries, closely followed by the imperial textile guild. The products of the imperial silk guilds, which included the state robes and garments for the ecclesiastic hierarchy, were regarded as treasures and not as commercial articles to be traded. This direct relationship of the imperial silk guilds with the emperor and the court accorded them high social status among all the guilds (Lopez, 1945). Next came the various types of silk guilds, then those of the clothiers and jewelers, which outranked the spinners' and weavers' guilds. The guild of the perfumers, whose members enjoyed the privilege of placing their stalls in the immediate vicinity of the palace gate, occupied a position in the middle of the rank order, higher than the soapmakers' guild but lower than the candlemakers'. The bakers' guild headed those of the food providers; those of the tanners and tavernkeepers were at the lowest level. It can be concluded therefore that the major influence was exercised by the level of services to the emperor and the court. The forces of centrality thus played a decisive role in shaping the organizational structure of the manufacturing, trade, and services activities in Constantinople.

The introduction to the Book of the Prefect indicates that the dominant motive for the minute regulations controlling the manufacturing and commercial guilds was the keeping of order and harmony among the citizens in their daily activities. The guild system was aimed at reinforcing the divinely appointed harmony. Members of the guilds and the Prefect's officers were responsible for maintaining the highest professional standards of the guilds and the meticulous preservation of their membership. Recruitment into the guilds was strongly regulated; once admitted to a guild, the member was forbidden to change his trade and join another guild. All these regulations and restrictions, enforced by the Prefect in the name of the throne, were aimed at achieving order and harmony, which were the basic imperatives of the imperial regime.

The state's purpose in devising and enforcing the guild system was to establish a mercantile community in which competition would be limited and whose prime aim would be service to the throne and the public rather than profit (Miller, 1969:73). Given the basic social orientations and political traditions of the Byzantine Empire, however, the power of the guilds was, in fact, much stronger (Vryonis, 1963).

G
FROM CONSTANTINOPLE TO ISTANBUL

Constantinople was taken over in 1453 by the forces of the Ottoman Sultan Mehmet II ("El Fatih," the Conqueror) after fierce battles and a harsh siege that brought an end to the long and illustrious history of the Byzantine Empire. The last emperor, Constantine XI, was killed beneath the walls of the city, marking the termination of the Byzantine imperial rule of Constantinople. Sultan Mehmet II transferred his government there from Edirne (Andrinople) and made the city the capital of the Ottoman Empire. The imperial character of the city was thus preserved, although Constantinople changed from a Christian to a Moslem city. It is nevertheless important, within the context of this study of urbanization in the Byzantine Empire, to follow the transformations that occurred in the structure of the city as the imperial regime and the basic cultural orientation changed.

When Mehmet II entered the conquered city, he found it in ruin as a result of the long process of decline and decay, aggravated by the ordeals of siege and battles. Most of the surviving (some 50,000) inhabitants were enslaved and taken to the slave market of Edirne, then the Ottoman capital. Constantinople was thus almost deserted, creating a void, in human terms, between the Byzantine and Ottoman occupation of the city. In order to bolster the city's development, Mehmet II launched a large-scale policy of resettlement. By offering housing and land and by establishing new urban facilities, the sultan attracted population from all parts of his empire into Istanbul, as the capital was now called. This policy of repopulating the capital city was not limited to Turks or Moslems alone: Greeks and other Christians, as well as Jews, were permitted and in some cases even encouraged to settle in Istanbul. Through this active resettlement process, the capital revived rapidly. By the time of Suleyman the Magnificent, in the mid-16th century, its population had again risen to about a half-million, the city regaining its primate position within the urban hierarchy of the Ottoman

Empire. The process of resettlement was characterized by the sultan's liberal and tolerant attitude and assembled a most diversified and dynamic population. The Christian and Jewish communities enjoyed freedom of worship and were allowed a large measure of communal autonomy. On the northern bank of the Golden Horn, the communities of Western merchants were reestablished, thus adding a European element to the mosaic of the urban population. Yet the most important process, instrumental in shaping the character of this population resettlement, was the steady growth of the Turkish-speaking Moslem majority, recruited either through conversion to Islam, by assimilation, or by transfer to the city from the Islamic parts of the empire (Lewis, 1963:103).

Making Istanbul the capital of the Ottoman Empire and resettling population in it brought about changes in the structure of the city and in its basic design and architecture. The simplest of these changes, which had the most visible effect, was the substitution of Islamic religious buildings—above all mosques—for the Byzantine churches. This was sometimes accomplished by the direct conversion of churches into mosques, but more often it came about by building new mosques. Such a substitution, although visible and impressive, was not merely external but was connected to far-reaching changes in some crucial aspects of the city's internal structure, especially in the degree of centralization of its public spaces and central control over them, as well as in the structure of these public spaces and their interrelation with private ones. These transformations were not accidental but were closely linked to the basic Islamic religious and sociopolitical conceptions, to the structure of the elites and the modes of control they exercised, which indeed contrasted with those of the Byzantine Empire. We will analyze these in greater detail in Chapter 8.

Given the simple facts that, ecologically, we are speaking here of the same city and that it continued to be an imperial capital, the change from Constantinople to Istanbul provides one of the most interesting and clear comparative illustrations of the impact of different cultural orientations and structures of elites in shaping the internal contours of cities. Some of these differences between Constantinople and Istanbul will now be examined.

A rather far-reaching transformation of the whole pattern of control over public lands and institutions took place. This change was fully evident in the fact that most of the mosques and *madrassah* (institutions of higher education) became *waqf* (public or private) foundations established (in most cases by the sultan) in perpetuity for these specific purposes. These foundations were administered by special notables and were no longer under the central bureaucracy's direct control. They became semiautonomous, combining both the public and private functions in a way typical of most Islamic lands (Baer, 1969; Gibb and Bowen, 1957). The establishment of these mosques and *madrassah* thus gave rise not only to a growing decentralization of urban life in general and public urban life in particular, but also changed the relations between public and private land, weakening the distinction between them, the maintenance of which had at least been attempted by the Byzantine emperors.

The most visible change in the structure of the city occurred through building a large number of mosques in all parts of the city. In urban terms, the mosques were much more than a place of worship; they constituted a focus for the major public institutions, mainly those of higher education. In many ways, the mosques and the various public institutions clustered around them fulfilled the role of large-scale and well-organized community centers, thus diminishing to a large extent the urban population's reliance on the court and imperial service outlets. Compared to the relatively unicentered Constantinople, which had been focused on the imperial palace and had been controlled by it in many ways, Istanbul was much more multicentered. The most outstanding symbol of the city's transformation from a Christian to an Islamic capital is evident in Mehmet II's order to convert Haghia Sophia into a mosque to be called Aya Sofya Camii. In his attempts to enhance the Islamic character of his new capital, Mehmet II built the sultan's mosque on the fourth hill of Istanbul—a new monumental structure whose size was comparable to that of Aya Sofya. The location was chosen for its central position within Istanbul and also for its distance from the former center of the city, on the first hill. The Conqueror Mosque was not merely a splendid place of worship. It was surrounded by eight *madrassah* and their dormitories and it constituted a "university city" that remained one of the chief centers of learning in Istanbul for hundreds of years. This urban concept of building a mosque surrounded by *madrassah* was also adopted by other sultans and their *vizirs* (ministers) and added several new educational and community centers to the city structure.

The first three sultans made significant contributions to the revival, growth, and restructuring of Istanbul. These urban processes culminated in the efforts of the fourth sultan Suleyman the Magnificent (1520-1566), during whose reign the city reached the peak of its splendor. Among the many attempts made during Suleyman's reign to restructure the imperial capital into a Moslem one, the most important was the building of the mosque of Suleymaniye, designed by Sinan, the second largest but by far the finest and most magnificent of the imperial mosque complexes in the city. It was located on top of the highest hill in the center of Istanbul, quite distant from the site of the former imperial center at the eastern corner of the triangle. This particular structure changed the skyline of the city forever, creating a counterbalance to the conspicuous building of Aya Sofya.

In this case as well, from an urban point of view, the buildings surrounding the mosque were of great significance, indicating some of the specific Islamic urban construction works. Some of them were *madrassah*, schools for the study of the Koran, law, and medicine. Another building in this vast complex was the Dar-us-sifa, a large public hospital, with a special section for the care of the insane. The hospital was well known for its size, organization, and the level of care and charity provided. Adjacent to the hospital was the Imaret, which housed an enormous kitchen as well as refectories supplying food for the several thousand dependents of the Suleymaniye: the clergy of the mosque, the faculty and students of the seven *madrassah*, the staff and patients of the hospital, and the travelers using the caravansary, which also formed part of the complex. Furthermore, the Imaret fulfilled a most important communal function in providing meals for the poor of the surrounding neighborhood. The Suleymaniye imperial mosque and all the related public facilities thus actually transformed the basic structure of the city: The main center of activity was moved from Constantinople's forums to the complex of this imperial mosque of Istanbul.

In a parallel manner, the functional structure of the city underwent a significant change with the transfer of commerce and shopping activities from the former forums and the *Mese* to the covered bazaars situated north of the former Constantine Forum. These bazaars formed a unique concentration of thousands of shops and workshops offering an immense variety of goods, unparalleled elsewhere in the Ottoman Empire and perhaps in the world. The Suleymaniye complex fulfilled the central need for public services; the bazaars fulfilled the role of major center of commercial and economic activities.

The last change brought to the city structure was the transformation of the street network. While a relatively clear distinction existed in the Byzantine period between the upper level of streets—mainly the *Mese* and the wide avenues branching off it—and the alleys of the lower level, this distinction became blurred in the Ottoman period because the building of private houses encroached on public space and narrowed the major arteries. With the deterioration of the broad avenues, the city lost its monumental axis. The emergence of the multicenters—the various imperial mosques and bazaars—rendered the pattern of accessibility much more uniform; the network of alleys expanded and encompassed the whole city.

Despite the continuity of the imperial center and the spatial locations, the transition from Constantinople to Istanbul caused the far-reaching transformation of several crucial aspects of the city's internal structure, mainly changes in the structure of public spaces and the relations between public and private spaces. First, there was the change in the imperial city's pattern of centralization, which transformed the concentrated core adjacent to the palace into a decentralized pattern of public spaces around the mosques and sociocultural complexes surrounding them. The imperial palace thus lost its central controlling position.

Second, the nature of the public spaces was modified. From their overall broad character in the Byzantine Empire, rooted in the combination of classical city-state and the centralizing imperial traditions encompassing, in principle, all the strata of the population and giving them the possibility of participating in public life, the public places changed to a more "communal"-religious pattern. This was characteristic of many Islamic cities and closely connected to the different relations existing between the political and religious elites in these two civilizations.

Third, a change occurred in relations between public and private spaces, as seen by the transformation of the street pattern as well as the public spaces in close connection with the general pattern of decentralization and establishment of *waqfs*.

These changes indicated a general inward-turning tendency toward the closed family dwelling places on the one hand and toward the communal-religious pattern on the other. Little space was left for specific overall urban, municipal—and imperial—public spaces.

The last important changes were in the structure of the city's commercial pattern. From a linear one, with a marked concentration in the forums along the main monumental axis, this pattern was transformed into a few bazaars whose location was related to the main mosques. Each of these bazaars included a large number of commercial and artisan enterprises evincing high specialization.

On the whole, these changes were in line with the general Islamic pattern of construction of urban life. At the same time, among all the Islamic capitals, Istanbul probably stands out as the most imperial and relatively the most centralized one. This is probably closely connected to a tighter control over both the *ulemas,* unparalleled in any other Islamic state, and the merchant and artisan groups that were mostly foreign.

CHAPTER 8

Urbanization in the Early Periods of Islam

We will now analyze the patterns of cities and urban systems developing within the realm of Islamic civilizations. We will see a great diversity of political-ecological settings and historical variabilities. These will enable us to examine the effect of several variables on the structure of centrality and concentration forces and on shaping the contours of cities and urban systems.

A
THE URBAN LEGACY OF THE PRE-ISLAMIC PERIOD: THE CITY UNDER BYZANTINE AND SASSANIAN RULE

The Islamic city emerged out of the legacy bequeathed by the pre-Islamic city and the specifically Islamic patterns of urbanism. On the eve of the Arab-Muslim conquests, the Middle East was an area already possessing a well-developed urban structure, and where large-scale urban settlements had been in existence for several millennia. The Arabian nomads who swept over the region in the 7th century established Islamic rule over the whole Middle East within a very short period, and many of their military and administrative centers were based on the system of urban settlements then existing. The Middle East of that time had been divided between the (Christian) Roman Empire, west of the Euphrates, and the mainly Zoroaster Sassanian Empire, to the east of it.

At the beginning of the 7th century, the cities under Byzantine rule had reached the end of a long, continuous process, as a result of which they gradually lost their political independence and autonomous civic institutions (Jones, 1940:85). The Greek and Hellenistic cities of pre-Roman days were often politically autonomous, but under Roman rule they relinquished their internal autonomy to some degree, still preserving the institutional framework of an urban assembly, council, and magistrates. Under Byzantine rule, however, the local urban government declined and was increasingly replaced by the direct control of the central government. The municipal services also depended to a great extent on the initiative of the imperial governor. By the end of the 3rd century, the administration of justice had passed from the city courts to the provincial governors (Stern, 1970:28). Alongside the imperial government, during this period the Christian Church was instrumental in taking over many of the functions formerly assumed by the city authorities. Finally, the municipal government was altogether eliminated and a military provincial government (the "themes" system) was introduced at the beginning of the 7th century. The major towns of the Byzantine provinces now became the seats of military governors with their military and civil organizations (Ostrogorsky, 1959:65). In the words of S. M. Stern, "Islamic civilization did not inherit the municipal institutions of antiquity because, owing to their gradual decline, there was, by the time of the Muslim conquest of the provinces of the Roman Empire, nothing left to inherit" (1970:26).

Furthermore, if the former Byzantine provinces were unable to pass on the ancient municipal traditions to the new Muslim rulers, the provinces conquered from Sassanid Persia, which were outside the sphere of Byzantine influence, were even less able to do so. Even if some cities in Persia possessed a few urban institutions inspired by the Hellenistic and Roman traditions, the majority of the cities in the Sassanid Empire had no autonomous institutions.

The urban legacy that the ancient world bequeathed to Muslim civilization was composed of two elements. Thousands of settlements existed in the Middle East on the eve of the Arab conquest and until the 10th century and later. Their vast majority originated from ancient times, and only a small number was established in the Muslim era. Until that time, the existence and growth of these towns had depended entirely upon imperial governments, whether Sassanid or Byzantine. City administration was ultimately in the hands of officials appointed by, and responsible to, the emperors. Imperial governors and tax collectors, backed by imperial garrisons, dominated the towns in both empires. The second aspect of the ancient urban legacy was a particular urban social structure that was a combination of this imperial administration with religious, community, and fluid market institutions, all based upon family or small clannish groups (Lapidus, 1973).

B
EMERGENCE OF THE ISLAMIC CITY

The Arabs, however, did more than merely take over these old settlements. Indeed, with the Arab-Muslim conquest a new era of urbanization processes began in the Middle East. The nomadic tribes, spreading outward from Arabia, settled in many different locations: in existing towns and cities, in suburbs of existing cities, in newly established towns, in villages, in Syrian desert oases, and on fertile land along the major rivers. Despite their nomadic origin, many of the conquerors preferred urban to rural existence, and the cities and urban way of life thus became a dominant feature of Muslim society.

The first cities erected by the nomads were semipermanent camps at the edge of the desert. It was from this kind of settlement that the *Ansar* evolved: a sort of fortified camp used by soldiers taking part in the military-religious expansionism, often termed *jihad* (holy war).

The second type of city built was the royal one, or capital, usually established when a new dynasty ascended to power. New capitals were founded when the caliphs required administrative centers that would isolate elite regiments and administrative staff from the mass of the subject population. The construction of a political-administrative center usually generated the growth of a large city because the very construction of palaces, mosques, administrative offices, and army camps attracted a large labor force including craftsmen, servants and the like to service the new imperial or sultanic center (Lapidus, 1969). The overcrowding of an existing capital or the ascendency of a new dynasty would lead to the establishment of a new capital, often in the suburbs of existing cities. Sequential construction of capitals thus caused expanded urbanization of already favored sites.

Apart from imperial-administrative capitals, the Arabs founded numerous frontier fortresses and fortified garrison towns. Along the Byzantine frontiers in Syria, Mesopotamia, and Armenia, as well as along the Khazar frontier in Azerbaijan, large numbers of fortresses were built to defend the Muslim core area in the Middle East and to prepare for further conquests to the north and east. In this respect, Muslim rule urbanized the frontiers of the empire; in all these ways, and mainly through the establishment of new urban settlements outside the network of ancient cities, the Arabs gave a strong impetus to urbanization in the Middle East. These new towns and cities differed from those inherited from antiquity by their ethnic and social structure, their specialized functions, and cultural traditions. In the early Muslim period, the two types of urban places—the old non-Arab city and the new Arab one—existed side by side in the Middle East for a considerable period. It took several centuries for these two types of cities to merge into a single Middle Eastern urban society, with a common religion, common institutional structure, and common culture. Only toward the 11th century did this "typical" Middle Eastern Muslim city become the dominant feature of the settlement system of the region (Marçais, 1956).

C
THE IDEOLOGICAL IMPORTANCE OF THE CITY IN ISLAM

It is important, in this context, to stress that a very strong ideological impetus to the creation of cities evolved within Islam.

Some scholars have claimed that the Islamic civilization was primarily an urban one. We have mentioned that the Arab conqueror favored city life over a rural existence, although the ideal of the tribal nomad inherent in the very origin of Islam has constituted a continuous element in the history of Islam. In the words of an authority, "Islam prefers the sedentary to the nomad, the city-dweller to the villager. It accepts the artisan but respects the merchant" (Von Grunebaum, 1955).

Two characteristics of the Islamic outlook have been singled out as particularly favorable to the development of urban life: the urban character of the later, post-Koranic, Islamic legislation—the *Sha'aria*—and the factual identification of settlement in a city with the adoption of the faith—that is, the fact that most of the actual conversions to Islam took place at least in the initial period of expansion. Both characteristics are related to the actual urban environment. The view gradually became widespread that the way of life prescribed by the Koran was most appropriate to an urban milieu such as the Hijaz trading centers existing at the time of the Prophet. In the first decades of Islamic expansion, Arabian cities were isolated pockets of urban society in vast nomadic areas. There the contrast between urban and nonurban modes of life was stark and uncompromising. The Islamic propagation, aimed as it was at the commercially oriented urban communities, expressed low esteem for the Beduin nomads, although the latter viewed themselves as superior and continued to constitute, as attested by the words of Ibn Khaldun (among others), a constant element in Islam's history (Benet, 1964).

To these characteristics should be added the fact that in the first three centuries of Islam, the city became (at least de facto) the main or only proper setting for communal worship. At that period communal worship was not regarded as entirely valid unless conducted in the presence of the caliph or his representative. This took place only at the *masjid-as-jami*, the principal mosque in a city, where, on Fridays, a Muslim was required to fulfill his religious duty and participate in communal prayer with his fellow believers and his political duty of hearing his sovereign's name mentioned in the sermon *(Ahubbeh).* Until the end of the 10th century, the term *medina* (city) was interchangeable with *masjid-as-jami*, and was restricted to settlements of substantial size, their number being controlled by law (Wheatley, 1981).

The mosque was not only the center of religious life but the major manifestation of urban life. The association of *jami* with the urban status of localities recurs in various sources of that period, particularly in Al Mukaddasi's geographical works. Later, however, large cities acquired more than one *jami,* and this practice eventually spread to smaller towns and large villages.

By the end of the 10th century, the *jami* began to lose its powerful significance as the place where the close connection between the political authority of the caliphate and the religious authority of the *ummah* was manifested. The mosque then became associated with local religious communities. Large cities acquired two *jami*, and gradually this practice became quite common. Some *jami* became identified with certain schools of law or with religious sects, thus representing the aims and aspirations of particular religious groups (Von Grunebaum, 1962).

D
BASIC INSTITUTIONAL PREMISES: THE NATURE OF ISLAMIC SOCIETY

To understand the development of Islamic urbanization, it is not enough to stress the general positive evolution of city life in Islam. It is also necessary to take into account the political-religious aspects of the Islamic culture, its basic premises, the social structure of its elites, and the main types of social and political institutions that developed within it (Gibb, 1962; Hodgson, 1974; Holt et al., 1970; Lewis, 1950, 1973; Von Grunebaum, 1946, 1954).

The most important cultural orientations predominating in the Islamic civilization were the following:

(1) the distinction between the "cosmic" transcendental realm and the mundane one, and the importance given to overcoming the tension inherent in this distinction by total submission to God and through this-worldly (above all political and military) activities;
(2) the strong universalistic element in the concept of the Islamic community;
(3) the autonomous access of all members of the community to the transcendental sphere and the possibility of salvation through submission to God;
(4) the ideal of the *ummah*: the political-religious community of all believers, distinct from any ascriptive primordial community; and
(5) the vision of the ideal ruler as upholder of the ideals of Islam, the purity of the *ummah,* and of the life of the community.

The fact that the original concept of the *ummah*, in the Islamic realm, assumed complete identity with the sociopolitical and religious community, is of special importance. The Islamic state emerged out of the conquests of nomadic tribes and was inspired by a new universalistic religion, in the initial stages of which the identity between polity and religion was very great. Similarly, many of the caliphs and later rulers (such as the Abbassids or Fatimids) came to power on the crest of religious movements, legitimized themselves in religious terms, and sought to retain popular support by stressing the religious aspect of their authority, wooing the religious leaders, and encouraging the religious feelings of the community.

Owing to the very extension of the Muslim conquests, however, and to the tensions existing between tribal conquerors and their conquered peoples, a state of unrest developed between the universal Islamic community and the primordial local or ethnic groups. The ideal of a common political and religious community was never realized, and a growing dissociation developed in Islam among the political, religious, and local institutional spheres, albeit with a strong latent ideological orientation toward their unification (Gibb, 1962:3-33; Lapidus, 1973; Turner, 1974).

The *ulemas* generally had little autonomous political power and were politically "domesticated" by the rulers, even to the extent of becoming organized in a distinct bureaucratic structure under the ultimate supervision of the rulers in Ottoman Europe. The identity of the Islamic religious community depended on observance of the Holy Law (*sha'aria*) as interpreted by the *ulemas* and enforced by the rulers. Between the religious leaders and the rulers, a very peculiar relationship developed. The *ulemas* became politically passive or subjected to these rulers, even though they were relatively autonomous in the performance of their legal-religious functions (Schacht, 1970).

The consequences of the spread of Islam in general and the severe ideological tension existing between the universal Islamic community and the various primordial ones in particular weakened the solidarity between the political and religious leadership of Islam. This led to a high degree of symbolic and organizational autonomy among the political elites, but it also resulted in their growing dissociation from the religious elites and the broader strata. These tensions also gave rise to the relatively high symbolic (but not organizational) autonomy of the religious elites and to a growing separation between the two. As has been mentioned, the religious leadership was very dependent on the rulers and did not develop into a broad, independent, cohesive organization. Religious groups and functionaries neither formed a separate entity nor constituted a tight body except when organized by the state, as in the Ottoman Empire (Gibb and Bowen, 1957, chaps. 8-12; Itzkowitz, 1972). At the same time, they tended to become increasingly associated locally with political elites, and local political notables—the *Ayah*—(Bulliett, 1972) maintaining, or at least attempting to maintain, a distance and dissociation from the central power. Growing dissociation also tended to evolve between the higher echelons of the *ulema*, which were more closely linked with the central or local political elites, and their lower echelons, which were more closely associated with the local masses.

The relationship between the political and religious elites and the fact that they did not form a cohesive, fully organized body resulted in a lack of autonomous political participation of the religious leaders. Neither the clergy nor the vast population played any part in political life, which was mainly confined to court cliques and the bureaucracy. Usually initiated by dissident elements of the *ulemas* and political-religious sects, revolt constituted the only possible form of political expression (Laoust, 1965; Lewis, 1973: 217-266).

Although the initial basic cultural orientations of Islam seemed to create conditions favorable to the establishment of imperial systems after the first caliphates, these systems were not the most common in the Islamic realm, and polities tended to move in the direction of a patrimonial society (Turner, 1974).

The regimes most frequently found in the Islamic realm were not therefore imperial, which occurred during the Abbassid, Fatimid, and Ottoman periods, but patrimonial and/or sultanic, found in the post-Abbassid period and tribal ones in Arabia and the Maghreb. The patrimonial tendency was often reinforced by the elite structure that developed within the later Islamic framework and counteracted the tendency toward more imperial patterns.

E
STRUCTURE OF MARKETS AND CONTROL

Owing to these basic cultural orientations, the structure of elites, their institutional derivations, and the oscillations occurring in them that characterized the history of Islam, a rather peculiar structure of markets and modes of control over production and flow of resources developed within the realm of Islamic civilization. This structure was closely related to the relative importance of more imperial as against more patrimonial regimes, in any given period or place.

The basic cultural orientations and heterogeneity of elites generated a tendency toward a relatively wide scope of rather differentiated institutional-religious, political, and economic markets and close interweaving between them and a rather flexible but strong pattern of control by the coalition of heterogeneous elites.

The development of the Islamic polity in a sultanic and patrimonial direction greatly narrowed the scope of the major institutional markets and their mutual interweaving. It also changed the mode of control into a more coercive but less intensive one in the later imperial systems—especially the Ottoman and the more patrimonial or sultanic regimes. Moreover, it almost entirely weakened or negated the possibility of autonomous access of the broader strata to the centers of control, which in principle was inherent in the Islamic credo. All this was evident in the structure of the stratification evolving in the realm of Islamic civilizations. With respect to the group basis of that structuring, a multiplicity of status groups tended to develop in these societies—ethnic, religious, local-regional, sectorial, or tribal—as well as a high degree of status segregation among them. Their rulers attempted to weaken or suppress whatever tendencies might have developed within the groups and to coalesce them into broader, country-wide strata. Rulers in patrimonial societies were more successful in their attempts than those in imperial societies.

The major criteria of status here were membership in the family and proximity to the basic attributes of social and religious orders. But these attributes were controlled to a large degree by the major political and religious centers of these societies, which minimized the autonomous access of different status units to such controlling positions and the major attributes of status. The combination of these structuring principles of hierarchies with those of the group basis of such hierarchies generated the following tendencies in these societies: enhancing the aristocracy and bureaucracy (often of a service nature) and, from the 10th century on, the local wealthy groups as the apex of the higher classes and channels of social mobility; sharpening somewhat the division between upper groups in the center and wider groups in the periphery, as well as between the upper and lower urban groups; and arranging the status sets and units vertically rather than horizontally.

Within this context, the development of urban society is of special interest to us. Von Sievers (1979) summarized this development well, as related above all to Syrian cities until the 10th century:

> In the course of the two centuries, the principle of social stratification was modified from one dividing society according to military privileges into one ranking society according to economic employment. During the same time, urban society differentiated itself into the upper stratum of nobles, elders or merchants on the one hand, and the *ahdâth*, representing the lower urban stratum of craftsmen, artisans and laborers, on the other hand. . . . As far as the rural populations are concerned, they descended during the two hundred years from a militarily and economically superior position to one of poverty and (after the Qarmatian revolt) political weakness [1979:240-241].

We should note that these developments took place under conditions of relative decline of imperial power, its growing dissociation from the different strata of the population—particularly the urban population—and the emergence of foreign rulers.

The process of urban differentiation and the emergence of a nascent urban autonomy were thus intermittent and eventually came to an abrupt end. Reestablishment of an imperial system over Syria by the Fatimids resulted, therefore, in the termination of social differentiation in the Syrian city. Von Sievers concludes that "in Syria, the history of urban autonomy movements did not begin with the tenth century; it essentially ended at this time" (1979:244).

F
URBAN SYSTEMS

These cultural orientations and the Islamic world's political structure shaped the combination of the centrality and concentration forces as they developed in the realm of Islamic Middle Eastern urbanization, and through it they greatly influenced the patterns of Islamic urbanization.

Several kinds of urban systems evolved in the Islamic world: a more rank-size order in the imperial Abbassid and Ottoman periods and a primacy pattern in the more patrimonial or sultanic societies. In addition, they varied according to the political and economic situation of individual cities and representative centers of the urban systems and according to the influence of a separate urban or semi-urban Islamic religious hierarchy.

An illuminating example of an Islamic urban hierarchy may be found in Saari's work (1971) dealing with urban systems in Egypt in the Middle Ages. According to this author, until the 10th century, when most of the population was still Coptic, the regional model of the Egyptian cities was created by administrative needs and did not present the characteristics of an economic hierarchy. With the new administrative organization and the growing Islamization of the masses, however, the character of the regional model of urban development began to change. By the 13th century, an economic hierarchy can already be seen in Egypt.

According to Saari's urban grading, based on the information provided by the Arab geographer Yaqut (13th century), there were in Egypt 1 capital city *(Qasba)*, 6 large cities *(Medina kebira)*, 29 medium cities *(Medina*—plural: *Mudun*), and 62 small cities.

Ismail (1972) has presented a theoretical typology of Arab-Islamic settlements and their regional organization based on the work of the 10th-century Arab geographer, Al Mukaddasi. This scholar mentioned that large areas were divided into subareas in which the settlements were graded according to a system whereby the *Ansar* was compared to a king, the *Qasba* to ministers, the *Mudun* to cavalry, and the *Qura* (villages) to infantry soldiers. Al Mukaddasi used two hierarchies, one for grading regional units and the second for urban settlements, and proceeded to develop the spatial regional model for each unit.

According to Tekely, (1971), however, there existed in the Ottoman Empire of the 16th century a hierarchic urban system consisting of a capital, regional centers, market towns, villages, and nomadic groups. In his opinion, this hierarchy depended on a differentiation of the functions of cities, and on their relationship with the hinterland and other cities. This interregional equilibrium was violated in periods of a weak ruling regime.

The Islamic world possessed an additional, completely different form of urban system: the Islamic Holy Places, organized and maintained by the activities of the religious elites. In this case, the hierarchical order depended on the degree of holiness attributed to each place (Wheatley, 1981) and was directly related to events in the lives of the Prophet and his close disciples. Thus Mecca held first rank in the religious hierarchy, Medina second, and Jerusalem third; a score of other places had a lower level of holiness. It should be emphasized that this order of cities was not hierarchical in the strict geographical sense of the term, as the Holy Places did not constitute complete levels of hierarchy in terms of mutual flow of resources and interrelation of functions, and their sphere of influence was not related to their rank in the religious hierarchy. The same applied to the numerous local cult places and shrines that had emerged throughout the realm of Islam.

The most striking fact about the urban system of Holy Places is its noncongruence with the urban system, based on a political and economic hierarchy, which attests to the dissociation between the religious and political orders. Despite the fact that the Holy Places of Islam were fixed and immutable, they gradually lost ground to the political-administrative urban system because they were located in the periphery of the great Arab empires, centered in Damascus, Baghdad, and Cairo. Not situated on the major transportation routes, they failed to play a significant role in the economic urban system. Despite Islam's great contribution to the history of urbanization and its attitude favoring the urban way of life, its Holy Places lay outside the Muslim empires' major urban systems.

G
PATTERNS OF URBAN GROWTH: *VILLES CRÉÉES* AND *VILLES SPONTANÉES*

The constellation of political, religious, and economic concentration and centrality forces at work in the early Islamic period also greatly influenced the internal structure of Islamic cities in the Middle East.

The internal structure of most Islamic cities was the product of pre-Islamic urban heritage, modified and remolded by Islam, with its particular social structure, religious-cultural orientations, and characteristic lack of political urban autonomy. In order to identify the *ideal* structure of Islamic urban design, however, we must examine those cities that were established on unoccupied sites, either as frontier garrison towns or, more significantly, as capital cities for caliphs or princes. The building of these new towns was a manifestation of central political power, imagination, and resources. These newly established cities—designated *villes créées* by the French Arabists (De Planhol, 1959, 1970; Pauty, 1951)—were the creation of a small power group of imperial origin and were often determined by the caliph or his delegate, who selected the site, defined its extension, and laid out the city plan.

These cities were built around a central area characterized by a regular design, which was intended to give the impression to inhabitants and visitors alike of orderliness, organization, power, and discipline. As we have mentioned, these cities appear to have begun as imperial retreats of the Umayyad caliphs, and then evolved into large-scale palace complexes that attracted large numbers of builders and laborers, thus becoming full-fledged capital cities in a short time.

Baghdad was built by the Abbassid Caliph Al-Mansur in 762-767 on a circular plan and is a good example of an imperial city established according to a regular plan (Hitti, 1973; Lassner, 1970; Le Strange, 1924). The imperial palace and mosque were at the center of the circle and a complete ring of residential quarters surrounded them. The area within the circle, between the residential ring and the palace and mosque, was kept empty, thus separating the imperial center from the populace. There was no direct connection between the residential quarters and the central complex, and the four gates in the inner wall encircling it could be sealed off in case of riots. The physical centrality of the royal palace and mosque, the accentuated separation of the imperial court from the capital's population, and the perfect geometric shape of the circular city were all manifestations of imperial centrality, power, orderliness, and discipline. It should be pointed out, however, that the city's geometric pattern had no cosmological significance, and neither the site nor the plan of the city carried any religious or symbolic meaning. The circular city and its central complex—the imperial palace and the mosque built in splendid isolation from the residential quarters—can be regarded as an excellent example of the centrality principle shaping the internal structure of a city.

The vast majority of Islamic cities—the *villes spontanées* (Pauty, 1951)—developed from no preconceived plan; a small number of these cities came into being during the Muslim era. Most, however, were based on preexisting urban forms adapted and modified under Muslim rule.

From the beginning of the 7th century to the 10th century, the cities under Islamic rule underwent radical changes. Despite significant regional differences, a characteristic form of city emerged by the end of that period. This can be described as the typically "Islamic" city because the common features of the cities' internal structure were more pronounced than were the regional and historical differences between them. It is highly significant that cities of diverse origins—Hellenistic, Sassanian, or North African—were all ultimately modified and transformed according to a common urban pattern. This consisted of a centripetally organized aggregation of intra- and extramural areas arranged in roughly concentric zones around a central complex, ideally combining citadel, administrative enclave, congregational mosque, and principal market (Wheatley, 1981). This concentric pattern had a hierarchical base, as residences or activities of higher social prestige were situated nearer the complex at the city center.

H
SOCIAL AND INSTITUTIONAL STRUCTURE OF THE ISLAMIC CITY

1 Background

The internal ecological and social structure of the Islamic city presented some very definite features that were shaped, beyond the specific local ecological conditions, by the development of certain characteristics of the centrality and concentration forces within the realm of Islamic polities. The most important of these characteristics were (a) decline of the institutional reality of the ideal of the *ummah*, the combined political religious community; (b) weakening of the religious legitimacy of the caliphate as the *ummah's* representative; (c) increasing weakness of relatively more autonomous urban economic strata; (d) dissociation between the central political elites and the local communities, their growing inward-looking tendencies, the minimization of their access to the center, the consequent weakening of the imperial orientation of the center and its tendency toward a more patrimonial direction; and (e) consequent weakening of public institutions and commitment to them and the relegation of many of their functions to semiprivate institutions, the best illustration of which was the *waqf* (Baer, 1969).

2 Religious Institutions: The *Ulema* and the Schools of Law

The basic concept of the religious community in the Muslim world was the *ummah,* the community of believers who were united by their faith in Muhammed and the Koran and, at least in the early period, by their allegiance to the caliphs who symbolized this faith, as well as by their collective adherence to Islam. Within the community of believers, however, a body of scholars and students, the *ulema,* began to emerge who devoted themselves to the study and teaching of the Koran and the practice of Muslim ritual and law. By the 8th century, these activities had become of central importance in the constitution of Muslim communities. In the early periods of Islam, the religious scholars were often self-employed as laborers, craftsmen, or merchants; they did not hold office, received no institutional means of support, and had no sacerdotal or priestly status. They upheld the idea of the *ummah* but were unable to implement it.

With the breakdown of the central imperial government, new military regimes arose, dominated by slave-soldiers or nomad chieftains who were unfamiliar with local traditions and learning. This situation gave new importance to the *ulema.* The combination of the religious prestige and judicial authority they enjoyed made them one of the major components of the social structure. They developed therefore from a purely scholarly and religious leadership with some judicial functions into a social and political elite, becoming fully embedded in the local patriciates. By diffusing the normative principles of Islam among the general population, the *ulema,* as also the Sufi order (Bulliett, 1972), generated a high degree of cohesion among the inhabitants of Islamic cities.

Finally, another institution should be mentioned, one that played an important part in Muslim urban life: the *waqf,* a religious or semireligious, politically accepted endowment, which provided some of the most important public services (such as education, health, and even minimal municipal services) and was often based on private or family nuclei.

I
INTERNAL STRUCTURE OF CITIES

1 Basic Components of Islamic Cities

The operation of all these forces greatly influenced the internal ecology as well as the social structure of the Islamic city and the structuring of its basic components, the first of which related to the areas of the city connected with central government affairs. In the early period of Islam, the administration was concentrated in the "house of government" (*deir-al-imara*); this accommodated officials of the central government and the offices (mostly fiscal and financial) they needed to exercise their functions (Grabar, 1976:96). At this early stage, the house of government was not of monumental scale and was not spatially separated from the rest of the city.

A change occurred, however, after the foundation of Baghdad. The administrative offices were separated from the officials' living quarters and became the governmental palace. The sultan's private palace was adjacent to it, and both were detached from the main built-up area of the city. To the inhabitants of Baghdad, the imperial palaces became a myth, the seat of fearful power and legendary wealth.

A new type of government quarters meanwhile emerged in the provincial capitals. This was the citadel (*Qal'a*) that, in most instances, was the largest and most impressive monumental complex in the town. Citadels were quite numerous in frontier areas but relatively rare elsewhere in the first centuries of the spread of Islam. They began to proliferate in the 10th century with the weakening of imperial power and the emergence of local dynasties, usually supported by soldiers of ethnic origin different from that of the local population. This political situation transformed almost every provincial capital into a center of power. Quite often a citadel evolved into a "mini-city" with its own mosque, palaces, cisterns, houses, and the like (Grabar, 1976:97). In many cities the citadel was built outside the city area but contiguous to it. This spatial pattern occurred in Aleppo, Damascus, and Cairo. The creation of a separate urban entity, controlling the main city but differing from it, reflected the evolution of the relationship between military rulers and urban community.

The second component of the Islamic city relates to the religious community and its physical manifestation. As already mentioned, the focus of the Muslim community was the congregational mosque, the *masjid-as-jami,* the mosque constituting not only the center of religious life but, until the 10th century, the major manifestation of urban status. By the end of that century, however, the *jami* began to lose its powerful significance and to be associated with local religious communities. Large cities acquired two *jami*, and gradually this practice became quite common. From the ecological point of view, however, the *jami* retained its preeminence and remained the central point of the city, having the highest prestige and attraction. Indeed, other areas of the city defined their relative importance by their proximity to the congregational mosque. It should be emphasized that the site of that mosque did not carry a cosmological meaning; it could be located anywhere in the urban area, bestowing, wherever it was erected, prestige and respectability upon its surroundings. It was the pivot of urban activities, the pole of attraction around which most of the public institutions and commercial activities were spatially laid out.

Visually, the skyline of the early Islamic cities was accentuated by the minaret of the mosque, which was the tallest structure in the city, the major point of orientation for the population. There was a sharp contrast between the spacious courtyard of the mosque and the extremely crowded built-up area of the city. In many cases this was the only large public open space, and it provided a site for large public gatherings, a focal meeting place for the community on solemn or special occasions. In contrast to the inward-looking residential dwellings, the mosque projected toward the outer world with its prominent entrance portal, which was often massive and ornamented, inviting believers in and at the same time separating the mosque area from the city.

In the early period of Islam, the *jami* fulfilled a wide variety of religious roles. In addition to being the place of public prayer, it was the court of justice, the meeting place for religious scholars (*ulema*), an intellectual and educational center. As the cities expanded and the needs for various public and religious services multiplied, some of the *jami's* functions were transferred to other buildings and institutions, which were always built in proximity to the main mosque. The most common of these religious bodies was the *madrassah*, the place of learning, which was a privately and locally sponsored institution aimed at strengthening the learning and scholarship of the local religious leadership. It grew out of the need to provide the local community with legal and spiritual leaders and practitioners. At the same time, the *madrassah* was the place where the aspirations of the local bourgeoisie, which endowed such establishments, found their physical expression (Grabar, 1976:94).

The various religious institutions (the *jami*, the neighborhood mosque, the *madrassah*, the various sanctuaries) where people congregated for prayer and learning were the places in which religious, intellectual, and social gatherings were held. It was through these institutions that the population of a city, if divided into factions, developed its specific identities.

The *souq* (market) was an assemblage of instruments for the execution of market exchange. Accompanying *souq* transactions were simple processing and manufacturing activities, mostly concerned with the production of foodstuffs, textiles and cloth, small hardware items, jewelry, repairs, and a variety of service trades. It may be concluded that many of the traders were producers-retailers who fulfilled production, distribution, and sale functions simultaneously. Trading in the markets was highly competitive and intensive, dealing mainly in small bulk commodities. Yet despite the intensity of competition and "haggling," transactions were carried out in a leisurely and quasi-ritual manner. This fact reflects the important social function assumed by the market in the city dwellers' way of life. It was the major meeting place for people, integrating the various factions and the various levels of urban communities through the market exchange (Wheatley, 1981).

The various commercial activities carried out in the *souqs* were highly differentiated spatially, each type of commodity being concentrated in one corner or street of the market. This spatial differentiation allowed for maximum availability and choice for the customer and provided an efficient milieu for the sliding-price system that was the basic modus operandi of the *souq*. The order of spatial concentration of the trade in various commodities reflected a hierarchical principle that was mentioned previously: The more valuable the class of goods stocked by a shop, the closer the latter's location to the congregational mosque—the main urban focus. A ring of shops selling books, incense, candles, and the like could thus be located in the immediate vicinity of the mosque. This first ring could be surrounded by a wider one, sheltering shops dealing in textiles, rugs, and jewelry—all items of high value, allowing those tradesmen to hold a high position in the commercial hierarchy. A third ring, furthest from the mosque, could be occupied by shops selling food and hardware—items of low value. This ideal concentric structure of the market area around the mosque seldom occurred in reality, but its governing principles had a decisive influence in shaping the actual structure of the central markets in most Islamic cities (Kark, 1978).

This spatial differentiation was relevant only on a micro scale. On the urban scale, however, religious and commercial activities were intimately interwoven. This spatial intermingling reflected the undifferentiated lifestyle of the townsmen in Islamic cities. Prayer, learning, and public discussion did not constitute specialized activities that were segregated in special quarters, but formed an integral part of the daily life of all classes of the urban society. A scholar could be a part-time merchant; a craftsman could regularly participate in theological discussion; a shopkeeper could serve as notary. The mixed spatial structure of the city allowed easy access to all functions and encouraged integration of the various spheres of urban activity (Wheatley, 1976).

2 Irregular Street Plan

Beyond those differences, the various components of internal ecological structure blended in most Middle Eastern Islamic cities of this period—and even later—into a very specific ecological pattern: the irregular street plan. This was greatly influenced by the irregular basic constellation of centrality and concentration forces we analyzed earlier.

The streets were narrow and winding in all directions. When walking in the maze of lanes of an Islamic city, one's sense of direction was soon lost. Many streets and lanes were not part of a road network but were unconnected and of the cul-de-sac type. Orientation was almost impossible in the Islamic city as the streets were bordered by high walls, or even totally roofed, thus making it impossible to obtain any perspective or view of urban landmarks. Open spaces or public gardens were very few and of limited area. The highly irregular and confusing street pattern was characteristic of all the *villes spontanées* of Islamic culture—even in those cities that, in the pre-Islamic period, had a regular street pattern, such as the rectangular grid pattern of Hellenistic origin (English, 1973).

The transformation of a highly ordered street pattern into an irregular, formless one was investigated in detail in Aleppo by Sauvaget (1941), who traced the stages through which the well-organized Hellenistic city-plan of Berrhoea evolved into the irregular pattern of the *souqs* of Islamic Aleppo. The pre-Islamic streets were wide and paved, lined by colonnades under which large shops were located. By gradually extending their walls forward, the shops then occupied the colonnades and even encroached on the public way. The paving stones were then removed and used for building. The public way was often completely blocked by shops or dwellings, thus creating a cul-de-sac and diversion of traffic. This process of encroachment on public space—particularly streets—by private interests was most pronounced at the city center. The demand for land for commercial activities was highest in the center—the bazaar area adjacent to the mosque—and competition between public and private use of space therefore was most acute there. This process also occurred in residential quarters, albeit at a slower pace. In all these areas, privately owned buildings encroached on the streets and increasingly distorted the street pattern. This phenomenon was undoubtedly encouraged by Islamic law, which decreed that the open space around and along a building constituted part of the property. The precise extent of the private property within the public space of the street is a matter for discussion by legal specialists, but its immediate impact on the Islamic city's urban design is not left in any doubt (Brunschwig, 1947).

The transformation of the original regular street pattern into a formless maze of winding lanes, which occurred in nearly all Islamic cities, even in imperial capitals that initially had been built according to a preconceived plan, was closely related to the basic characteristic of the Islamic city—the absence of urban autonomy and lack of corporate identity and municipal administration. Although municipal authority was lacking, the imperial government did not create in its place a bureaucratic body that would allow orderly governing of the cities. The only urban official, the *muhtasib,* began to function as inspector of markets and gradually became the supervisor of the religious and moral welfare of the urban population. The *muhtasib* was guardian of the community's morals and protected the faithful from exposure to undesirable influences (Marçais, 1955; Von Grunebaum, 1955: 152). Initially the *muhtasib* had been responsible for various urban functions such as street cleaning, practice of fair trade, safety of the buildings, and so forth, but later his religious functions and control of morals were more emphasized. His jurisdiction was limited to uncontested abuses and the rectification of obvious wrongs; he could not hear evidence or administer an oath (Wheatley, 1981). The development of the *waqf* reinforced these tendencies. In the ongoing tension and conflict between the public milieu and private interests, the private householder and the commercial entrepreneur held the upper hand. The relation of the Islamic city's irregular form to its lack of political autonomy and to the weakness of its urban administration was thus fully manifest in all these ways.

J
URBAN AUTONOMY

The special cultural orientations prevalent in Islam and the structure of the concentration and centrality forces that evolved within Middle Eastern societies have greatly influenced the structure of urban social and corporate organizations and the concept of urban autonomy. The most fundamental characteristic of Islamic cities, in this respect, was the low level of development of urban organizations and consciousness. Certain types of urban organizations (particularly economic ones such as artisan and merchant guilds) emerged, however, in a number of Islamic cities. Their growth usually coincided with a period of economic development and a relatively strong political regime; they declined in periods of weakened economy and waning of imperial power.

On the whole, especially in the later, more patrimonial, stages of Islamic regimes, the cities could not be considered distinct social and political entities. None of the social organizations in Islam—the quarter, the fraternity, the religious community, and the nation—was specifically an urban organization, nor was it identified with the city as a physical unit (Lapidus, 1967). The groups to which the population belonged were either subcommunities within cities or supercommunities with affiliations that extended beyond the city walls. The cities were mainly foci of population that intermingled with a larger society; they were concentrations of people, activities, and services. Consequently, with the possible exception of certain periods in Abbassid history, corporate urban organization and identity were extremely weak (Ashtor, 1956; Cahen, 1959, 1970).

Islamic urban society operated on three levels, none of which was specifically of municipal character: the parochial, the religious, and the imperial. The parochial groups were confined to the locality, the religious organizations transcended local or regional boundaries, and the imperial regimes were either centralized caliphates or sanctioned the rule of nearly independent regional governors supported by slaves and foreign troops. The organization of urban life and government depended on the relationships among individuals, groups, and institutions on these three levels. The Islamic city thus evolved into a conglomeration of distinct neighborhoods, each of which was characterized by a combination of ethnic origins, religious affiliations, and occupational status (Landay, 1971). Daily activities and personal and family affairs all functioned within the framework of these local groups.

The urban neighborhood (*mahallah* or *harah*) was a spatially defined, village-like residential community, furnished with local markets and sometimes workshops. It consisted of people—notables, tradesmen, and common folk—recognizing mutual ties of kinship, origin, ethnicity, religious affiliations, and occupation. Class distinctions, expressed by unequal access to strategic resources, did not form the basis of neighborhood organization (Lapidus, 1973:63).

The cities were divided into quarters, each of which enjoyed a certain limited autonomy and the right of decision on routine matters. The citizen, therefore, pledged his loyalty first to his family and then to his ethnic group.

Whereas the imperial government concerned itself with the overriding interests of military security, taxation, and general economic welfare, neighborhood groups had a comparatively narrower sphere of interest. Schools of law played a major role in city life by defining beliefs and values, preserving the sense of Muslim solidarity transcending the city's boundaries, practicing judicial and local administration, providing education and ensuring the continuity of religious and intellectual traditions.

We see, then, that the separation evolving between the political and religious elites and existing between the ruling groups and society as a whole, as well as the population's nonparticipation in the process of government, led to a lack of cohesion that hampered the development of urban and municipal organizations. The city was therefore not an autonomous entity but a conglomeration of sectors, none of which represented the city as a whole.

The Moslem cities thus failed to generate administrative bodies. The administrative framework, including the appointment of important functionaries, was imposed on them by the state. But when the central political authority was weakened, most scholars agree that the cities were given a de facto political authority that was seldom used.

To sum up, there were situations in the Islamic world when the inhabitants of certain towns tried, with greater or lesser success, to free themselves from their ruler's control and to gain full political autonomy. These aspirations sometimes resembled those of Western European cities, which gave rise to great communal movements (Ashtor, 1975; Cahen, 1959; Goitein, 1966). This attempt, however, never assumed institutional form in the Islamic world and perhaps for this very reason remained abortive.

CHAPTER 9

Urbanization in India

A
HISTORICAL BACKGROUND

The history of India is made up of short periods of flourishing civilizational development separated by long eras of disintegration and political weakness. Ancient India (2500-1500 B.C.) appears to have fostered a highly developed urban civilization. This very early primary urbanization came to light at the excavations of Harappa and Mohenjo-Daro in the Indus Valley. Both sites show clear evidence of large urban centers: They were carefully laid out cities with a grid pattern of streets intersecting at right angles; the houses, several storeys high, were built around courtyards and were notable for their careful drainage, linked to a citywide sewer system. Each city had a citadel built on a mound and included palaces, granaries, and bathhouses (Wheeler, 1968).

Harappan cities were characterized by great regularity of design. For example, the grid street network, of the same width and right-angle intersections, was repeated at various locations, often imposed over an earlier irregular street pattern. Even the size and shape of the bricks were standardized as were weights and measures. The copper tools, vessels, and pottery also displayed a marked uniformity (Possehl, 1979).

This great urban civilization came to an end at about 1500 B.C., possibly as a result of the invasion of the Indus Valley by Aryans originating from the great plains of Central Asia. This first large-scale invasion heralds one of the main forces that shaped Indian history: great masses of people flowing into the country mainly from the northwest, conquering large areas and eventually assimilating into the indigenous society and culture (Thapar, 1966).

The seminomadic Indo-Aryans conquered parts of northwest India during the 14th or 13th century B.C. The penetration in the central Ganga-Yamuna plain and the eastern valley of the Ganges by Aryan tribes and clans and their stepwise settlement took more than half a millennium. This process can easily be divided into three major phases, each of which is associated with a particular area. The earliest period, up to the 11th and 10th centuries B.C., is known from early Vedic literature and depicts a seminomadic life of the predominantly pastoral Aryans during their conquest of the area, which is roughly the present North Pakistan and the Indian Punjab (Piggot, 1950).

Since about 1000 B.C., climatic changes and increasing economic tensions may have caused the further movement of Aryan tribes into the Ganga-Yamuna Doab between Delhi and Allahabad, which became the future Midland of the Brahminic culture. This late Vedic period is known from the ritualistic Brahmana texts, which clearly depict a process of detribalization and increasing social stratification. Agriculture now became the people's main occupation. The major social change that took place during this period was the emergence of the Brahmins as highly qualified and dominant ritual specialists and the Kshatriyas as dominant warriors and landholder caste. This period of social stratification and territorialization of former Aryan tribes engendered the emergence of strong chieftaincies and small kingdoms—a process that, of course, was not limited only to the Aryan tribes.

It was in this last period, from the 10th to the 6th century B.C., that the basic contours of the Hindu-Brahminic civilization started to crystallize (Thapar, 1983). It was also during the millennium following the Aryan invasion that the major contours of Indian society were shaped. Above all, the Brahminic system of caste organization and division of labor gradually crystallized. The Brahminic system was ideally composed of four strata: Brahmin ("ritual specialists," inaccurately designated by European scholars as priests), warriors, merchants and cultivators, and outcasts.

These social cultural processes were concomitant with a territorial shift toward the east. The geographical focus of India at that period was the eastern portion of the central Gangetic plain, the ancient Pataliputra area that coincides vaguely with present-day Bihar and its capital, Patna. The 6th and 5th centuries B.C. witnessed a dramatic, coincidental emergence of several socioeconomic, political, and intellectual factors that formed the basis and background of India's Great Civilizations. These are the more or less simultaneous and sudden development of the Gangetic towns and interregional kingdoms and (on the intellectual level) the emergence of a new type of religious movement, best known through the teachings of Buddha and Mahavira, the founder of Jainism.

None of these factors is completely new. Most of the towns that flourished around 500 B.C. had grown out of the seats of extended patrimonial household courts of the clan chiefs and early tribal rajas. The new kingdoms themselves had emerged in the whole Gangetic valley from the process of territorialization of former tribes and lineages and the extension of the more powerful among them. This development of the tribes *(jana)* and their area *(pada)* to *janapadas* and, finally, to the *mahajanapadas* of the late Vedic period, was linked with a process of urbanization (Thapar, 1983). The emergence of a new intellectual elite had its origin in the ascetic renouncer of the Upanisad period who withdrew from the life of householder and its Brahmin rituals in order to seek the principles of life and to strive for salvation (*moksa*) from the cycle of rebirth. What was new during this post-Vedic period, however, was the sudden acceleration of the overall material and social developmental and an awareness of change that led to new and often radical, innovative thinking. This general development centered increasingly in the eastern part of the central Gangetic valley.

Furthermore, a seemingly separate factor of this development, the conquest of large parts of the Indus valley by the Achaemenidian Empire, under Darius the Great, should be mentioned. For the first time in Indian history, the south Asian subcontinent became directly linked with a major center of world history.

The basic structure of India's settlement system must have been laid out during this period; it was composed of village settlements of different scope, with different degrees of interrelations among them, through the monuments, temples, and urban or semiurban networks.

B
MAJOR CHARACTERISTICS OF INDIAN CIVILIZATION

Two starting points are of special importance for understanding this civilization. One was the combination of political decentralization (or multicentrism) and continuously changing political boundaries and economic structures together with a relatively (although never fully) unified civilizational framework. From this point of view, it resembled the (Western) European framework in many ways. The second point, however, was that this civilization was characterized by distinct types of cultural orientation and structure of elites.

Hinduism (Biardeau, 1972; Brown, 1961; Dumont, 1966), most fully articulated in the Brahminic ideology and symbolism, was based on the recognition of tension between the transcendental and the mundane orders—a tension deriving from the perception that the mundane order is polluted in cosmic terms. This pollution can be overcome through ascriptive ritual activities that identify social with cosmic purity or pollution, and through adherence to the arrangement of social ritual activity in a hierarchical manner that reflects an individual's standing in the cosmic order.

Accordingly, Hinduism emphasizes the differential ritual standing of wide ascriptive social units called castes and the occupations or tasks tied to these units. Mundane activities were arranged in a ritual hierarchy based on their other-worldly significance vis-à-vis elimination of pollution in the mundane order. Such activities ensured the transmission of differential ritual standing through the basic, primordial, kinship units. In all these ways it had a more direct relation to worldly activities than did Buddhism (Cohn, 1971; Dumont, 1970; Heesterman, 1964; Mandelbaum, 1970; Singer and Cohn, 1968). At the same time, however, the very stress on the pollution of the world also gave rise to attempts to reach beyond it, to renounce it. The institution of the renouncer (*Sannyasa*) has been a complementary pole of the Brahminic tradition, at least since the postclassical period (Heesterman, 1985).

Given this strong articulation of the tension between the cosmic and mundane orders, a distinctive center developed within Hindu civilization, the ideological core of which was Brahminic ideology and symbolism. But because of its other-worldly emphasis, wide ecological spread, and strong embedment in ascriptive primordial units, this center was not organized as a homogeneous, unified setting. Rather, it consisted of a series of networks and organizational-ritual subcenters—pilgrimages, temples, sects, schools—spreading throughout the subcontinent and often cutting across political boundaries (Cohn, 1971; Singer, 1959; Singer and Cohn, 1968).

The religious center or centers became closely associated with the broad, ethnic Hindu identity (even more closely than the religious symbols and symbols of political community in Buddhist societies). The vague, general, yet resilient boundaries of Hindu ethnic identity constituted the broadest ascriptive framework within which the Brahminic ideology was transposed.

At the same time, however, the major center of Hinduism was not political. Accordingly, center-periphery relations in most Indian principalities and kingdoms did not, themselves, greatly differ from such relations in other patrimonial regimes, city-states, or tribal federations. These various political centers, although organizationally more compact than the ritual centers, were not continuous (regimes and kingdoms arose and fell), nor did they serve as major foci of Indian cultural identity. The political units and centers that developed in India were relatively weak in terms of the major orientations of the cultural system and commitments they could thereby command. Hindu India's essential religious and cultural orientations, then, were not necessarily tied to any particular political framework. This was not only true in the last centuries of Muslim and, later, English rule, but it also applied earlier.

Although states of different scope arose in India, from semi-imperial to small patrimonial centers, not a single one developed with which the entire Indian cultural tradition was identified. Classical Indian religions had a lot to say on the problem of policy, the behavior of princes, and the duties and rights of subjects. But to a much higher degree than in many other historical civilizations, politics was viewed in secular terms that emphasized its distance from the ideological center of the civilization, its tradition and identity (Heesterman, 1971).

The relative independence of the cultural traditions, centers, and symbols of identity from the political center was paralleled by the social structure's relative autonomy, the complex of castes and villages, and the networks of cultural communication (Beteille, 1965; Ishwaran, 1971; Mandelbaum, 1970). These castes and caste networks were not, however, simple primordial or territorial units of the kind known in many primitive societies. They were much more elaborate ideological constructions that raised primordial givens or attributes to a higher level of symbolization, thus giving rise to a wide definition of communities, markets, and networks (Rowe, 1973).

It was within these groupings and networks that the major types of institutional entrepreneurs and elites emerged. These were political and economic entrepreneurs and articulators of models of cultural order and ascriptive solidarity whose activities were structured by the two fundamental aspects of Indian social life. On the one hand, these activities were rooted in and defined by the combination of ascriptive primordial and ritual characteristics. On the other hand, such definitions placed very strong emphasis on the proper performance of mundane activities.

In most of these groups some combination of ownership of resources and control over their use and conversion developed. In the macrosocietal setting, such conversion was effected primarily through the interrelations among different caste groups and the networks (Morrison, 1970; Neale, 1962; Rudolph et al., 1975). A peculiar characteristic of markets thus developed in India.

Relatively wide institutional markets emerged that were embedded in broader ascriptive units, mainly in the local and regional caste networks. These were relatively broad constructions that were continuously being reconstructed. The markets, the widest of which were religious, centered on temple centers and fairs and to some degree were cross-cutting. That is, the religious markets (temple festivals and fairs) were controlled in a relatively flexible, yet not unstructured way by the association between the Brahmins, the kings, and the different caste networks (Conlon, 1970).

Throughout its long history, India has witnessed the continuous rise of new organizational settings, many religious movements, redefinition of the boundaries of political units, changes in technology and levels of social differentiation, some restructuring of the economic sphere, and transformation of social and economic policies, all directed by these coalitions and set within the basic premises of this civilization.

C
MAJOR PERIODS IN THE HISTORY OF INDIA

1 The Mauryas—First Period of Centralization

These basic characteristics of Indian civilization have crystallized and recrystallized at the same time as, in the major periods of Indian history, the different political, social, and economic traits were being shaped. The processes of migration and conquest have played a central role in their formation.

While the Aryan invasion was a gradual extension of seminomadic tribes, a new kind of intervention into Indian life occurred with the military conquest of northwest India by Darius I in 518 B.C. This imperial operation heralded similar successive invasions by imperial powers, culminating in Alexander the Great's conquest of northwest India in 326 B.C.

The Greek incursion stimulated the emergence in India of political powers that crystallized into the Maurya Empire, the first India-wide power, which persisted from the 4th to the 1st century B.C. and which, at its zenith, ruled the whole of India. In many aspects, it resembled the Persian Empire and was held together with the assistance of a strong army and efficient police force (Spear, 1972:55). The Mauryan rulers established a far-flung empire and created an elaborate administrative machinery to govern the subject population. They had an implicit belief in the virtues of a universal world order supported by an expansionist imperial drive. The Mauryas were able to mobilize their material resources more effectively than any other previous regime in the subcontinent, probably with the help of international caravan trade. The bulk of the imperial revenue came from the land as well as from commerce and trade. The Mauryas' administrative machinery was remarkably effective. The Mauryan Empire was divided into provinces whose governors were generally experienced officials appointed by the capital.

To ensure adequate surveillance of the provincial capitals and the collection of resources from various parts of the empire, the Mauryas established at substantial cost an extensive road network throughout the empire. These well-maintained highways facilitated administrative control, allowed fast movement of troops, and enabled the transport of goods as well as the migration of population to various parts of the empire. They also allowed the spatial diffusion of new ideas and beliefs. It was along these highways that Buddhism eventually spread across the Punjab and reached China through Afghanistan. It seems that the road network persisted long after the Mauryan Empire faded away and constituted the most substantial and enduring part of the Mauryan legacy. The Mauryan administrative structure, which was based on controlling land and encouraging trade and commerce, led to the emergence of a hierarchical urban system topped by the capital city. Its provincial capitals constituted its middle stratum; small market centers formed its lower level. The Mauryan period was indeed an age of vigorous urban growth exemplified by the prosperity of such cities as Pataliputra (the capital), Ayodhya, Benares, Ujjain, and Taxila (Ghosh, 1973).

The 6th century B.C. was also an age of intellectual and moral revitalization, evident in the emergence of new sects, cults, and religions, the two most important being Jainism and Buddhism, the latter having a larger following and a strong impact on the Indian way of life (Farmer et al., 1977; Spear, 1972:97). For a short period, through the conversion of Emperor Asoka (274-237 B.C.) to

Buddhism, it became the state's religion. But at the end of Asoka's dynasty, Buddhism declined and never regained its position against Hinduism. It was under Asoka that the Mauryan Empire expanded and reached its farthest boundaries, and Buddhism was thus allowed, together with the political power, to attain its peak. This phase of an imperial regime imposed on a rural society was short-lived, however, and did not often recur in Indian history (Thapar, 1961). The second appearance of such a regime took place in the 4th century A.D., when the Gupta Empire was founded. Buddhism's third recurrence occurred ten centuries later, during the Moghul Empire.

2 Scythian and Sassanian Invasions—The Gupta Empire

These phases of imperial rule were short-lived and separated by long periods of successive tribal invasions of northern India. The 2nd century B.C. witnessed the Parthian invasion, followed by the Scythian (Sakas) incursion deep into India, which reached the area later known as Gujarat and western Deccan. The largest tribal movement was that of the Kushans, a branch of the central Asian Mongols, who embraced Buddhism and dominated northern India and Afghanistan from the 1st century B.C. to the 3rd century A.D. During this period, peace and prosperity reigned in northern India and the foundations were laid for the second flourishing of Indian civilization under the succeeding Gupta Empire.

The rise of the Sassanian Persian Empire put a stop to tribal incursions into India from the north and west, thus enabling the revival of the imperial regime by the Gupta dynasty. Like the Mauryas, the Guptas had their roots in Magada, in southern India, which remained under Kushan domination. Gradually the Guptas encroached upon parts of central and northern India until, by the middle of the 4th century A.D., almost the whole of Hindustan was under their rule. The Gupta Empire endured until mid-5th century, when it was beset by the invading Huns. In its heyday, the Gupta Empire extended over northern and central India and most of Afghanistan, reaching as far south as the Vindhya Mountains. The period of the Gupta Empire can be characterized as the Golden Age of Rebirth of the Sanskrit culture; Sanskrit replaced Pali as the official and literary language. Economic prosperity and culture flourished and gave rise to a marked urban development, a somewhat unusual feature in the Indian rural structure (Basham, 1959).

With the decline and disintegration of the Gupta Empire, under the pressure of forceful invasions, imperial rule disappeared for about ten centuries, until the ascendance of the Moghuls.

3 Tribal Invasions and Periods of Decentralization—The Rajputs

Another great wave of invasions occurred in the 5th century A.D. First, the Gujaras penetrated northern India, settling in the area that later became known as Gujarat. Then the Huns stormed into northern India, devastating the Punjab and destroying its flourishing Buddhist civilization. This left an everlasting mark on Indian society, above all due to the incorporation of tribal elements.

One of the outcomes of the assimilation of tribal elements into Hindu society was the emergence of new military aristocracies, among the most important of whom were the Rajputs. These warriors were originally barbarians whom the Brahmins admitted into Hindu society on their own terms. The Rajputs, ruling a large number of small principalities, constituted the dominant political and social feature of India from the 7th to the 12th century. During these five centuries, the apex of a decentralized system of government was reached, with no imperial authority and organization. The subcontinent was split up into many small political units, each dominated by a military ruler and made up of the members of his clan.

The Rajput principalities clashed constantly, uniting and splitting in varied ways. Struggle for succession further added to the segmentation of this political system; each Rajput chief ruled over an area that varied in size and was settled by members of his own kin group. Sections of the domain were controlled by the ruler's kinsmen, and ranking of the elite depended on their lineal connection to the chief. Subsidiary governors enjoyed a great measure of autonomy in political and economic matters and often attempted to be independent of the ruler, thus adding to the political fragmentation.

Against this fluid political background the relative continuity of the social structure organized in the caste system loomed large. Tribal elements, which joined the population throughout the invasions, assimilated and adapted to the existing caste system, thus reinforcing the everlasting power of the social order. Throughout the Rajput period, Buddhism declined to almost complete disappearance, to the benefit of Brahminism, again attesting to the fact that Brahminism did well in a situation of political fragmentation.

4 Muslim Invasion—Centralization Under the Delhi Sultanate

With the advent of Islam in the west, a new cultural and political element began to impinge on Indian polity and society. India's first contact with Islam occurred in 712 upon the conquest of Sind by an Arab tribe, and for the following three centuries Muslim advance was confined to this part of the country. Toward the end of the 10th century, however, new tribal movements brought Muslim troops from Central Asia into Afghanistan and Punjab, which adopted the Islamic faith and has remained Muslim ever since. The major Islamic invasion occured only at the end of the 12th century when the Turk Muhammad of Ghor defeated Prithvi Raj, the Rajput ruler, and captured Delhi in 1191. This marks the start of a major new period in Indian history.

Although the new invasion was of a military nature, the invading armies were accompanied by a steady stream of clans that settled in India in the wake of the military conquest. The Muslim invaders gradually formed a military aristocracy spread thinly over the country. This military invasion and establishment of the Delhi Sultanate at the end of the 12th century had a profound influence on Indian society and culture. Contrasting with the prevailing pattern of invaders being assimilated by Hindu society and religion, the Muslims were able to impose their faith on significant segments of the local population without changing many basic characteristics of the Indian social system (Akram, 1964).

The Delhi Sultanate constituted the political center of Muslim rule for two centuries (1192-1398). During this period, Delhi became the capital of a military kingdom that united most of northern India under its control, a feat that had not been accomplished since the Gupta Empire. For a short period in the middle of the 14th century the Delhi Sultanate controlled almost all of India. This deep penetration into the south stimulated strong political opposition that brought about the rise of the Hindu Empire of Vijananagar. In time this empire controlled the whole peninsula south of the Kistna and, from mid-14th to mid-16th century, remained the chief ruling power. Throughout this period the core of this Hindu empire flourished through the channeling of wealth that flowed into it from all corners of southern India. The firm control over the flow of resources within the empire made its capital, Vijananagar, one of the greatest cities of that time, with its sumptuous design and architecture and very large population—the "Wonder City" of the first Portuguese who described India.

5 The Moghul Empire

The era of Moslem rule in India reached its peak with the Moghul Empire, founded by Babur (1483-1530). He defeated the Sultan of Delhi, crowned himself emperor, and gained control over most of Hindustan by overcoming the two chief opponents to his supremacy: externally, the Afghans and, internally, the Rajputs. The Moghul Empire remained in power and flourished for seven successive generations of the same ruling family, maintaining very high standards of statesmanship.

The actual firm establishment of the Moghul Empire was reached under the rule of Akbar, who reigned for almost half a century (1556-1605) and extended the imperial realm as far as the Indian Ocean and Central Deccan. Even more important was the institution of a well-organized central administration, which became the cornerstone of India's governance over the next centuries. Akbar based his rule on a foundation of understanding with the Hindus in general and the Rajputs in particular, granting them participation in the empire's administration and exhibiting a general tendency toward religious tolerance. Hindus were among his most important and trusted counsellors, while Rajput Rajahs governed provinces and commanded armies.

Akbar's successors extended imperial rule and, toward the end of the 17th century, controlled almost all of India. This territorial expansion, however, also contributed toward the Moghul decline. Aurangzeb became involved in the fight against the southern Moslem kingdoms and simultaneously provoked a rebellion of the Marathas of western India. The continuous internal tension and drain on resources due to overextension heralded the end of the Moghul Empire in the middle of the 18th century.

Under the Moghuls, India enjoyed a relatively high degree of affluence and prosperity for extended periods, due largely to expanding the country's settlement through fostering villages. This enabled the construction of large-scale urban centers comprising buildings and structures of magnificent design. The strong centralized rule allowed an extensive increase of trade, the income from which, coupled with land revenue, enabled the Moghuls to maintain their sumptuous courts, build splendid palaces, and patronize arts and crafts. Yet despite the outstanding urban growth and its unique architecture, this imperial rule left intact the more local social organization of villages into networks, the economic base remaining agricultural. The entire country was split into many small, separate networks that, because of sheer territorial size and poor communications, could not be integrated into a single national market having a strong influence over the emergence of an urban system.

D
CITIES AND URBAN SYSTEMS IN INDIA

1 Introduction

The development pattern of towns and urban hierarchies that emerged in Classical (especially Medieval and early Modern) India has been influenced by specific concrete characteristics of concentration and centrality forces as they evolved in India. The processes of concentration were shaped by demographic and economic development and expansion, both agricultural and commercial (internal and external and, above all, caravan and sea trade) and by the diverse forces of centrality that evolved in India—by the various political forces, the continuous recrystallization of caste formations and those of religious expansion, and the establishment of religious networks and frameworks.

The concrete combinations between these forces greatly varied in the different periods of Indian history and in different parts of the subcontinent, and they are very unevenly documented and researched. We will therefore illustrate such different combinations and their impact on the development of cities and urban hierarchies in India by only a few select examples—namely, the Rajput, the Cola, and the Moghul systems. Our analysis will provide comparative indications about the relative importance of centralized and decentralized political systems for the development of cities and urban systems in general, as well as about the relations between the economic and political urban systems, particularly the religious systems.

Before analyzing the urban systems at different periods, we will elaborate on some of their common characteristics. Throughout the greater part of Indian history, the urban system was fragmented and did not develop into a well-structured hierarchical organization. This relates directly to the regional role of towns and interurban relationships. The traditional Indian town has frequently been characterized as having served as fortress headquarters of a local political military ruler and center for the local or regional trade, often fostered by the ruler. The fortress, the bazaar, and the temple or mosque formed the basic components of the Indian town *qasba*, as they were later called. These head-quarter towns linked the villages with the regional level of government in a predominantly agrarian society and were the nexus of cash flow within that society.

The local ruler, who was a major landowner, tended to invite merchants, artisans, administrators, and professionals to settle in his fortress headquarter town. These nonlanded tradesmen and professionals depended on the ruler for protection. Not being tied to the land, they had great mobility that they could use as leverage against the ruler in case of oppression or excessive taxation. With the expansion of commerce and accumulation of economic power in the tradesmen's hands, conflicts often arose between the military ruler and the business and professional classes. In the historical context of security and political instability, the military ruler usually had the upper hand, as he provided the indispensable protection for these fragmented polities.

Whereas the towns constituted hinges linking vertically the lower levels of the settlement hierarchy with the higher ones, their main role was to act as military headquarters in the basic antagonistic relations with neighboring towns, most of which were on the same hierarchical level. Except for the short periods of central imperial government—occurring under the Mauryas, Guptas, and Moghuls—the numerous towns were antagonistic toward one another. This hostility and competition resulted in constant warfare, inhibiting full regional and market integration (Weber, 1958).

The urban system in pre-British India was, accordingly, critically affected by two types of conflict and tension: between opposing neighboring towns that served as headquarters for the local military rulers and between the landholding ruler and the nonlanded merchants and professionals.

The political-geographical pattern of the warriors-rulers—who dominated their local kingdoms, were in constant strife with their neighbors, and struggled with competitive internal economic powers—was well suited to a great variety of regions across a wide span of time. It has developed in many periods and areas of India, the best studied of which have been the following: Eastern Uttar Pradesh, 13th-14th centuries (Fox, 1970); the Vijananagar Empire of south India, 14th-16th centuries (Stein, 1976); Bengal, 17th-18th centuries; and Saurashtra, 18th-19th centuries (Spodek, 1973). In all these cases, the towns were the locus of militant antagonism between local neighboring small princedoms, each of which was in turn afflicted by the inner tension existing between the landed military ruler and the mobile traders and professional groups. Thus the towns reflected the fragmentary structure of Indian society that persisted in most periods of its history.

2 Religious Urban Centers

Side by side and often simultaneous with these various urban centers, many cult or religious centers developed in India. Some evolved characteristics of their own and became interwoven, in different ways at different periods, with other urban centers.

As is well known, Hinduism comprises various forms of rituals with household gods at the base (actually the ancient Vedic religion had no other form of worship). The second level of ritual activities was conducted at the site of inconspicuous village shrines that were distributed in various parts of the rural areas. The most important places of worship were the temples, which were monuments of remarkable architectural design and housed one or more gods; worshippers may also have frequented other temples. It was these places that had some potential of becoming urban centers. The role of the temples as stimuli for urban growth was defined by two major factors: First, because the temple "officials" did not have to provide ritual services to a specific surrounding area, the temples did not constitute congregations. Second, of crucial importance was the fact that although regionwide (and sometimes even countrywide) networks of Brahmins and religious specialists indeed developed, they were never structured in a churchlike organization and were usually embedded in local or regional settings (Appadorai, 1977).

The Hindu temple was the house of a god, and its architectural and spatial features distinguished it from the places of worship of other religions. As it did not serve a congregation, it did not need a large enclosed space. The god or gods (made of stone or wood) were often concealed in a dimly lit inner shrine. An open space, for the use of ritual processions and public circulation, surrounded the shrine. These open grounds were usually walled to demarcate the sacred space; in south India these gateways were elaborate and conspicuous and they dominated the landscape. The grounds surrounding the temple housed the Brahmin caretakers of the resident deity as well as the artisans employed in beautifying and maintaining the temples. The temple compound always included a large pool for ritual baths for both gods and worshippers. Thus in many ways Hindu temples constituted almost self-contained units, quite autonomous in their religious functions and not necessarily connected with other urban functions.

Hindu temples were usually autonomous institutions endowed by wealthy people through whose patronage they remained free from dependency on large fixed communities. However, urban communities often grew around temples and sacred places, mostly to cater to pilgrims' needs, but only rarely did they constitute a major stimulus for urban growth. The size and opulence of these temples and their proximity to other temples were related directly to the region's economic level and productivity. Local rulers often drafted regional resources for the construction and maintenance of impressive temples (Spencer, 1969).

A specific type of contact developed among the different worship places. They were influenced by the great importance placed on pilgrimage to centers of religious learning and ritual purification and by the lack of any overall, church-like religious organization (Bhardwaj, 1973).

The spatial structure of the Hindu religious urban systems was very near the pole of a self-sufficient local autonomous structure but relatively distant from a fully integrated centralized hierarchy. The informal circulation network among the various holy places and temples generated the most important mechanism of spatial integration. The great pilgrim assemblages provided a regular opportunity for the exchange of religious ideas among the many different Hindu sects. The travels of holy men among holy places provided another integrating mechanism and allowed the diffusion of ideas among the various religious centers. This informal circulation network substituted for spatial integration, which in other religions was provided by a centralized hierarchical structure.

A temple complex or a sacred site, such as a confluence of the Ganges River, attracted pilgrims often from great distances to perform rituals and be blessed and purified. The flow of pilgrims created a circulation network of a religious nature, promoting secondary flows of traders and artisans who catered to the pilgrims' physical and spiritual needs. The number or pilgrims and their itineraries were partly related to the ease of transportation and to the distribution and density of the population in the areas surrounding the temple or sacred site. But more important was the informal hierarchy of the level of the holy places' sanctity, allowing some to become only local or regional foci of pilgrimage whereas others became pan-Indian pilgrimage centers. Thus over the centuries, the major holy places have drawn a significant part of their pilgrim traffic from all over the Hindu realm. The pan-Indian centers of pilgrimage were situated in all parts of the country, defining, in some way, the Indian religious realm, from Kedarnath and Badrinath in the Himalayas (near the sources of the Ganges) to Rameswaram and Cape Comorin at the southernmost extremity of the Indian Peninsula.

Many of the temples did have some networks of subsidiary temples, but usually these did not develop into a full-fledged hierarchical system. Only insofar as they became interrelated with the military or trade centers did they become part of fully evolved urban hierarchies.

Not having a formal hierarchical structure, holy places contributed to the circuit pattern of the pilgrimages. This included the most important pan-Indian religious centers, beginning with holy places at the sources of the Ganges (Kedarnath, Badrinath, and Hardwar), continuing with those in the mid-Gangetic plain (Varanasi, Allahabad, Gaya) and along the Bay of Bengal (Jaganath Puri, Kanchipuram, and Rameswaram), then northward at Nasik-Tryambak and Dwarka, on the shores of the Arabian Sea, and completing the pilgrimage circle through Mathura, returning to the sources of the Ganges (El Faruqui and Sopher, 1974:72).

The degree to which temple towns developed into full-fledged cities and major urban centers varied greatly. In the south, many important temple towns emerged, such as Madurai or Tangore; in many ways the temples, with their artisans and merchants, constituted major urban centers. Otherwise, it seems that only Varanasi (Benares) and Allahabad became major urban centers; all the other pilgrimage centers remained religious centers—that is, part of the autonomous religious network—which did not coincide with the urban network and did not develop into large cities. This could be explained by the fact that Varanasi and Allahabad had significant locational advantages and their central regional locations encouraged their urban development, much beyond their role as pilgrimage centers (Eck, 1982).

3 Major Components of the Indian Urban Systems

The following elements constituted the permanent components of the Indian urban system: the military-political town, serving as a center for the flow of cash nexus in the society and often for the redistributive system, and the temple or the full-fledged temple town. The great variations existing among the different periods and areas developed with respect to (a) the degree of existence of a more centralized hierarchy—or, at least, attempts at maintaining such hierarchy—and hence also the relative importance of the town at the upper levels of such hierarchy; (b) the relative importance of coastal towns (littoral) in relation to those of the hinterland (Heesterman, 1980); and (c) the importance of temple centers and networks in relation to the more political and commercial towns.

These variations (above all, the first two) were influenced by the degree of political centralization and the relative importance of the internal and international trade. Common to all these systems were continuous shifts in the relational importance of different towns, mainly related to the political fortunes of their rulers and the shifts and vagaries of international trade (Das Gupta, 1970; Heesterman, 1980).

E
COMPARATIVE HISTORICAL INDICATIONS: CITIES AND URBAN SYSTEMS AMONG THE RAJPUTS, COLAS, AND MOGHULS

We will now analyze the structure of the urban system during the Rajput, Cola, and Moghul periods. The first two exemplify a decentralized structure of government; the third is a fine example of a centralized system. The Maurya and the Gupta empires are other examples of centralized systems of government but are not discussed here due to lack of detailed information.

1 The Rajput Period

The Rajput period is a very good illustration of the development of urban systems in a decentralized political framework in Indian history. As we have seen, the Rajputs made up a class of local rulers. They formed many small states that were decentralized polities in which political power was dispersed and no single state was economically outstanding. Each Rajput ruler dominated an area of varying size inhabited by his own kinsmen and the local people he had subjugated. Under the main ruler, lower-ranking kinsmen governed sections of the state. These subsidiary governors were granted great autonomy in political and economic matters and, quite often, tried to gain independence from the main authority. In the Rajput political system, the rank of the elite depended upon caste standing and kinship closeness to the state's founder or its present chief (Fox, 1977:44).

The urban settlements of the Rajputs' principalities were greatly influenced by the political and social system. No general urban hierarchy, embracing the entire country or large parts of it, developed. Instead, a small-scale local urban hierarchy crystallized in each Rajput state, not strongly related to the other states. The size and complexity of the urban hierarchies in each state varied with the status of the ruler and extent of his domain. The structure of urban India under the Rajputs can thus be characterized as being composed of numerous small-scale and local hierarchies of roughly the same order of magnitude, lacking a dominant urban center and having minimal political or economic integration.

A typical settlement system in a Rajput state—not dissimilar from what would later also be found in the Moghul period—consisted of three levels: the capital city, with its immediate hinterland (the *haveti* and *pargana*); the urban center of the second level (the *tappa, taif*), with their secondary center; and the basic local unit (the *gaon*). The capital cities (the *pargana* and *haveti* centers) possessed a distinct urban character. The *gaon* centers were well-defined villages. The *tappa* centers and some of the small *pargana* centers presented a mixture of urban and rural characteristics—"rurban" settlements, to use Fox's definition (1970:181).

The *pargana* was the local ruler's seat and was also the location of the clan deity and temple. Being capitals of the Rajput states, these towns originated as strongholds to withstand the forays of other chiefs or dissident kinsmen, or even centralized empires that periodically arose in this part of India and tried to infringe on the local ruler's authority. Because of the military and security functions of these towns, a strong defense element was always present in their urban structure. The smallest or plainest capital was built around a mud fort, often surrounded by moats and protective hedges. In their largest and most developed form, the Rajput capitals were surrounded by massive walls with ornate castles and towers and complex military installations (Fox, 1977:50).

The population of the Rajput cities was made up mainly of the ruler's kinsmen and different groups of artisans and merchants attracted by the rulers. In general, the population of a Rajput capital consisted of the chief, his close kinsmen, and other members of his kin group, as well as the retainers and agricultural tenants who lived in the city and worked the neighboring fields. The Rajput capital was often also the religious center of the kin group and contained the deity worshipped by the chief and his kinsmen. Thus Rajput capitals were characterized by temples (elaborate in various degrees), sacred artificial ponds, and other ritual structures. Sumptuous Brahmin rituals emphasized the religious function of the Rajput cities and legitimized their rulers.

These cities—especially their bazaars—usually managed the flow of the cash nexus. It was there that the land revenue was concentrated, fed into, and was exchanged for commercial products and handicrafts. The specific requirements of the ruler attracted craftsmen, traders, and ritual specialists to the city. The chief usually endowed a market by the side of his compound, guaranteed safe trading, and even provided inducement for merchants trading mainly in luxury goods and specialized services to settle in the city. The markets were established by the chief in order to maintain his control over the cash nexus and to benefit from the different levels of commerce. City size and prosperity were directly related to the chief's status and power among the other Rajput rulers. When the ruler's power and that of his kin group declined, so did the markets they had created.

The *tappa* centers were offshoots of diversified kin groups. In function and urban structure, these secondary centers were quite similar to the major center in the capital city but were much less complex and elaborate. The *tappa* was governed by a close kinsmen of the ruler who, due to his link with the ruler's lineage, could maintain authority. The *tappa* center was smaller than the capital but was equally fortified and well defended and sometimes surrounded by walls and moats. A small market was sometimes created to supply the needs of the chief's local kinsmen and also as a central place of low order to serve the rural population around the *tappa*. Genetically and structurally, the *tappa* center was a small-scale replica of the capital city of a Rajput principality (Singh, 1968:73).

The most basic units of the settlement system were the *gaon* centers, which were villages of a special kind fulfilling the economic functions of rural settlements. Because of political fluctuations and numerous military forays of Rajput rulers, many of the larger villages were fortified, moated, and walled. This also caused the *gaon* centers to assume considerable size and unite in clusters of villages (Singh, 1968:73).

2 The Cola Period

A different pattern of urban and marketing system in a decentralized patrimonial regime can be found in south India during the Cola period (A.D. 850-1279), the golden age of Early Medieval south Indian civilization.

Somewhat stronger but not very successful tendencies toward centralization developed within the Cola realm. It was mainly the international trade, however, that became especially important in this period, together with the emergence of littoral cities and their connections with internal cities. International traders used the shores of southern India as their regular ports of call. This entailed a regular, ongoing relationship with the hinterland to allow locally produced goods to reach the coastal centers and networks, connecting the external and internal trade that emerged to supply the requested commodities. The revenue derived from port duties on various luxury goods transformed the basic inward-looking orientation of the south Indian kings; they became promoters of the seagoing trade and supported the organizations that made possible the continuous flow of products in and out of the ports. The Cola kings in particular, whose lifestyle demanded a conspicuous display of grandeur, perceived the benefits to be drawn from a positive orientation to commerce and shaped their relationship with the people involved in the trade accordingly.

The support and encouragement of the Cola rulers in the 11th century enhanced the development of regional commercial centers of considerable magnitude and wealth. In the late 12th and in the 13th centuries, however, these centers began to challenge the authority of the Cola kings. The process culminated in formal alliances among the regional centers under the 14th-century Vijananagar Empire.

During the Cola period, the major element in the urban panorama was thus constituted by coastal cities. Their relative growth, especially at times of decentralization of middle regional cities, was similar to their equivalent in the Rajput realm, except for the somewhat greater emphasis on commercial activities. The spatial-political landscape of south India during this period was made up of administratively autonomous villages and regional assemblies known as *Ur, Nadu,* and *Nagaram,* which remained stable despite the rise and fall of several imperial dynasties (Hall, 1977:2). Of special interest were the *Nagaram*, the regional commercial centers that fulfilled a major role in the area's trade and marketing system. Each traditional unit of regional administration (*Nadu*) had one *Nagaram* controlling the area's entire commercial activities and also, to some degree, its relations with the coastal towns.

The commercial regional centers (*Nagaram*) were centrally located and closely linked to several village clusters. They were basically commercial towns that fulfilled all the functions of a central place in the lower level of the urban hierarchy. The local *Nagaram* assembly of merchants was the most important organization in the towns. *Nagaram* towns functioned as points of redistribution, from which local products were transferred through the town's market to a wide network of commercial exchange with links to coastal ports and international markets; they also made specialized commodities from external sources available to local villages.

These regional commercial centers began to strengthen their autonomy simultaneous with the gradual weakening and decline of the Cola kingdom and formed the nucleus of a new structure of political authority. The growth and economic strength of the *Nagaram* centers exemplify the influence of concentration forces on the evolution of the urban hierarchy's lower levels. With the decline of imperial authority, regional commercial centers became the backbone of the decentralized political structure.

Another major feature of the urban complex of the Cola period was the growth and increasing conspicuousness of temple towns. This confirmed that the Brahminic power flourished in periods of decentralization and that there existed, beyond the lower levels of the urban hierarchy, some contradiction between the forces generating a strong religious urban hierarchy as opposed to the political-commercial one.

3 The Moghul Period

The Moghul Empire provides the best example in the history of India of the development of urban systems under a centralized government. The Moghul period witnessed important changes in the settlement system as a result of major transformations in the political administrative realm. These changes affected the upper levels of the urban hierarchy while the structure and role of the lower levels, the small local communities, remained almost unchanged. Evolution of the urban system was then directly related to the transformation of northern India from a decentralized, semifeudal system into a more bureaucratic imperial regime. This political administrative transformation, which occurred under the reign of Akbar (1542-1605), attempted to centralize the government on the national and regional levels but left the local level fragmented and quite autonomous under the hegemony of the dominant lineages. The administrative structure comprised three levels: the central government, the provinces, and the local lineage territorial units (Gumperz, 1974:581). In fact, as B. C. Cohn has shown (1961), these levels often tended to collapse upon each other, especially as the rulers tended to intervene in local conflicts.

The central government was the top level and was located in the capital city, mostly in Delhi but sometimes transferring to Agra and Lahore. The next level comprised the provinces (*Subah*), which totaled 15 at the end of Akbar's reign, each ruled by a governor (*Subadar*) who was a senior imperial officer responsible for security and order and commanding a body of troops. Many *Subadars* were

Rajput chiefs who had become loyal officials of the Moghul emperor (Spear, 1970, 1972:133) and were left to control their territories as Moghul agents. By becoming Moghul agents they were able to ascend the hierarchy of imperial services. The authority of the provincial governor was counterbalanced by that of the *Diwan,* also an imperial official, who was responsible for collecting revenue and paying the troops, thus ensuring imperial control also at the regional level. The seats of the provincial governors were located in capitals of the provinces, which in the Moghul Empire made up the second level of the urban hierarchy. The size of these second-level cities varied according to the extent of the province and its economic situation, but they all fulfilled the duties of political administrative control and represented the imperial government—a prime example of the operation of centrality forces in urbanization processes (Ali, 1966).

At the base of the administrative structure were the local lineage territorial units. Land was usually controlled by the clan, which constituted the major social framework of these local units. The territorial hegemony of the dominant lineage, which was the essence of the third level of settlement hierarchy, actually meant that the central government interfered little in local affairs; these were left to the authority of the local lineage chiefs, although the rulers often meddled in local politics. But in principle, the lineage chiefs were responsible to the upper levels of the imperial government for maintaining peace within the area, collecting taxes, and other obligations. This political and economic recognition of the local lineage chiefs by the central authority enhanced the self-definition, cohesion, and corporate activities of the local lineages, tending to give the greatest behavioral, political, and economic significance to clan membership (Fox, 1970:171). As in the earlier period, the basic lineal unit was itself divided territorially into three levels—the *pargana*, the *tappa,* and the *gaon*—as a result of which a hierarchy of settlement evolved within the local level of centers. The clan chief settled in the *pargana* and the head of families in the *tappa* level. The *pargana* and the *tappa* thus evolved into small urban centers due to their respective territorial and subterritorial command of leadership and resources, whereas the *gaon* evolved as a strictly basic rural settlement, subsisting on local resources. The lineage chief living in the *pargana,* therefore, assumed the major function of linking his lineage to larger urban political units (Fox, 1970:172).

This territorial system at the local level persisted for a long time and withstood numerous political upheavals; the land was owned by the clan, and the caste social structure, acting as a cohesive force, did not allow the dependent noncorporate groups to break away from the prevailing territorial dominance pattern. It should be emphasized that the Moslem rulers did not tamper with the deep-rooted social systems, and the official hierarchy of the Moghul administration stopped at the *pargana* level and did not effectively penetrate the lower levels (Naqvi, 1972).

The mechanism connecting the various levels of urban hierarchy was provided by the contracts drawn and alliances entered into between the central rulers and different local groups, organizations, and other chiefs, whether headman or council *(panchayat)* of the dominant caste, heads of guilds, or different "strong men" or local *mahajans*. Thus the distinct features of the Moghul urban system were the relatively greater importance of the primate city and the close interweaving of the political and commercial (but not the religious) urban systems (Chaudhuri, 1978).

CHAPTER 10

Urbanization in Japan

A
HISTORICAL BACKGROUND

The development of urbanization in Japan reflects the course of Japanese history as a continuous but intermittent expansion of central control. Each attempt at unity was immediately followed by a prolonged period of breakdown, decentralization, and gradual reconstitution of a new and more effective base of centralized power.

Japanese history can be said to start around the 4th century, when some of the Yamato tribes or clans (later designated "imperial") succeeded in establishing tentative sovereignty over the main inhabited parts of the country (Reischauer and Craig, 1978). This rule was so strongly inhibited by powerful territorial, tribal, and clan chieftains, however, that the imperial clan could only be considered *primus inter pares.*

In the 7th century another effort was made to strengthen central control, this time under the influence of China's T'ang dynasty. The Taika reforms of 645 and the subsequent elaborations in the Taihō Code of 701 and the Yōrō Code of 718 attempted to establish the T'ang principle that all land belonged to the emperor and could be granted only by him. In order to reduce the independent power of tribal or territorial chieftains and landed magnates, a centralized bureaucracy was created by the emperor, from whom authority was considered to emanate. The Chinese model inspired this system of centralized court bureaucracy, appointment by the political center of provincial and local governors, development of a national tax system, and improvements in communications.

The Yamato state was thus seemingly transformed from a loose tribal coalition with a poorly defined center of sovereignty into a centralized state ruled by an emperor and court bureaucracy. Its authority was increased by the establishment of a fixed capital (until then the capital had moved from one temporary site to another upon the death of each emperor)—first in Nara (in 710) and, at the end of the 8th century, in Kyoto (Heian).

The long era of cultural expansion known as the Heian period began in 794 and continued until 1185. In spite of bold centralizing moves, however, the forces of decentralization were continuously active. Gradually throughout the Heian period, and with gathering force after the 11th century, powerful provincial military clans controlling lands, arms, and men built up their own political alliances, land tenure, and local administrations. By the end of the 12th century, under the leadership of Minamoto Yoritomo, they were able to displace the central court and asssume effective power over the country, giving rise to the Kamakura period (1192-1333). This was an initial step toward the development of a provincial warrior class, its growing predominance, and the crystallization of a feudal society.

The Kamakura period was characterized by the predominance of a central group, originally led by Yoritomo, and by intermittent attempts to unify the country and reinstate imperial control. Yet the central court and the military aristocracies maintained uneasy and shifting "dual power" until the 14th century, when another breakdown of the central power occurred with an attempt at imperial restoration. This was followed by a new centralized military power, the Ashikaga *shogunate*, which heralded the Muromachi era (1336-1573). This new centralized power itself broke down under the pressure of dissident local forces, giving rise to misleading and shifting coalitions and a period of civil war (sometimes quiescent, sometimes in violent eruption) before new centralizing forces appeared. From the latter part of the 16th century, three great centralizers appeared in succession: first, Oda Nobunaga, then Toyotomi Hideyoshi, and finally the most important, Tokugawa Ieyasu.

1 The Tokugawa Regime

Ieyasu's military victory in 1600 at Sekigahara marked the beginning of the final premodern phase of centralization. This regime was far more effective than earlier ones and lasted until 1868, when it was overthrown by a modernizing coalition, the so-called Meji Imperial restoration. Although the Tokugawa unification was the most effective in Japanese history until that time, it was still far from complete and perhaps could best be characterized as centralized feudalism (Reischauer and Craig, 1978). The shogunal government permitted substantial power over almost three-fourths of the country to remain in the hands of mighty territorial barons, the *daimyo*. The *daimyo* were vassals of the *shogun,* who maintained complete control over them by several means: the system of hostages and compulsory residence in the court city, called the *sankin-kōtai*; spying; the sequestration of strategic areas for the *shogun*'s direct domains (*chokkatsu-chi*); and careful placement of personal vassals and family members to control the movement of potential enemy lords.

B
MODE OF DECENTRALIZATION: FAMILY AND GROUP IDENTITY

Japanese history has thus been characterized by the continuous expansion and crystallization of a relatively homogeneous national community, with a strong symbolic center that emerged relatively early. Another feature was a tendency toward recurring oscillations between political centralization and decentralization and the concomitant development of distinct types of decentralization based on a sort of dual center with varying degrees of power held by the effective political center.

This decentralization, which Bloch (1964) and others (Baechler, 1975) have called the closest approximation to European feudalism, differed in many ways from the other decentralized regimes we have analyzed (European, Indian, and Islamic systems), as well as from European feudalism itself (Hall, 1962). Unlike in those regimes, decentralization in Japan did not evolve within the framework of a civilization based on some universal premises and comprising many different ethnic and political units and communities whose primordial attributes differed greatly from those of the civilizational collectivity. Rather, they took place within the framework of a relatively homogeneous national collectivity in which a close relationship developed very early between the national, the political, and the ethnic regional or family symbols of collective identity. In this collectivity, the center (the emperor) constituted the major pivot of such an identity, even when this center was relatively weak (Duus, 1969).

From the early periods of Japanese history some specific structural characteristics became apparent. Of special interest from the point of view of our analysis was the structure of group and intergroup solidarity, identity, and loyalties that evolved as one of the basic features (although certainly not the only one) of Japanese society and that became relatively predominant, probably as early as in the period of late feudalism. The nature of these characteristics could be best observed in the structure of the Japanese family and in its development, mostly from that period, or as a special type of extension of family relations beyond the family and kin group.

This extension of familial relations was characterized by a combination of strong internal organization and solidarity of the small nuclear family or groups of such families, bearing the weakness of broader kinship categories or units as well as their obligations. To a large degree this facilitated the outward transfer of manpower and resources from within families and a great capacity for continuous reorganization. In Herbert Passin's words (1968:241):

> We know from history that the Japanese familial form was present before Japan became a feudal state and continued after it ceased to be. The ideal Japanese family has always been hierarchical, and internal family relations have provided a model for authority relations in non-kinship groupings. The terms *oya* (parent) and *ko* (child), for example, have been extended in meaning to indicate superiors and inferiors: lord and vassal, boss and henchman, employer and employee, leader and follower.

Closely related to this extension of familial principles beyond the family was the special conception of rank and hierarchy, based on relations among leader, followers, and intermediaries, with little consciousness of equal status (Passin, 1968: 245). This conception became predominant from relatively early times (Ledyord, 1975; Edwards, 1983), although it was certainly not the only one in Japanese society.

This mode of extension of family or kin relations competed with another mode that probably predominated in the earlier periods of Japanese history in the Kinai (central Japan) and in regions that were based more on the extension of horizontal clan principles, of the *uji* or *meta-uji* society (Murakami, 1984: 11-13). Such conceptions of hierarchical rank were extremely important in structuring the relations between family groups and the broader social structure, and creating the *ie* and proto-*ie* social structure (Murakami, 1984). This structure was not mainly an extended kinship or clan unit as the *uji* organization, but was based on a flexible extension of "kin-tract-ship" principles beyond the family and on a wide use of the practice of adoption. To follow Murakami (1984: 11, 31):

> The loose interpretation of kinship as well as the practice of adoption suggest that each proto-*ie* strongly demanded core members with military and managerial talents in order to compete and expand against other proto-*ie,* and also that each proto-*ie* group *per se* was established as an indomitable entity so firmly that it did not need to rely on blood ties. . . .
>
> We would like to emphasize here that the proto-*ie* broke through the limits of kinship group, thereby becoming a more "achievement-oriented" organization that could expand its members and territory. However, of at least equal importance is the fact that the proto-*ie* could maintain a high degree of group cohesion in spite of the practical renunciation of kinship principle. . . .
>
> The ultimate collective goal of the proto-*ie* was to guarantee the very-long-run subsistence of its members and their posterity [and it] . . . can be defined not as the attainment of a *specific* target, but only as the *diffuse* goal of eternal continuance

> and expansion of the group. . . . The hierarchy which characterized the *ie* system aimed at collectively fulfilling some function by assigning a specific part of this function to each individual stratum within the hierarchy. . . . Taking a slightly different angle, the *ie* structure may be said to stand on a balance of two almost contradictory principles, homogeneity and functional efficiency.

The differences between the *uji* and *ie* types of societies were closely related to the diverse patterns of settlement that developed in Japan. The Kinai region in central Japan had some characteristics that differed significantly from those of the rest of the country (the holdings in that area were substantially smaller). In the Kinai and other economically advanced areas, management and cultivation of excess land was entrusted to tenants, and the application of this practice went so far as to reduce some holdings to almost mere legal entities; in these cases the *uji* system tended to evolve. In the rest of the country large holdings were worked mainly by landowners and their dependents (Smith, 1959: 1-5), and the *ie* system developed there more frequently.

The social relations and inheritance patterns that emerged from the two forms of landholding differed greatly. The smaller holdings produced a special kind of landlord-tenant relationship with the nuclear family (a couple and their unmarried children) at its center. These families were not the sole inhabitants of the holding but formed a distinct unit within the larger group because the family's rights to the land and its ceremonial position were well maintained.

In areas where land was plentiful, it belonged to the village and was partitioned exclusively among its members. The purpose of partitioning the land was not to establish new familial branches; thus a family was never forced to divide its property, and this practice was discouraged. The number of families in each village was fixed by custom, and a new family could not be added unless another died out or moved away. The eldest son therefore inherited the property and became the new head of the family (Smith, 1959: 36-40).

The inheritance system thus generated an economic hierarchy among the landholders, shaping the village's class structure in the image of the hierarchical family. In large holdings it was the extended kin unit of the *ie* type that formed the class system; in the small holdings the nuclear family constituted the building blocks of society. The combination of demographic expansion and the tendency not to divide the family holding constituted an important factor in the continuous settlement in Japan (Befu, 1981).

C
CULTURAL ORIENTATIONS

Closely related to these characteristics was the relatively continuous predominance, from early times, of certain basic cultural orientations and principles of institutional organization that were closely related to these orientations. They were important in shaping some of the major institutional features and dynamics of Japanese society, the modes of conflict and conflict resolution that developed within it. The most important among these orientations were the following: (1) the stress on a relatively low level of tension between the transcendental and the mundane orders; (2) in close relation to the former, a strong combination of this- and other-worldly orientations and emphasis on ritual activities; (3) a strong inclination toward obligation and commitment to the social (and cosmic) orders, extending from the family to various wider circles—in principle, although not always in practice—to the center or to the whole collectivity; and (4) as developed in Japan, the strong emphasis on group identity generally and particularly on special combinations of vertical and horizontal group loyalties.

Of special importance here was the fact that Japanese tradition did not distinguish between the societal and cultural orders represented by the center and those represented by various types of collectivities at the periphery. A close relationship developed between the symbols of the center and those of peripheral groups, the orientation of the center being a basic component of these groups. The collective identity of the major local and occupational groups was expressed in terms defined by the center's ideology.

As the cosmic order's representative, the emperor occupied a strong mediatory position connected to potentially open access to the national symbolic center through the various vertical networks and orientations of all groups and sectors of the population.

D
THE GROUP MODEL AND MODES OF CONFLICT REGULATION

It was the relative predominance of the combination of group solidarity and such rank and hierarchical principles that gave rise to the model of Japanese society formulated by Chie Nakane (1970). This was based on the vertical interlocking of groups, networks, and group loyalties—a model that has been subject to much criticism, especially recently (Mourer and Sugimoto, 1980; Najita and Koshman, 1982). But these criticisms did not necessarily obviate the importance of the principles of Japanese social structure stressed by the model. Rather, they have pointed to the need to take into account several aspects of social relations in Japanese society not sufficiently considered by this model. Thus the existence of such principles does not imply that far-ranging conflicts did not develop either within the groups themselves or among them.

Japan's turbulent history, on the central as well as the local levels (manifest in peasant litigations and rebellions), fully attests to the ubiquity and intensity of such conflicts.

Indeed, several bases of conflict were built into the very institutional application of these orientations. The first basic conflict existed between the hierarchical principles of any group represented by its designated (ascriptive or elected) leaders and the more egalitarian, horizontal tendencies within it. The second was between the concrete implications of such principles and the interests of various subgroups within the family or village group. The third conflict occurred between the internal solidarity and interests of the family group and their extension—mostly in terms of a hierarchical rank order—to broader settings that necessarily extracted resources from the family or the village (Hall, 1966). The fourth conflict focused on specification of the exact locus of vertical contacts and the mutual obligations of lower and higher echelons. The overt ideology of such obligations tended to stress mutual harmony and benevolence—themes that became predominant with the infusion of Neo-Confucian orientations into Japanese thought (Scheiner, 1978)—yet many acute dissensions often developed.

The resolution of such conflicts tended, however, to reestablish some of the vertical hierarchical principles even if in different concrete organizational institutional configurations. Other—horizontal or egalitarian—orientations were more evident in peasant rebellions and their millenarian components (Scheiner, 1978).

The nature of the concrete constellation of such forces, the institutional arrangements, the relative degree of centralization, depended on the composition of the forces predominating in a specific period of Japanese history or in a specific region of Japan. Yet such differences, variations, and changes notwithstanding, these different constellations tended to be restructured according to the above-mentioned principles.

Continuous reestablishment of some combination of group solidarity and vertical hierarchical orientations took place despite the fact that many such conflicts and movements were often organized by seemingly unrelated individuals. They could not, however, establish their effective positions without acting in accordance with these principles. Even the more horizontal orientations tended to develop in the direction of a sort of populist participation and much less in terms of horizontal class or sector identity based on autonomous access to the major attributes of status and the center of the different groups.

The relative predominance of all these institutional tendencies was related to two closely interconnected facts. The first was that these orientations were carried by the major collectivities—the family, the village, the feudal, or some regional sector and their representatives, be they heads of families or more specialized elites such as warriors, religious specialists, and the nuclei of administrative groups—all of whom were embedded in these ascriptive groups. The second fact was the concomitant, almost total absence of autonomous elites (except in such limited spheres of activity as the artistic or intellectual fields) whose criteria of recruitment and activity were not embedded in such primordial ascriptive groups. The elites' lack of autonomy was closely related to the absence of universalistic criteria based on a transcendental vision stressing a chasm between the transcendental and mundane orders.

E
ELITE STRUCTURE AND SYSTEM OF CONTROL

The combination of these cultural orientations and their institutional repercussions epitomized in the relative predominance of the *ie* type of society since the late middle ages had serious effects on many aspects of Japanese social structure and on the modes of social control that developed in Japan. They also particularly affected the relations between the self-government of the different social groups—families, villages, or domains—and their control by overlords or the central government.

Due to the strong cohesion among the basic groups in Japanese society, relatively far-reaching self-regulation developed within them. This was vested above all in their respective leaders, whether ascribed or elected, representing such collectivities, as members did not possess independent rights of access to those groups. At the same time, the leaders of these groups were themselves supervised by superiors in their hierarchy, potentially up to the emperor, and did not enjoy full, legal rights of access to the higher centers.

This mode of control was best illustrated in Japanese villages that had achieved some degree of controlled self-regulation (Befu, 1968). Two bodies of rules—those originating from the central government and those established by the villages themselves—regulated the peasant's life, existed side by side, and had different objectives. The state's aim was to maximize taxation and to preserve a static and stratified society; the village wished primarily to maintain internal peace and protect itself. Compliance with state laws governing fiscal matters and regulations was paramount. The village codes emphasized policing and other aspects of village life not stressed by the state (Hall and Jansen, 1968: 310). The state's centralized power was based on the definition of the social hierarchy (warrior-farmer-artisan-merchant) and on officials' ability to govern their own status group (fief, town, village, family). Local matters were dealt with by traditional law without interference by national law or courts, but state law was applied on such sensitive matters as taxation, status relations, and policing the village.

The state's control of the villages and self-control of the villages were implemented through a series of devices. The state representatives at the village level were the headman, elders, delegates, and five-man groups. The headman was by far the most important official, in terms of both his delegation of power and the responsibilities assigned to him by the government. The headman and other officials were elite members of the village community, not administrators sent by the center. To enforce its laws, the government thus relied on individuals who were locally influential. Selection of the village's own leader as government representative meant that the government's unchallenged authority would be accepted positively. Moreover, the fact that the village officials were chosen by and from among the propertied peasant elite meant that the authority of the state would be accepted and enforced in the village through its existing political structure (Hall and Jansen, 1968: 312).

One of the main reasons for this coherent organization was that the villagers felt strong group responsibility. The village was united by deep solidarity; mutual cooperation and labor exchange were indigenous patterns of the social organization. The government was therefore capitalizing on an accepted concept when it levied tax on the village as a whole, and held the five-man group responsible for villagers' infractions. This strong degree of commitment to the village as a unit and of solidarity enabled the villagers to enforce their self-made laws and impose sanctions on transgressing members (Hall and Jansen, 1968: 314). Yet such headmen or other intermediaries in middle echelons became the focal points of conflict and ambivalence (Scheiner, 1978).

As in most sectors of Japanese society, the control of such solidarity was vested in the hands of the respective leaders who, as mentioned, were organized in vertical lines. Although these vertical lines converged on the center or on the leader, access to the center was not based on an autonomous approach but rather on strong commitment to the central authority, a commitment that, in principle, was controlled by the center. Hence even in the feudal period, no contractual relations developed in Japan between lord and vassal, nor did countrywide autonomous organization of vassals develop as occurred in Europe.

F
THE POLITICAL SYSTEM, STRUCTURING SOCIAL HIERARCHIES, AND MODE OF ABSORPTION OF EXTERNAL INFLUENCES

The basic cultural orientations and the modes of self-regulation and control we have analyzed had a far-reaching impact on several basic aspects of Japanese macrosocietal order. The political system appears similar to the various patrimonial systems analyzed in the other case studies in which little distinction existed between center and periphery and neither impinged significantly on the other. In Japan, however, the center or centers continuously attempted to permeate the periphery. This permeation was much less oriented toward restructuring the periphery in comparison to the imperial systems already discussed. Rather, according to a transcendental vision, it focused on endeavors to mobilize the economic, political, and military resources, as well as the loyalty of the different groups in the periphery.

Such permeation and attempts by the center or by different ("decentralized") centers to mobilize the resources of the periphery were based on the strong emphasis laid in Japan on group commitment as well as on the identity of the center and periphery's basic cultural parameters focused on the center.

It was this combination that distinguished the Japanese social system from most of the patrimonial systems and gave rise to the special combination of centralized semi-imperial or decentralized, feudal-like regimes that characterized

Japanese society with the special type of structuring of social hierarchies and systems of control, which will be analyzed shortly.

The orientation and structure of these elites had repercussions on several crucial aspects of the structuring of social hierarchies and on the control wielded by Japanese society. Given the special importance of at least some centers, it was natural that they should attempt to become the source of prestige and the regulators of the social hierarchy. Indeed, under Chinese influence a relatively elaborate yet constantly changing class structure emerged in the Nara period that was composed of *samurai*—defined in Confucian terms as an elite of *bushi* (warriors and gentry)—farmers (peasants), and artisans and merchants (in order of descending importance, according to Confucianism). In this class structure the proper performance of the duties and responsibilities of each class was emphasized (Hall, 1970:179). Based on the prevalent cultural orientations marked stress was laid on the this-worldly duties and activities of these classes, which in principle were oriented toward the center.

The concrete outlines of this complicated class structure were greatly influenced by the changing fortunes of the center and by the continuous oscillation between the centralization and decentralization policies. Yet they were focused, at least in principle, on the idea of the center.

Only in periods of centralization was the center able to control the land, cities, and class system through such rather elaborate mechanisms as the conferring of prestige and honor on elite groups and the control and distribution of land. In cases in which the center lost control over the land, its controlling power also became extinct and other ways of gaining access to power were sought. Thus during the 14th and 15th centuries, in the period of so-called feudalism, it was the feudal lord and not the individual knight who increasingly became the dominant figure in the political system as the land became more secure and was divided among the *daimyo* (lords) into a large number of feudal domains (Reischauer, 1981:71). In times of war, however, soldiers as well as members of the lower classes were able to push their way into the upper ranks of the feudal aristocracy, while many members of the old warrior class sank to the status of peasants (Rozman, 1973: 68). Until about 1640, even without owning and cultivating land peasants were able to trade with the Chinese, Dutch, Koreans, and Russians and could thus amass wealth. Yet although merchants and artisans could become wealthy, in most periods they largely remained outside aristocratic society, not only because of the Confucian stigma attached to their classes but also because of the strong control exercised over them by the center. Only in periods of far-reaching decentralization could they achieve a certain autonomy.

The advantages of such groups were thus suppressed by the Tokugawa rulers. Fearing that the Japanese traders who traveled to foreign parts might bring Christianity or dangerous foreign ideas back to Japan, and fearing the emergence of a rural independent, rich, and intelligent elite group, the Tokugawa rulers banned trade with other countries. Similarly, although the peasants had great mobility, constituted the backbone of the concentration forces, and officially were highly evaluated, they could gain no significant autonomous access to the center.

It was not just the structure of the social hierarchies that was more diversified than the official designation of such hierarchies—a situation common to all societies. In close relation to some of the major orientations, the entire Japanese social hierarchy was structured in a very specific way. Despite the use of the Confucian nomenclature, it differed greatly from the class structure of the Chinese and other imperial systems.

To begin with, the class structure was much more complex than represented by the Confucian image. Each class tended to be split into several subdivisions. The *bushi,* for example, could be divided into different levels in accordance with their stipends. The farmer or peasant class included very rich as well as very poor people. Toward the end of the Tokugawa period, wealthy peasants were gaining power, literacy, and culture (Smith, 1959: 177). The two classes at the bottom of the basic hierarchy, the merchants and the artisans (particularly the merchants), had numerous opportunities for accumulating wealth. Interclass mobility was prohibited; wealthy merchants were not allowed to adopt consumption patterns similar to those of the *bushi*, and laws were enacted to keep the merchants in an inferior position.

Beyond such complexity, social hierarchies were structured according to specific principles of hierarchical organization, with special conceptions of rank, leader-follower relations, and orientations of commitment. Thus the structuring of social hierarchies and classes was organized more on vertical than on horizontal lines, which could form the basis for the organization of groups or strata with autonomous access to the attributes of status and the center.

Most tendencies toward horizontal organization were, as already mentioned, manifest in millenarian and populist images and movements.

The combination of cultural orientations, structure of elites (especially the level of their embedment in various ascriptive settings), also had far-reaching effects on the pattern of response to external impingement.

Whatever were the changes occurring in the fortunes of classes and groups in Japan during the course of its history, they were all characterized by a high level of embedment in the major ascriptive groups, by the relative weakness of the autonomous (nonembedded) political or religious elites, and by their own embedment in their respective collectivities and classes.

The major indigenous religious specialists—priests of the different cults that crystallized into what was later called the Shinto—thus formed part of and were fully embedded in such elites. The intellectual and religious elites who came later from abroad—Buddhists and Confucians—were often organized in distinct communities, monasteries, schools, and the like. They were involved in many autonomous activities—artistic, intellectual, and even educational—but they did not become autonomous in the sense of restructuring the society's basic premises and breaking through the primordial ascriptive framework of the major social units.

At the same time, openness to various external intellectual influences—above all those of Confucianism and Buddhism—also emerged in Japan. Yet the mode of incorporation of these traditions was rather peculiar and closely related to the structure of the predominant elites. These traditions were incorporated in the realm of artistic and educational activities and within these spheres gave rise to far-reaching creativity and new types of organizations. They had only a minimal effect, however, on structuring the sociopolitical order's basic premises (Nakamura, 1964). The Confucian educational and moral ethos was indeed incorporated into the ideology of commitment; yet, unlike in China, it did not generate autonomous elites independent of the wider ascriptive groups. Rather, it served as a stimulus to educate the existing embedded elites.

This mode of incorporating traditions is perhaps best illustrated by the way in which the Chinese saying "the rulers should be educated" was adapted by the Japanese. To them, this implied that their existing rulers—lords, *shogun,* and the emperor—were to be educated in the Confucian classics, unlike China where, according to the criteria of Confucian learning, a new ruling class should be created or recruited by autonomous means (Dore, 1965).

G
STRUCTURE OF FORCES OF CENTRALITY AND CONCENTRATION AND PATTERN OF SETTLEMENT AND URBANIZATION IN JAPAN

1 Introduction

Within this framework of major cultural orientations, the structure of elites, their major institutional implications and systems of control, cities and urban systems developed in Japan. As in all other societies, the patterns of urbanization, of structure of cities and urban systems, were shaped in Japan by the specific combination of the concentration and centrality forces as they evolved in its history. The continuous confrontation between these forces was largely influenced by their respective basic characteristics, which were greatly shaped by the prevalent cultural orientations and their institutional implications and the relative strength of these forces. It was the combination of the basic characteristics and orientations of the concentration and centrality forces and their relative strength—whose particularities changed throughout Japanese history—that has determined the development of cities and urban hierarchies in Japan at different periods.

A crucial aspect of Japanese rural and urban history was the ongoing conflict between the population's tendency to grow and migrate and attempts by the center or various lords to control such growth and migration in rural and urban settings. The forces of concentration, characterized by their internal cohesion and special features, brought about a strong tendency toward demographic expansion and continuous settlement in rural and urban settings alike. The demographic expansion was closely related to the process of continuous settlement typical of Japanese history from its beginnings, which was in turn greatly influenced by the basic cultural orientations and characteristics of the family structure.

The traditional patterns of Japanese settlement, both rural and urban, have largely been determined by the distribution of cultivable plains. The main islands have great mountainous spines and the limited lowland areas contain small, isolated patches of river sediments that form a series of relatively small sea-bordered plains, separated by mountains and crossed by numerous rivers. Villages and towns were built on locations dominating these plains. Only about 20% of the islands was level enough to allow cultivation, and a variety of factors determined the population distribution among the habitable areas. The earliest records indicate settlements in northern Kyushu and along the borders of the protected sea area; there were also concentrations of population in the alluvial plains (Reischauer, 1981: 6). This settlement pattern was the starting point for the development of the urbanization process in Japan.

The structure of the rural society was evidently of crucial importance in this settlement pattern and in its relation to the process of urbanization, as from very early times, the Japanese economic structure was based on agriculture, land, and peasants. This constituted the base of military and economic power and was controlled and regulated by the government; each village comprised a few large holdings, a greater number of middle-sized ones, and numerous small ones. As has been mentioned, these villages were organized in two distinct molds that influenced the country's patterns of urbanization (Beardsley et al., 1959).

In most periods of Japanese history, with the exception of the later medieval period, the processes of urbanization were strongly influenced by forces of centrality of a political-administrative nature. The number, location, and size of urban centers were largely determined by the territorial orientation that allowed the central authorities—or, in periods of a decentralized regime, the local rulers—to control the different areas efficiently. Economic considerations of production and distribution of goods were often subordinate to the political imperatives of the imperial or military centrality. Only during periods of feudal fragmentation did the forces of concentration shape the urbanization processes more significantly (Smith, 1960).

2 Ancient Urbanization: Kofun and Early Yamato Periods

The earliest stage in the urbanization process took place in the Kofun and early Yamato periods, from the 3rd to the 7th centuries A.D. At the beginning the basic layer of settlement was established; villages were dwelling and working places for extended families and had an autonomous organization. The Kofun clan leaders wielded great political and social power; their functions were increasingly sanctioned by religious power.

The social system established by the Yamato engendered a kind of social stratification that became the basis for the rigid status system that was to characterize Japan for centuries to come. This social system was based on three classes: the clan heads who became local elite rulers, legitimized by the prevalent religious beliefs; the peasants and artisans who provided food and material needs of the clan heads; and the slaves who were kept by the latter. As weapons became available and agriculture more extensive, the clan leaders acquired vast political and military authority. Some became stronger and began to subordinate and control the weaker leaders and their village units (Ledyord, 1975; Edwards, 1983).

This development signaled the first appearance of urbanization when the strong Yamato clan heads rose to power in the 5th century and formed a national oligarchy under which the former system of small independent villages was controlled by a single court sanctioned by religious beliefs. The early capitals of the Yamato state were identified as the principal residence of the ruler (Wheatley and See, 1978).

At this early stage of the urbanization process there existed little differentiation between the court village and the large number of ordinary villages. The court village—the residence of the Yamato clan heads—exercised control over the local clan head, but this control mainly consisted of an expression of personal allegiance and had not yet produced a bureaucratic administration.

3 First Imperial Urbanization (7th-11th Centuries)—Taika Reform

The first administrative cities in Japan developed as a result of the Taika reform (645), which was introduced by the Yamato's descendants to unify the country under a single political rule. At this time two major Yamato clans having opposing religious orientations clashed. The Soga family, favoring the introduction of Buddhism, eventually subdued the Mononobe clan; a major change in the country's administration thus occurred, the central aspect of which was the *Ritsuryo* system of land allocation and management.

The *Ritsuryo* system, the first systematic attempt at land distribution established after 645, was based on Chinese techniques of land distribution and administration and created two new local divisions: the *gun,* at the lowest level, headed by prominent local families and responsible to officials of the next administrative level, the *kokufu* center (Rozman, 1973:27). The purpose of this system was to improve the efficiency of land management, increase the state tax revenue, and provide central authority over the provincial level (Hall, 1970:52-54). This laid the basis for political organization and the centralization of power in the hands of a governing elite headed by the emperor, who ruled by divine right. The reforms were instituted mainly to free the peasants from the clan head's control by centralizing the ownership of land in the hands of the emperor and by an equal allotment of land. The new system introduced an even stricter form of political organization based on a stratification system as fixed as the former one but dominated by the imperial court rather than by the clan heads.

The country was divided into various administrative districts that formed a hierarchy system of settlement categorized as follows: The land was first divided into eight administrative districts, one of which constituted the imperial center. Each district was divided into *kuni,* or provincial government units, and the *kuni* were subdivided into groups of *gun*—townships or villages, each comprising about 50 families. The entire system was controlled by a Grand Council of State composed of eight ministers (residing in the administrative capital) who supervised the regional administrators. From their seats in the *kokufu* or provincial capitals, these ministers directly controlled the provinces and local administrators who headed the villages in which they resided. Urban settlements thus mainly fulfilled the role of administrative centers.

From Nara, and later from Kyoto, representatives of the imperial bureaucracy traveled to the domains of vassal clans. Together with the local provincial administrators, these representatives formed cores around which numerous secondary centers developed. The spread of Buddhism and the building of temples further added to the concentration of population. The centralization of power in the hands of the imperial government brought about the establishment in 695 of Nara, the imperial capital, to be the national administrative center as well as the residence of the emperor and his retainers. Nara was soon followed by Heian (Kyoto) in A.D. 770 (Hall, 1970:30-32; Reischauer, 1981:25).

During the first imperial phase of Japan's urbanization, cities were characterized primarily by forces of centrality shaping the administrative centers of a unified imperial rule. The capital city and secondary centers were functionally organs of political and administrative control. Geographically, the cities and towns were concentrated in the periphery of the inland sea. Nara, and later Kyoto, topped the hierarchy of cities that overwhelmingly constituted political and religious centers. Important handicraft production and commercial activities developed there, especially in Kyoto, but their political-religious functions still predominated (Wilkinson, 1965:14).

This pattern changed with the disintegration of the centralized state and the beginning of the feudal period from the 10th to 12th century. The *Ritsuryo* system of land division and allotments deteriorated, as did the flow of manpower, grain, and other commodities to the capital city (Yamamura, 1974). Titled officials and the representatives of temples and shrines competed to acquire increasing acreage of tax-free land on which to build manors (*shoen*). Because many peasants had been impoverished by the high taxation, the *Ritsuryo* system began to break down, and consequently it became easy for the nobility, priests, and richer peasants to buy land from poor villagers. Manors were then built from which the rich landlords controlled the surrounding area.

In order to wield their control, local landlords required military assistance and this led to the increasing importance of the *bushi* warrior class. However, in order to thwart this group's accumulation of power, the landed nobility incorporated the *bushi* into the administrative hierarchy as their retainers. The growth of tax-free estates in the hands of private landlords backed by strong military and religious factions diminished imperial authority and led to the emergence of decentralized feudalism.

This process had far-reaching effects on the development of the urban system. Because the central government was not strong enough to impose the payment of taxes on important landlords, it also lost its ability to control the provincial authorities. *Kokufu* cities soon disappeared; lords of the *shoen* no longer supplied goods to these outposts of the central government, and the storehouses, public buildings, and other structures of the *kokufu* became redundant. The distribution of *shoen* was so fragmented, however, that they could not replace the *kokufu*. Those functions that had previously been assumed by and concentrated in the hands of the *kokufu*, as well as new duties relating to the expanding commerce and handicrafts, were now dispersed over the countryside. Only in the 13th century did the *kokufu* begin to consolidate into new types of cities (Rozman, 1973:29-30).

4 Urbanization in the Feudal Period (12th-16th Centuries)

With the appearance of military leaders (*shogun*), the effective imperial-aristocratic rule quickly declined. Beginning with Yoritomo, political control was transferred to a military government (*bakufu*). The Hojo regents ruled in the name of the *shogunate* from 1205 to 1333, and the Ashikaga *shogunate* was in control from 1333 to 1573. Disintegration of the Ashikaga control marked the beginning of a period of civil wars.

From the 12th to the middle of the 14th century, the absentee landlords became increasingly divorced from local areas, some of them finding new sources of revenues by taxing the periodic markets and patronizing guilds of merchants and craftsmen in the cities. These landlords were gradually squeezed out during the 14th and 15 centuries in favor of local *bushi* who consolidated their control over the previously fragmented *shoen*. Castle building then flourished, and *bushi* lords competed among themselves for the control of local lands, constructing networks of branch castles controlled by a central one (Rozman, 1973:39).

This medieval period was characterized not only by increasing decentralization but also by the growth of relatively new autonomous social forces, the breakdown of some of the *meta-uji* society, and the continuous spread of the *meta-ie* principles of social organization.

It was during this long period, spanning about four centuries, that Japanese feudalism, with its basic characteristics, evolved. Compared with classical European feudalism, Japan's feudalism was characterized by the lack of specific contractual relations between lords and vassals, lack of countrywide and even regionwide strata organization of vassals, predominance in the feudal context of vertical relations, and absence of the vassals' direct access to the centers of national or regional power.

The major changes in this system, to follow Ryosuku (1978:28-29), were those from the middle ages when the feudal system resulted from the combination of the

> *shoen* system and the patriarchal lord-retainer system, both of which were primarily of an ancient period nature; . . . [to] early modern feudalism, . . . [when the feudal system] was a combination of the village-based support system, featuring villages as the basis of the fiefs, and the individual lord and retainer system, both of which were distinct in nature from the ancient systems.

It was also during this period that the most far-reaching urban developments occurred, generated by the forces of concentration, mainly by merchants and artisans, strongly connected with the intensive economic and social developments in the country, especially the emergence of a commodity economy (Wakita, 1980). These developments gave rise to a wide variety of new rural and urban forms and the emergence of tendencies toward autonomy.

The cities that emerged between the decline of the aristocracy and the end of the civil wars (12th to 16th century) reflected the changes in political-military leadership, the evolution of strong religious factions, and the economy of a decentralized feudal order. The administrative capital was moved to the fishing town of Kamakura, and Kyoto was retained only as a symbolic cultural religious center without administrative authority.

Major changes occurred in Japan during the four centuries between the failure of imperial control in the 12th century and the rise of the Tokugawa regime in the 16th century. These changes considerably influenced the process of urbanization in the direction of greater diversity and even, to some degree, of autonomy. These consisted of the expansion of domestic and foreign trade, leading to the emergence of the genuine commercial city, the shift of political-military control to the northeast away from imperial and aristocratic rule, evidenced in the establishment of Kamakura and the growth of important religious communities, manipulating a significant amount of economic power in the form of feudal fiefs (Yazaki, 1963).

During this period the largest cities were no longer exclusively the domain of political administrative centers. The stimulus provided by foreign trade, together with the highly organized domestic trade, made the commercial city a strong rival for urban dominance. As mentioned previously, the merchant classes enjoyed a measure of political and economic autonomy previously absent in the history of Japanese urban life. Wholesale and retail monopoly systems, guilds, and other economic devices were utilized to generate and maintain effective control by the merchant groups (Wakita, 1980).

The most important urban development that occurred between the 12th and 16th centuries was the rise of the commercial city, reflecting the growing importance of concentration forces in urban development. Unstable national leadership and the turmoil caused by the civil wars resulted in an atmosphere conducive to the spread of social mobility and freer commercial activity. The growing trade with China and its extension to the Philippines and Indochina, as well as the arrival of European traders and missionaries in the 16th century, also contributed to this process. Commercial activities were supplemented by handicraft and manufacture. Weapons and luxury items produced in these cities increasingly met the demand of local concentrations of *bushi,* as consumption decentralized from the big cities' lords to the *bushi* in the countryside. The new consumers still depended on the cities of the Kinki area for better-quality products, and this led to an unprecented growth of trade in the provincial areas. Intermediate and smaller ports also emerged as crafts centers to meet the needs of local *bushi* (Rozman, 1973:40). The inhabitants of most of the commercial cities exercised a hitherto unknown degree of self-government and judicial autonomy, which extended even to the maintenance of their own military forces.

In addition to these trade cities, new centers of local political-military control flourished. Seeking to avoid pressure from the older imperial and aristocratic factions, the first *shogun*, Yoritomo, made Kamakura his capital and thus enlarged the effective boundaries of Japan to the north and east, into the Kanto region. This created a new urban focus removed from the traditional area of city life dominated by Kyoto and Nara. Except for the period under the Ashikaga *shogunate* (1336-1573), when the political bureaucracy returned to the Kyoto area, the Kanto region remained the seat of political administration from the 12th century. Present-day Japan's national capital and largest city, Tokyo, commands the Kanto region and owes its dominance to this early shift of national control to the northeast.

Influential religious centers such as Ujiyamada and Ishiyama also arose during this period. The administration of large rice-producing fiefs by religious factions led to the growth of towns that served as centers of secular commercial activities in addition to their more sacred functions.

5 Castle Towns

A new type of city, the castle town *(jokamachi)*, also appeared around the 16th century and became the major type of urban development in the Tokugawa period.

From fortified strongholds, initially built to resist enemy attacks during the period of the civil wars (14th and 15th centuries), the castle towns gradually became centers of control and of resource mobilization in the lord's domain. The authorities joined forces with the local military power; and urban activities, previously scattered over the domain, were increasingly concentrated in the *jokamachi* of the primary lord. The *jokamachi* thereby steadily increased in size and, by concentrating the soldiers (*bushi*) in the castle cities, the local military lord (*daimyo*) had greater control over his troops. For a time the top retainers of the *daimyo* lived in residence near the central castle, whereas the others and most lower *bushi* lived in branch castles or in villages under the lord's control. A *daimyo* who succeeded in expanding his domains required increasing numbers of *bushi* in his central castle. The abolishment of branch castles was a continuous process that contributed to the growth of the main castles.

With the emergence of castle cities, the control of the absentee lords waned, and their former supporters strengthened independent local bases of power. Following the late 15th-century civil wars, Sengoku *daimyo* emerged, crushed the interests of the absentee lords, and competed among themselves to dominate compact areas (Wakita, 1980:325). The various institutional innovative measures, providing the *daimyo* with a stronger hold on their vassals, assisted the development of the castle towns, and by the end of that century the *daimyo* held direct control over entire domains. The lands of absentee lords were largely confiscated and *bushi* were gradually removed from the villages and restricted to a fixed stipend (Rozman, 1973:46), thus coming more under the *daimyo's* control. The concentration of authority in the hands of the *daimyo* made the presence of *bushi* necessary in the castle center, and their removal from the villages brought about the emergence of some sort of self-organization of the village communities. This process led the *daimyo* to collect taxes from the villages on a joint quota basis for each entire village community (Hall, 1970:132; Wakita and Hanley, 1981).

Although the rise of castle towns was initiated, to no small degree, by the political authorities, they were not the sole source of this development. The castle town was indeed strongly based on earlier, diversified processes of more autonomous social forces. James McClain (1980:267) describes the situation as follows:

> Many types of communities shared in Japan's seventeenth century urban growth . . . but undisputedly, the most significant type of seventeenth century urban settlement was the *jokamachi,* or castle town. . . . Between the late sixteenth and the middle of the seventeenth century, perhaps as many as two hundred castle towns came into existence, most of which were entirely new communities. By the early 1700's, Edo, a mere village a century earlier, had become the world's largest city, with a population well in excess of one million. Kanazawa and Nagoya, two castle towns in western Japan, were among the twenty largest European and Japanese cities. With populations of more than 100,000 each, those two castle towns rivaled in size such cities as Rome, Amsterdam, Madrid and Milan.

McClain's work on Kanazawa further provides detailed insight into the growth process of one of the new castle towns, its social structure, and the political and economic forces shaping the town (McClain, 1982).

6 Urbanization in the Tokugawa Period (17th-19th Centuries)

The economic strength and the extent of internal autonomy of the commercial centers in the 16th century were, however, insufficient to withstand the pressures of a reunified national political authority. Once this authority was directed to the cities, as potential centers of independent power, the latter either declined rapidly or were absorbed into the new administrative structure.

The period of civil wars in Japan ended in the latter part of the 16th century, when reunification of the nation took place under the successive leadership of Nobunaga, Hideyoshi, and Tokugawa Ieyasu. During the late 15th and early 16th centuries, the feudal lords consolidated and stabilized their holdings. The multiple feudal rivalries of the civil war period evolved into relatively stable alliances under the leadership of the more powerful feudal lords. The culmination of this process was the reestablishment of effective national unity under Tokugawa Ieyasu, in the early 17th century. The imperial court, the powerful Buddhist hierarchy, and the merchant class of the cities all came under the authority of the powerful Tokugawa *shogun* (Webb, 1968).

The economic base of Tokugawa society rested firmly on peasant agricultural productivity. Feudal lords were allocated fiefs on the basis of an area's rice yield. Tokugawa Ieyasu chose Edo (Tokyo) as his administrative capital and was thus able to control access to the Kanto region, the most productive of Japan's alluvial plains and the dominant source of economic strength. In order to maintain the elaborate political machinery and standard of living of the ruling hierarchy, the Tokugawa rulers and their vassal lords secured, in taxes and tribute, approximately one-half of the peasants' harvest. For nearly 300 years, the peasantry was kept at a near-subsistence level, while a relatively small political hierarchy controlled the total resources produced over and beyond the subsistence needs of the peasants.

The eventuality of fluctuations occurring in rice production or in the exchange value of the crop constituted an essential weakness in the Tokugawa economic system. The funds necessary to maintain the extensive bureaucratic staffs and to meet the frequent levies of the Edo *shogunate* made it imperative for the lords to rely on a consistent income level. In times of crop failure or low exchange value, the lords had to mortgage their future rice income to the merchants. This practice became so widespread that in the late Tokugawa period, the *shogunate* itself came under the financial domination of merchants-townsmen. The merchants were thus gradually transformed into bankers, and only secondarily were they storekeepers and traders.

The pattern of urbanization in Tokugawa Japan, which reflected a consolidation of the political and military power in the castle towns, also led to changes in the nation's economic structure and created opportunities for the development of an influential merchant class. The castle towns rapidly assumed importance as commercial and consumption centers when population concentrated around the permanent residences of the lords and their retainers. Then, serving as the commercial link between the city and its hinterland, an increasing number of merchants were attracted to move to the castle towns rather than continuing to reside in older trade settlements. This attraction induced the emergence of a climate basically antagonistic to the land-centered feudal regime as conceived by the Tokugawa.

The numerous castles built at strategic points during the civil wars and the period of unification on the orders of the Tokugawa rulers became nuclei around which cities and towns sprang up. In order to tighten administrative control, the Tokugawa rulers decreed, in 1615, that all but one castle in each fief should be destroyed. The impetus to urban growth, inherent in the Tokugawa political structure, was then mainly focused on a limited number of locations. The consequent expansion of these castle centers meant the decline of the towns previously existing except where they could be incorporated into new defense or communication systems. The castle towns can thus be defined as the dominant urban form throughout the Tokugawa era (Hall, 1955).

The process of absorbing local castles and markets into one central castle town changed the spatial organization of the central places. From the 13th to the 16th century their number rose rapidly, and it is estimated that urban settlements totaled between 500 and 600, probably including the early *jokamachi*, major branch castles, ports, periodic marketing settlements, and temple-centered commercial cities. Apart from Kyoto, which recovered from its temporary decline in the 16th century, and some port towns, these urban centers had little continuity, and their rise or fall depended on the fluctuation of local conditions. The early *jokamachi* were generally abandoned, either with the defeat of their lords or when the lords' domains expanded together with their control over them.

In the middle of the 16th century, cities were still small. Except for Kyoto, only such large ports as Sakai and Hakata reached a population of more than 20,000. The inhabitants of few other cities exceeded 10,000, and the largest *jokamachi* of that period—including Odawara in the Kanto region, Fuchu in the Tokai region, and Yamaguchi in the Chugoku region—had populations numbering roughly 10,000. The growth of small urban settlements was just then beginning (Rozman, 1973:49). Consolidation of domains and growth of the national market progressed rapidly. These trends matured at the beginning of the 17th century following the establishment of a stable and centralized administration.

A number of secondary population centers developed, in addition to the castle towns. The constant travels of the lords and their retainers, demanded under the Tokugawa rule, required the construction and maintenance of a nationwide network of roads. The service, posting, and market towns that emerged along these routes grew to appreciable size, especially those that were also castle towns. Like the post stations, a few centers specialized in the production of textiles and handicrafts. Some, such as Osaka (the second largest city in Japan of that period), Nagasaki (the major port city, from which Japan's contacts with China were maintained for 200 years), or Sakai (which became almost an autonomous city) grew into important commercial centers, constituting a secondary but very prominent branch of urbanization alongside the major aim of building castle towns in Tokugawa Japan (Hauser, 1974, 1977-1978; Yazaki, 1968).

The two and a half centuries of Tokugawa rule were a period of relative peace, during which the *shogunate* made strenuous efforts to maintain Japan as an isolated feudal nation and as a society in which a hereditary military hierarchy ruled and administered, an agricultural peasantry produced, and a class of merchants and artisans was heavily circumscribed and accepted but not encouraged.

The Tokugawa controlled the whole country through their feudal lords and the administrative centers were therefore built in areas where the rise of urban settlements would not otherwise have been encouraged. The concentration of city population around one castle in each fief led to a system of widely dispersed, relatively large urban centers. This did not mean that these centers were isolated; the rulers' requirement of alternate residence in Edo of the local lord entailed an almost constant interchange of personnel and ideas between castle towns and capital. Although the geographical distribution of cities was such that no region was without an urban center, the urban population was not equally distributed throughout the country. Areas extending to the south and west of Edo, covering Nagoya and Kyoto and the regions bordering the Inland Sea, were the most heavily urbanized. Edo was the nation's largest city with the largest population, and Osaka and Kyoto were not far behind. With a peak of over a million inhabitants, Edo had not much more than twice the population of Osaka or Kyoto. Osaka, with its great merchant houses, and Kyoto, the imperial capital and handicraft center, were never completely overshadowed by Edo in either size or functions.

The relevance of these developments to Japanese urbanization lies in the creation of a social structure within which urban-oriented commercialism took precedence over feudal agriculture but still depended on it. The merchants were joined in many commercial activities by the lower ranks of the warrior class, who found it increasingly difficult to maintain themselves on fixed rice allocations. Furthermore, the peace and prosperity enjoyed during the Tokugawa period rendered superfluous the military services of the warrior class. The *shogun* and the local lords then attempted to preserve their large military forces by overstaffing their administrative bureaucracies, but they succeeded only in increasing their own economic obligations. The traditional way of life of the military aristocracy was further undermined when large numbers of *samurai* undertook commercial activities. The local lord often intentionally contributed to the alienation of the economically deprived *samurai* by granting them monopolies in the handicraft specialties of his domain, thus blurring the demarcation lines between warrior and merchant classes.

H
URBAN AUTONOMY

The major characteristics of the social political system and the social structure of Japan, mainly the combination of controlled self-regulation and strong internal solidarity of many groups—albeit a solidarity based on lines of vertical authority—can explain the special features of the urban autonomy that developed in Japan.

Perhaps even more than its rural sectors, urban Japan was characterized by a great distinction among the different strata and status groups, by the latter's relative self-regulation, usually vested in the leaders of the respective groups, and the severe control of such regulation by the higher echelons in the social hierarchy (Yazaki, 1963). In this context no concept of urban citizenship could emerge, and the city was never considered as an independent unit. The main administrative and economic cities were linked to their hinterland and to the rest of the urban hierarchy. Public law had created a power hierarchy but conferred on it or on its subjects no correlative juridical rights. Abuses of authority could be redressed in court, not on the basis of an inferior's "rights" but on the initiative and by order of a superior. In Tokugawa thinking, for instance, law and justice were synonymous. The law had not been made by humans, and people had to try to understand and conform to the existing laws of nature that encompassed all of society. Because of this way of thinking, no Japanese city would entertain the idea of rising against the central government to defend its rights. Although no city laws were enacted and Japanese cities did not possess their own charters, laws, and corporate identity, a certain degree of self-regulation of the urban sectors of the population developed, as well as some distinctive identity of such groups (Morris, 1981).

City laws (usually unwritten) and customs differed from locality to locality and were overlaid by a countrywide, political-legal tradition based on the Confucian social stratification, family hierarchy, feudal devices of ownership of tenants and servants, and the symbolism attached to the imperial court at Kyoto (Hall and Jansen, 1968:204). Except in cases in which disputes transcended feudal boundaries, the law of Edo did not reach deep into society, nor did the *shogunal* civil law extend beyond its own territory (Hall and Jansen, 1968:205). At the same time, such self-regulation was vested in different sectors in the respective, vertically organized leadership.

This self-regulation, however, did not ultimately create autonomous units, within which the members would have rights of access to internal centers of power or within which they themselves possessed such rights with respect to the hierarchy's higher echelons in general and the center in particular. From this point of view, the guilds' pattern of development in Japan is of special interest.

I
GUILDS

Guilds first developed in mid-13th century, when the Hojo regents assigned to artisans and merchants particular locations in which to reside and from which their businesses could be carried out. The establishment of guilds in Japan served the interests of both the government and the merchants and artisans. Whereas the traders defended their professional interests by organizing themselves into corporations, the political authorities benefited by dealing directly with guild representatives and, through their regulation, could exercise tight control over the entire merchant stratum. Having been at first restricted to certain sections of the commercial quarter (*cho*) according to the place of origin and occupation of their members, the guilds formed a strong solidarity group, based on common geographic origin, traditions, customs, and respect for authority. Later, as the *cho* grew in size, guild membership was no longer based on geographic origin but on occupation.

In the 15th century (Hall, 1970:121), the guilds' pattern of organization crystallized and they became a closed-membership community of merchants and artisans who claimed monopolistic rights to sell or manufacture certain commodities and were somewhat secured and protected by a patron. Because legal protection was unavailable to them at that time, the feeling of strength in numbers was the main reason for the merchant and artisan groups to rely on a patron. The guilds thus tended to cluster around the prestigious establishments of Kyoto, Nara, and Kamakura; with the emergence of the *shogun*, provincial guilds that had local patrons also made their appearance. The trading guilds, which specialized in dealing in certain commodities or local products, constituted the first extensive network of commercial distribution in a market economy (Hall, 1970:122-123).

The degree of self-regulation enjoyed by economic organizations varied at different periods—highest in the late medieval period—but remained quite low throughout the Tokugawa period. The control of the political over the economic order was prescribed by Confucian ideology, but the services of the merchants and artisans were needed by *shogun* and *daimyo*. The nobility and the center controlled merchants and artisans in two ways: by forbidding them to buy their way into an official title—thus preventing them from becoming a threat to the politicians in power—and by limiting the contacts of this class with other groups of the population, especially the peasants.

In the feudal and early Tokugawa periods, merchants residing in the castle towns were assured of patronage and protection, but within the domain itself, the *daimyo* tried to draw a boundary between the castle town merchants and countryside residents. On the whole, trade was limited to the central city and sometimes to the few urban centers existing before the emergence of the *daimyo*. Within the domains, the castle town merchants performed two prime tasks: accumulating produce from the hinterland and linking the domain economy to the national market. Fulfillment of the first function led to the emergence of monopoly associations under *daimyo* patronage; the second brought into existence the class of rice merchants and domain financiers who maintained the produce warehouses at Osaka and other centers. The *daimyo's* desire to render his domain self-sufficient and prosperous acted as a constant stimulus to the castle town merchants and to the continuous evolution of different unities and variegated city life (Rozman, 1973:119, McClain, 1980, 1982).

Administrative control was, however, quite strict in the city and especially evident in the restrictions on residence of the different classes, the unification of weights and measures, the limitations placed on guilds and Buddhist sects, and the legal decrees. All these measures were aimed at controlling the social strata. The military government (*bakufu*) and *daimyo* could thus direct and control taxation and the application of other policies.

These attempts at control were facilitated by the fact that the bulk of government revenue came from land taxes (Reischauer, 1981:96). There was therefore no firm basis for merchants' demands for concessions or for exerting pressure to obtain them. In rare cases when the central administration borrowed money from rich traders, this was considered as a compulsory loan, and the merchants' lender status did not give them any real leverage.

It was thus natural that during the growth of internal commerce and decline of central or regional power, mainly during the feudal stage, the center's attempts at control weakened and led to greater self-regulation of market centers, sometimes verging on a limited form of autonomy, and also to a pattern of intensive urban cultural life (McClain, 1982). Similarly, the marked growth of trade in the Tokugawa period produced a stronger merchant stratum. The "townsmen" (*chonin*) became steadily more prosperous, and the more affluent of them, working closely with the government as its economic agents, were sometimes given semi-*samurai* status, bore family names, and even wore swords (Reischauer, 1981:97). Merchants were able to gain some degree of self-regulation, mainly in managing the internal affairs of city commercial quarters. Independent commercial communities emerged in several ports and among them Sakai even had its own armed force, becoming essentially an autonomous town (Reischauer, 1981:72). Basically, the merchants' independence never approached the extent achieved by the "free" cities of feudal Europe, and the commercial sector remained largely dependent upon *shogunate*'s and *daimyo*'s support (Hall, 1970:123-124; Rozman, 1973:39).

Although it is sometimes maintained by Japanese historians that the merchants of the Tokugawa period lived under a system of local self-government, this actually consisted of little more than the privilege of managing certain private areas of their activities under their headmen (Hall and Jansen, 1968:180-181), who themselves operated under the control of a high *bushi* assisted by a low *bushi*. The privileged merchants and artisans who had been active in the original construction of the commercial blocks (*cho*) could pass their hereditary administrative posts in the *cho* on to their descendants, but most of the new headmen were usually elected (Rozman, 1973:96), thus preventing the entrenchment of political power in the hands of the more affluent and well-established merchants.

The Japanese city therefore never really became autonomous because no *daimyo* tolerated it, and the autonomous force of centrality, with its own specific characteristics, overcame the concentration forces. This could be observed in the fact that the city lacked specific civil or cultural consciousness. Even in periods of decentralization and decline of the central government, the cities did not rise in revolt. Even the widespread economic distress and dissatisfaction at the end of the Tokugawa period did not lead to overt or effective protest (Hall, 1970:237).

J
URBAN SYSTEMS

The preceding analysis of urbanization processes in Japan points to some continuous characteristics as well as to some important variations in the evolution of cities and urban systems. The process of urbanization was generally encouraged by the encounter between the powerful demographic impetus to settlement—reinforced by the special features of the Japanese family system—and strong attempts by various political elites (the emperor, the *shogun*, and the autocratic warlords) to control this process. Both factors were shaped by the interests of major social groups and the basic premises of Japanese cultural and institutional derivatives.

In order to maintain their roles, the emperor, feudal lords, *shogun*, and chief *samurai* established control points that formed the skeleton of the urban system. Such control was exercised through a variety of mechanisms, the most important being the following: accumulation of all surplus goods by political elites, thereby controlling the peasants; fixing market prices, thus controlling the merchants; and the center's application of a system of fixed status in society (i.e., nobility, peasants, artisans, merchants, outcasts) by enforcing strict codes of behavior that emphasized status levels and prohibited social mobility between the classes.

Throughout most of Japanese history the functions of the city were initially and predominantly administrative. Yet they included a strong element of commercial interest, with great variations in their relative importance. These variations in the structure of cities and urban system were shaped by the internal cohesion and strength of the forces of concentration and centrality and by their relative strength vis-à-vis each other.

The structure of the urban system—that is, the number of levels, the order of their appearance, and the relations between them—underwent significant stages, each one reflecting the changing relationships between the everlasting forces of centrality, the strong political control of central and local rulers, and the forces of concentration; this was expressed in the growing demographic pressure, in a well-developed agricultural base, and in a developing marketing system on regional and national levels.

Until the establishment of Heijo (Nara) as the imperial capital in A.D. 710, the urban system of Ancient Japan can be characterized as an equal-size system, composed of temporary capitals, built quickly, abandoned, and transferred to new sites with a new ruler. During the Yamato period, from the 3rd to the 5th century, the many *uji* were brought under the control of the Yamato court, but the location of the court changed with almost every new ruler and did not allow the development of a permanent large settlement in those locations. With the introduction of the *Ritsuryo* system and the Taika reform in A.D. 645, the upper level of the emerging urban system began to take shape. The establishment of Heijo (Nara) in 710 marked the transition to the second phase of the urban system's evolution. This was characterized by a strong tendency toward primacy, the primate city being the imperial capital—first Heijo (Nara) and then Heian (Kyoto). The population of the capital cities grew at a tremendous pace: 200,000 inhabitants in Heijo by the end of the 8th century (Yazaki, 1968:34) and 500,000 in Heian in the 10th century. This remarkable situation was the outcome of a strong centralized government in the early imperial period, whose extensive taxation of the rural population benefited the imperial court. Control of the entire country by a single primate center became possible due to Japan's relatively small size and the development of an extensive transportation network. The marked primacy situation was accentuated by the absence of middle-sized cities and the formal establishment of towns of a lower hierarchical level to fulfill the functions of administrative centers. These urban centers (*kokufu*) had quite small populations and were mainly outposts of the central government, with the task of collecting taxes in the form of grain and other goods for direct transfer to the capital city. The urban system thus consisted of only two hierarchical levels and lacked the middle and lowest levels of marketing centers. The flow of resources and interaction between the levels was unidirectional, from the villages to the *kokufu* administrative centers and from there to the capital city. It can therefore be concluded that in a situation of weak demographic pressure and scant agricultural surplus, the forces of concentration were insignificant and were overwhelmed by the centrality forces in the form of a strong centralized imperial government. This unequal confrontation occurring in a small territory with a well-developed transportation network brought about the crystallization of a marked two-level primacy structure.

Decline of the strong forces of centrality heralded the transformation of Japan into a feudal regime and subsequently led to a change in the urban hierarchy. From the 10th to the 12th century the system of land division and allotment deteriorated; titled officials and representatives of temples competed to acquire large tracts of tax-free land on which they built manors (*shoen*). This process undermined the bases of the flow of manpower, taxes, grain, and other commodities to the capital city, causing a breakdown in the administration of local areas. Private control gradually replaced the elaborate central bureaucracy in Heian. The most significant expression of the decline of the central government's power and the strengthening of local lords in their manors was the gradual disappearance of the administrative centers (*kokufu*). Thus at the beginning of the feudal period, the urban system lost its low level of administrative urban centers. The *kokufu* were not replaced by new urban centers due to the large-scale fragmentation of the manors (*shoen*) and landlords' tendency toward self-contained economies within their manors. The capital city declined in population and wealth due to the central government's weakening control on the flow of resources into it. Thus the urban system in the period between the 10th and the 12th centuries was transformed into a one-level, low primacy system characterizing relatively weak contrality forces and the almost complete absence of concentration forces.

From the end of the 12th to the end of the 16th century feudal Japan's urban system underwent significant changes; the trends that already operated in the last stages of the imperial period still persisted. The uppermost level of the urban hierarchy was altered by the adjunction of a second principal city to this level. Kyoto remained the largest city and seat of the imperial court and *shoen* lords, who received and consumed much of the surplus accumulated in their manors. Kamakura, which joined Kyoto on the upper level of the urban hierarchy, became a large city of about 200,000 people as a result of transferring the military government there. This separation between the remaining symbolic center in Kyoto and the emerging political-military center in Kamakura produced the bifurcated top of the urban system and heralded the future segmentation of the upper level of the urban hierarchy, which later took place in the Tokugawa period.

The decline in imperial authority caused the elaborate administrative machinery of provincial control to break down, and the *kokufu* centers almost disappeared. In this decentralized situation forces of concentration began to operate, building up the urban hierarchy's lowest level in the form of numerous small and temporary market centers. These reflected the gradual commercialization of *shoen* agriculture and the growing tendency of the manor lords to purchase locally goods for consumption in their Kyoto residence with cash received from the local sale of their estates' products. With rising agricultural production, however, periodic markets became more frequent and stable. For about three centuries of the feudal period this pattern of scattered urban activities in numerous small communities typified the urban system. Thus the lower level of administrative centers in the imperial period was replaced by a level of marketing centers of an even smaller size.

An important feature of feudal Japan's urban system was the emergence of cities in its middle level. These were mainly port towns largely concentrated in the Kinki region. The more commercialized the market centers of *shoen* agriculture became, the larger were the marketing and transshipment activities of the port cities. Such ports as Yoto, Otsu, and Hyogo (Kobe) developed into large cities, ranking next to Kyoto and Kamakura in size. Some of these primarily commercial cities were characterized by well-organized guilds and were able to secure a measure of autonomy and self-government, quite unique in the urban history of Japan. Thus in the later part of the feudal period (15th century) and in the Sengoku period (16th century) Japan reached the most decentralized state in its history. The urban system of this period can be characterized by a three-level, low primacy structure, the upper bifurcated level being shaped mainly by forces of centrality. But as these were quite weak and diffused, they were unable to support very large and prosperous primate cities; the outcome was a low primacy structure. The middle and lower levels of the hierarchy were primarily the outcome of concentration forces, reflecting a flourishing agriculture and extensive long-distance commerce uncontrolled and uncurtailed due to the weakness of the central government.

In the middle of the 16th century, cities were still small. Except for Kyoto, only such large ports as Sakai and Hakata reached a population of more than 20,000. The inhabitants of few other cities exceeded 10,000, and the largest *jokamachi* of that period—including Odawara in the Kanto region, Fuchu in the Tokai region, and Yamaguchi in the Chugoku region—had populations numbering roughly 10,000. The growth of small urban settlements was just then beginning (Rozman, 1973:49). The consolidation of domains and the growth of the national market were progressing rapidly. These trends matured at the beginning of the 17th century following the establishment of a stable and centralized administration.

A significant change in the structure of cities and urban systems took place under the Tokugawa, based on the combined effects of growing urbanization and commercialization and the strengthening of central control. The administrative urban hierarchical structure was formed in the Tokugawa period with a strong territorial-military orientation. This structure provided the framework for the commercial system, stimulating and ensuring the safety and security needed to implement the complex transactions and direct buyers' and sellers' movements among the various centers. The prosperity of the marketing system was further secured by the political stability of the Tokugawa government. This combination of administrative and economic systems was manifest in the fact that all *jokamachi* and, of course, all the cities higher in the administrative hierarchy were by decree also market towns, thus fulfilling a major role in each region's economic development.

The urban system under the Tokugawa thus represented the interaction of the forces of centrality and concentration, the centrality forces mostly dominating. The hierarchical pattern of the Japanese urbanization process was therefore expressed by a spatial organization resulting from the combination of the political-administrative system, evident in the well-organized bureaucratic structure, and a progressive local and national marketing system (Rozman, 1973:111).

K
INTERNAL STRUCTURE OF CITIES

The internal organization and ecological organization of Japanese cities were shaped by the various social forces that created the cities and predominated in them. It is natural, therefore, that the first periods of development were shaped largely by the forces of political centrality and their basic conceptions and orientations (Wheatley and See, 1978).

Following the establishment of the *Ritsuryo* system, aimed at achieving political unification in accordance with Chinese principles, political administrative orientations shifted from heads of clans to central organs of the imperial government. This shift necessitated building permanent headquarters in the capital city. Responding to the new political administrative demand, capital cities grew to unprecedented size, supported by heavy taxes levied on the peasant population.

During the 7th century, successive capital cities were established in the Asuka district. The process was initiated by building an imperial palace on a new site to be followed by the influx of nobility, officials, and monks settling in the vicinity. The frequent change of sites reflected attempts by the imperial power to lessen the influence of clans on the central authorities. In 645, an imperial capital was established at Naniwa, followed in 694 by Fujiwara (Kidder, 1977:72), which, like Naniwa, was patterned after the ecological structure of the Chinese T'ang capital of Ch'ang-an. Fujiwara was an outstanding example of the influence of Chinese institutions and culture upon Japanese society and of the shaping of the cities' ecological structure according to the Japanese elite's conceptions of social order.

Fujiwara was planned and built on a grid pattern forming square blocks. The main avenues were oriented in a north-south direction, the central avenue, which divided the city into two equal halves, leading to the imperial palace located at the northern end of the capital. The city blocks (*cho*) were created by the streets oriented east-west. Within city blocks lots were allocated to the nobility, officials, and the common people commensurate with their rank in the social hierarchy: The higher a family's rank, the closer their lot was to the imperial palace. Official markets were established in the two halves of the city, operating under the supervision of a market master. The Fujiwara's imperial complex was later replicated in Heijo and Heian. The imperial assembly halls were in the foreground; beyond them was the administration palace comprising the halls of the various ministries, and behind it, in a small cloistered and secluded area, was the imperial council hall, where the emperor conducted his activities. The imperial residence was in an adjacent section, usually to the east of the main administrative complex. Fujiwara served as capital city for only 16 years, after which the capital was again moved to Heijo (Nara) in 710.

Established in 710 and flourishing for about 80 years was the capital city of Heijokyo, the best and fullest illustration of the internal structure of Japanese cities patterned after the socioecological structure of the Chinese T'ang capitals. With an estimated population of 200,000 in its heyday, Heijo can be regarded as the first large-scale city in Japan's history of urbanization (Yazaki, 1968:34). Its basic structure was identical to that of Fujiwara but on a much larger scale, owing to the expansion of the administrative system. The city's layout followed the well-known Chinese grid system. The street pattern was determined by the main avenue leading from north to south, parallel or perpendicular to the main axis, creating neat, square city blocks. This meticulously designed symmetrical pattern represented forces of centrality in full operation; the land, controlled by the imperial court, imposing its political and cultural orientations upon the capital city's inhabitants. The precise geometric pattern, with its measurable and visible spatial relations to the focal points of the city, made it possible to define the social status and standing of each family within the imperial hierarchy by the location of each residence in relation to the focal points and street system. Despite the adoption of the Chinese urban model, it is noteworthy that Heijo, like other capitals of that period, was not surrounded by walls or ramparts, thus attesting to a strong sense of security and confidence in the imperial power and affinity to the population's broader strata.

As in other capitals, the city's core was the imperial palace, located at the northern end of the main avenue and forming an enclosed compound that included the various official courts and the emperor's residence. Due to the size of the buildings and their architectural style, the palace compound radiated authority, overshadowing the city's humbler residences (Johnston, 1969). The nobility and officials who moved to Heijo were granted land for their lifetime, the size of the lots varying according to their rank. Not only the size but also the location of those lots was a spatial indicator of the grantee's status, as only nobles of the first rank were allowed frontage on the main street. The lower the rank of the official, the farther away his residence was from the palace and main street and the more humble his house.

The shift of the imperial palace to Heijo was accompanied by that of many Asuka temples to the new capital. The transfer of those temples, supported by powerful clans of the Asuka region, was aimed at securing the continuous allegiance of those families to the emperor. As Buddhism constituted the moral base of the political regime and the Buddhist priests exercised extensive political influence, the imperial government provided substantial funds from the state budget for the construction of the official temples. Only a few years after the establishment of Heijo, about 50 temples were already constructed all over the city. In size and splendor they were comparable to the imperial complex buildings. The fact that the temples had not been concentrated in a single part of the city, thus potentially creating a physical counterbalance to the imperial palace complex, reflected the subordination of the monks to the imperial authority.

The imperial authority was delicately balanced, jointly supported by three groups: the imperial family, the nobility, and the priests. The emperor himself was regarded as the spiritual head of the national family, even when the latter was split and contentious. Accordingly, the imperial palace, which constituted the central symbol of national unity, was often relocated when shifts occurred in the balance of power among the main supporting groups. This motivated the transfer of the capital to Heian (Kyoto) in 794, the new city being established at a site on which no contending power had any claim, thus allowing the emperor's position to remain neutral in his relations with all rivaling clans. The diminishing influence of Buddhism within the imperial power structure played a major role in structuring the new capital. In Heijo, Buddhism was the official basis of religion, thought, and politics; the numerous temples acquired extensive landholdings in the city and around it and exercised great influence on the political and economic realms. The transfer of the capital to Heian was an attempt to eradicate this massive influence, especially in the political sphere. In the new capital, the construction of temples was strictly limited, and very few temples were permitted to move there. The emperor's direct control over all national affairs was thus strengthened and the urban scene was changed conspicuously.

Heian was planned and designed rigorously, its implementation taking many years to complete. The basic outline of the city was also molded on the Chinese T'ang model, closely following the structure of the former capital. The main difference was its scale, as Heian (Kyoto) was much bigger and extended over a larger area. The city, planned and built on the square grid system, was oriented around its main thoroughfare. The eastern and western parts of the city on both sides of the main axis gave it symmetry, and its precise subdivision produced an orderly appearance. Spatially, the various functions were arranged in descending order of importance from north to south; the all-important political sector, encompassing the imperial palace and the government complex, occupied the revered area at the city's northern end. Cultural and economic functions were located in the middle section; the lower but necessary activities were concentrated in the less desirable southern part of the city. The social status of the urban population was clearly expressed in the residential spatial organization; the higher the official's status in the imperial hierarchy, the nearer his residence to the main thoroughfare and government complex at the northern end.

The capital was planned as a balanced, symmetrical city. Yet the shifts in power and cultural orientations gave rise to changes that took place gradually in the city structure, causing an imbalance in the planned symmetry.

The major deviation from the original plan of the capital was the much faster and affluent growth of its left side (*sakyo*) as compared to its right side (*ukyo*). A specific cultural orientation emanating from the continent accorded greater merit to the left over the right side. Because of this, many members of the nobility who were originally allocated lots on the right side of the city transferred their mansions to the left side, as did numerous temples and shrines. This preference gradually led to denser population in the left side and also a marked imbalance in social status between the two sides.

This deviation from the symmetrical layout, reflected in the size and social status of the population, triggered an important change in the city's commercial structure. Originally a market was planned in the eastern and western parts of the city, to be strictly supervised by government officials. These two commercial centers were intended to be of equal size and to serve the population at successive periods of two weeks every month. With the decrease in population and purchasing power in the west side, the market there lost many customers, while the one on the eastern side gained greater economic importance and became the city's main commercial center. This phenomenon was further enhanced by the weakening of the *Ritsuryo* system, which led to an increase in the number and volume of commodities finding an outlet in the market. The asymmetric growth and affluence of the eastern market and the subsequent changes in the city's commercial pattern reflected the diminishing influence of the centrality forces and the growing impact of concentration forces in the last stages of the early imperial period (Yazaki, 1968:58).

The analysis of the evolution and structure of Kyoto and Kamakura has highlighted the different patterns evolving as an outcome of the imperial rule and the *bakufu* regime. Common to these two, however, was the dominance of centrality forces in shaping their development and internal structure. With the onset of the feudal period and its transformation in the later Middle Ages, structural changes took place in the existing cities and, as mentioned, new forms of urban settlement emerged: the commercial and religious towns and, above all, the castle town.

The location of the castles varied. Some were established adjacent to existing towns or near the lord's headquarters; elsewhere the castle and its attendant functional units were built in entirely new locations. Irrespective of the type of location, however, a common feature appeared in all castles at the earlier stage: Topography played a crucial role in site selection, and castles were built atop hills to ensure maximum security and control.

The castles located on hilltops were inhabited by the *daimyo* and their warriors. A new type of spatial differentiation evolved in the larger castles: The *daimyo* resided in the inner fortress and officers were allocated quarters surrounding the inner part of the castle. Soldiers were garrisoned in its perimeter. Outside the castle walls near its main gates were concentrations of merchants, artisans, and craftsmen (especially those engaged in arms manufacture), as well as innkeepers, all of whom were attracted by the wealth in the castle and the relative security it provided. In the early Middle Ages, however, the castle—the center of military and political authority—and the adjacent spontaneous concentration of commercial activity were not yet fully integrated into a unified urban settlement. During the civil wars, merchants' and artisans' quarters in front of the castle were often burnt to the ground in an attempt to obstruct an enemy's advance or to cut off the enemy's supply route, thus making these quarters most unstable.

Only toward the end of the civil wars and until the 16th century were the castles transformed from hilltop fortresses with auxiliary temporal commercial quarters into castle towns in the plains, and the military and commercial functions were included within their limits. At this stage the castle towns were carefully designed with a larger population and a sounder economic base, as compared with the haphazard evolution of castle settlements of the earlier period.

Maps of castle towns produced by Gutschow (1976) allow a detailed reconstruction of a typical castle town's internal structure at that time. At the core of every castle town always stood the castle complex. Taking advantage of natural features such as a promontory, a river bank, or the seashore, the castle was the most prominent structure of the town and the built-up area spread out over the open land surrounding it. In some castle towns the townsmen's residential quarters were included within protective walls or moats; in many others, defensive structures protected only the castle area that encompassed the *daimyo*'s mansion and his officers' quarters. Much attention was given to the internal structure of the castle area that was designed so that each part could be defended separately, even if the rest of the compound fell into an enemy's hands. The basic pattern of the castle complex comprised three traditional defense lines defined as the primary, secondary, and tertiary fortifications. Arising above the fortifications were overhanging white-walled towers that dominated the skyline. The main tower, usually five to eight storeys high, was the command center, the main storage facility, and the final defense position of the castle complex. It sometimes served as the residence of the lord and was built on an imposing and ornate scale, signifying the dignity and power of the lord. Within the castle complex were the residences of ministers, advisers, and high officers, as well as other storage areas and military training schools of the lord.

The ecological structure of the castle towns reflected the special type of combination and interaction, characteristic of forces of centrality and concentration, in contrast to both the earlier imperial and feudal periods. In general terms this structure reflected the feudal status system comprising four classes: warriors, farmers, artisans, and merchants, each of which, in principle, occupied a different quarter. Yet the special place accorded the artisans and merchants was of greater importance than in the earlier imperial system.

Following the official orientation, the most conspicuous separation existed between the warriors' quarters and those of the townsmen. The residential space allocated to the warriors varied according to their circumstances, as measured by their rice income. Senior officers of the *daimyo*'s army built spacious mansions; the higher their rank, the closer their homes were to the castle. Low-ranking warriors, on the other hand, resided in a concentrated circle at the town's periphery, quite often beyond the merchants' quarters, in order to ensure the town's defense (Yazaki, 1968:148). The area allocated to warriors in the castle town was much larger than that of the townsmen.

Within the townsmen's zone, several blocks were allocated to merchants for setting up shop. These blocks were along the main streets or along the national highways cutting across the castle towns. The merchants gathered, according to their occupation, for mutual benefit and safety. The numerous stores, together with the large inns also located along the main streets, provided castle towns with a lively hub of commerce, entertainment, and social gathering. These commercial centers, of elongated shape, were the product of forces of concentration and formed a counterbalance to the military-political center of the castle (McClain, 1980, 1982).

Similar to the internal structure of the quarters of warriors' residences arranged according to rank and rice income, the townsmen's ecological distribution also represented, in principle, the social standing of the various groups. Yet given the marked strength of the different forces of concentration at work, a discordance between the social status and the locational pattern of the artisans' and merchants' classes evolved and presents a point of major interest. Even though the artisans held a higher position in the social status scale than the merchants, the more affluent among the latter occupied the more central locations in the townsmen's zone. This reflected the influence of concentration forces in shaping the internal structure of the castle towns. These forces were strengthened by the accumulation of wealth and political clout in the hands of rich merchants. Although the artisans' economic power allowed them to choose central main street locations, their workshops and dwellings were kept to back streets, in much less prestigious and accessible positions. The craftsmen and artisans were spatially organized in these quarters according to their craft and products, thus creating distinct trade communities.

At the same time, the locational pattern of temples in castle towns strongly reflected their instrumental role and subordination to the military-political authority. In the process of establishing a major castle town in his domain, the *daimyo* either compelled temples to relocate in the new castle town or ordered the construction of new temples there. The location of the temples in the castle towns is of particular interest as it is indicative of the instrumental role the temples played in the town's defense. Because of their sturdy construction and the spacious grounds surrounding them, they constituted a prime defensive position. The temple sites were chosen with the view toward forming a defense line, especially in relation to weak points in the castle's fortifications. Another way of utilizing the temples for defense purposes was to build them in a circle around the town. The temples sometimes also served as a buffer zone between the castle complex proper and the town. In a few cases, the temples were aligned at points of entry into the castle town and served as sentries controlling traffic into the town. Apart from their crucial importance in defending castle towns, the temples fulfilled the primary urban function of public recreation in the many seasonal festivities. All these functions emphasize the equally important instrumental capacity of the religious institutions and buildings in addition to their spiritual and moral role.

Urbanization in Medieval Europe

A
INTRODUCTION: POLITICAL AND CULTURAL ORIENTATIONS

Within the framework of decentralized systems, Western European civilization, crystallizing since approximately the 8th century, was characterized by distinct features that have been of great importance in the development of cities and urban hierarchies. These features have been considered the pinnacle of urban society and the conscious or unconscious model for the study of cities.

European civilization was distinguished by several cross-cutting cultural orientations and structural settings. The symbolic pluralism or heterogeneity of European civilization originated in a multiplicity of traditions—Judeo-Christian, Greek, Roman, and various tribal ones—out of which its own cultural tradition crystallized. The most important among Europe's cultural orientations was the emphasis on the autonomy of the cosmic, cultural, and social orders and their interrelation. This autonomy can be defined in terms of the tension between the transcendental and mundane orders, and the ways of resolving this tension by combining this-worldly (political and economic) and other-worldly activities (Heer, 1968; O'Dea, O'Dea and Adams, 1972).

In close relation to these orientations, multiple elites evolved in Europe, possessing a large degree of internal autonomy and autonomous access to the center and characterized by growing specialization. These features generated the special traits of the centrality and concentration forces that emerged in Europe, mainly a distinct type of structural pluralism. This structural pluralism was characterized by a combination of the steadily increasing levels of structural differentiation and by the constantly changing boundaries of collectivities, units, and frameworks. Closely connected to these orientations and the concomitant heterogeneous autonomous elites, several institutional tendencies developed in Europe—namely, tendencies toward wide and cross-cutting institutional markets (their scope expanding with technological and economic progress); toward a low level of coalescence among the boundaries of the major collectivities (the broad civilizational—religious Christian—and political, ethnic, and national ones); and toward flexible modes of control exercised by the major elites constantly competing to gain access to this control.

These cultural and structural tendencies gave rise to flexible boundaries of ethnic and political collectivities, large multiple cross-cutting markets, multiple avenues of mutual conversions of different resources, and broad access of many of its groups to such markets. At the same time, the social field was perceived as a network of relatively broad, diverse, sometimes universalistic, open channels through which membership in the basic (kinship, territorial, political and other) collectivities could be gained, and also including criteria of justice according to which resources and positions were allocated.

In the first period of this emerging civilization, the combination of these orientations (the structure of elites and geopolitical features) gave rise to the classical European feudal system. Its most important characteristics were the multiplicity and cross-cutting of the centers, the contractual nature of relations between lords and vassals, and the markets' continually expanding scope and cross-cutting.

But even when the tendency toward more centralized monarchies and market economy became stronger and where the feudal system weakened or disappeared, several of the major structural characteristics that denoted the specific configuration of the centrality and concentration forces still persisted. The most important of these characteristics were the following: (1) the multiplicity of centers; (2) a high degree of permeation of the periphery by the center, impingement of the periphery on the center, and attempts by the periphery to mold the center in its image; (3) relatively little overlap of boundaries and of the restructuring of class, ethnic, religious, and political entities; (4) a comparatively high degree of autonomy of groups and strata and of their access to the centers of society; (5) considerable overlapping among various status units, together with a high level of countrywide status (class) consciousness and political activity; (6) a multiplicity of cultural and functional (economic and professional) elites mutually cross-cutting but enjoying a relatively high degree of autonomy and a close relationship with the broader, more ascriptive strata; and (7) a relatively high level of autonomy of the legal system in comparison with other integrative systems, mainly the political and religious spheres (Bloch, 1961; Brunner, 1968; Cam, 1954; Prawer and Eisenstadt, 1968).

B
PROCESS OF MEDIEVAL URBANIZATION

It was within the framework of these basic features that European cities and urban systems developed, exhibiting some very unique characteristics. As mentioned in Chapter 1, these traits have often constituted in the literature the yardstick against which the analysis of other cities and urban systems was based. The major characteristics of these cities and urban systems have been researched more than those of any other civilization, and it is not our intention to review, or even summarize these detailed studies. We shall concentrate only on some of the most salient aspects of European cities and urban systems that are of special interest to us and that underline the uniqueness of this urban scene, within the framework of our comparative analysis of cities and urban systems. This uniqueness is due to the special constellation of concentration and centrality forces in Europe and their interrelations, mainly to the specific features of the concentration forces, the high level of their autonomy and access to the centers, and to their concomitant attempts to mold the centers in their image.

Periodization is needed when analyzing medieval urbanization. Even if schematic, it is nevertheless a necessity, as this is a multifaceted and regionally overlapping phenomenon constantly changing and developing. First occurred the decline of the Roman Empire, lasting roughly from the 3rd to the 5th century. Then, the Early Middle Ages endured until the 9th to 10th century, when the first wave of urbanization started. This was followed by the High Middle Ages and the great movement of urbanization, in the 11th to 13th century. Finally came the Late Middle Ages, lasting until the 16th century, the period of the Reformation and the decline of urban independence. The first two periods pose the most enduring problems: the continuity from Roman times and the sources for the urban growth of the High Middle Ages.

As has been pointed out by C. T. Smith (1967), two aspects of these problems can be distinguished: whether there was a continuous tradition of trading activity and merchant settlement, and how spatial continuity persisted from the Roman to the medieval site and town plan. Even if we start from the assumption that Europe was divided into three great zones of continuity differing in kind and degree, as proposed by Edith Ennen (1953), we can safely conclude, along with Smith, that a "continuity of urban institutions and government is very rare; but a continuity of religious function is extremely significant all over continental Western Europe; topographical continuity of site is very evident" (Smith, 1967:309). In the zones of stronger continuity, some sort of urban core survived from Roman times, usually centered on an ecclesiastic establishment built on the site of a Roman town, surrounded by defensive walls, and concerned with religion and administration. It is generally agreed that the primary urbanization of the 10th and 11th centuries evolved when a merchant community settled in or near such a preurban core. Within a short time, by not later than the end of the 11th century, these merchant communities developed elaborate forms of self-government and attempted, usually successfully, to wrest from their ecclesiastic or lay lord, the right to tax, defend, and rule themselves. New problems arise here, however. The existence of such merchant communities depended on long-distance trade, yet new towns emerged in the 11th century that were not situated on the major trade routes. In addition, these small groups of traders, bellicose as they could be, were obviously not strong enough to challenge the feudal powers. But two new developments, which were to shape Medieval Europe decisively, were to the trading groups' advantage: The first was the wave of urbanization brought about by the demographic expansion of the 11th to the 13th century, which created a mass demand for manufactured goods and attracted into towns craftsmen and workers—the solid backbone of the urban population—from an already overcrowded countryside. Second, the towns began to grow and multiply in precisely the same period—the 11th century—when for the first time European monarchs attempted to reduce feudal power. Throughout the severe conflicts that erupted in the late 11th and 12th centuries, the towns prospered, constituting sought-after allies in the disputes among popes, kings, princes, and nobles.

The establishment of urban settlements was thus induced by the diffusion of international trade and the emergence of local demand—be it for goods, administration, defense, or culture. These two processes usually combined in different degrees: The growth of the regional urban centers, which was slow and on a modest scale, was given strong impetus by the settlement of merchant communities engaged in long-distance trade along the axis of the major routes that linked Italy with the Low Countries and Northern France through Southern Germany and the Rhine and Rhone valleys (Smith, 1967:331).

As a consequence of this general demographic and economic expansion, the towns of Medieval Europe underwent a continuous process of growth in number and size, which lasted from the 11th to the 14th century. The Black Death epidemics of mid-14th century caused a sharp drop in population, and the foundation of new towns then became rare. Interestingly, the losses in urban population were rapidly made up in the cities, mainly through migration, far more quickly than in the rural areas, and we find here a significant characteristic of the medieval town: its continuous dependence on rural migration. Its indigenous growth was never sufficient to counterbalance the high mortality rate, let alone furnish further expansion. Except for a small number of export and production centers and the residences of kings and princes, urban growth had slowed to a halt by the 14th century. In many places the areas enclosed by the walls remained gardens and fields quite often until the 19th century.

The question of the number and size of Medieval European towns has been given a lot of attention, and a fair amount of demographic information is available (Ammann, 1956; Mols, 1954-1956). Medieval towns ranged from the so-called dwarf towns, with a population of from about 300 to 2,000, to medium-sized towns with up to 10,000 inhabitants, to large cities with between 10,000 and 20,000 inhabitants, and to metropolises with populations of over 50,000. Of the total number of towns, naturally there were numerous small- and medium-sized ones, quite a few large ones, and very few metropolises. The interesting point, however, was the uneven spread of towns over different regions of Europe, the two most heavily urbanized areas being Northern Italy and the Low Countries. Venice, Milan, and Florence were the largest of all European towns, with populations ranging between 60,000 and 100,000 each. The cloth-producing cities of Ghent and Bruges, with their world markets, had over 50,000 inhabitants each and were, besides Paris, the largest cities north of the Alps.

Italy and the Netherlands possessed dozens of large- and medium-sized towns. France had only one really large city—Paris, the capital, with about 60,000 inhabitants—and some regional centers ranking as large towns, such as Avignon (which owed its growth to the papal occupation in the 14th century), Bordeaux, Toulouse, and Montpellier, each having 10,000 to 20,000 inhabitants. There were also a few large industrial towns in the north and a large number of medium- and small-sized towns. In Germany, Cologne was the largest city with a population of only 40,000, but there also were 9 large towns, each having 20,000 inhabitants, about 200 medium-sized towns with 2,000 to 10,000 inhabitants each, and a few hundred dwarf-towns. In England, London had a population of about 30,000, but there were also the large towns of York and Bristol and a number of medium-sized towns. Spain's structure was similar to that of France but with fewer towns. Northern and Eastern Europe had no large towns at all.

The total urban population of Medieval Europe was not impressive and only with the advent of the Industrial Revolution did the level of urbanization that had been reached until the 14th century prove insufficient to meet the needs of the expanding economy. It was then that the towns began to grow again. These processes of urban growth, operating within the framework of the constantly expanding economic developments and continually changing political boundaries, greatly influenced the pattern of the urban systems that evolved in Europe.

C
URBAN SYSTEMS

Urban systems began to develop very early in Europe, and the view that the medieval town was a self-sufficient, enclosed entity can no longer be completely sustained. Various studies of the flow of commodities, commercial and financial transactions, and movements of people clearly indicate that strong interaction took place among the rural and urban settlements, creating interconnected systems of settlement. More and more studies have shown lately that despite far-reaching political decentralization and inadequacy of communication during the medieval era, the economic relations stimulated the evolution of urban systems that overcame the political fragmentation and even became closely related to the numerous aspects of the decentralized political milieu.

Study of the structure of urban systems in Medieval Europe has become a major topic of research among historians and geographers. These studies encompass both micro and macro levels, ranging from the delimitation of the service area of one particular settlement to the analysis of entire urban hierarchies of a kingdom or region. The analysis of urban systems in Medieval Europe and of the regional relationships between the towns and the countryside they served was given an important impetus through the meticulous study of the urban population in Italy (Beloch, 1937-1941), France (Lot, 1945-1954), Germany (Keyser, 1941), and Britain (Russell, 1948, 1958). These and more recent works provided a basis for analyzing the size and spacing of towns all over Medieval Western Europe (Russell, 1972). Russell's study enabled to identify the levels of urban hierarchies in various parts of Medieval Europe. By applying a particular form of the rank-size rule, Russell established the size distribution of the urban settlements in each region, thus allowing us to compare the urban systems in the various regions and to define the general pattern of city-size distribution and the particular features of each region.

The study of urban systems through the application of rank-size distributions culminated in the major work, *European Urbanization, 1500-1800* (de Vries, 1984). This proves most convincingly the successful application of the urban system concept, as defined by city-size distributions in the study of urbanization processes.

European urban systems differed entirely in the two phases of the medieval period. In the early stage (up to the 11th century) the decline of cities was followed by the gradual disintegration of most urban systems. Increasing transport difficulties, together with prevalent insecurity, caused a dwindling of the economic interaction among cities, leading to its complete disappearance. And the breakdown of the central government led to the demise of administrative links between cities of various sizes and functions.

The populated areas of Western Europe were split up into large demesnes, each controlled by a local baron and becoming almost a self-contained economic unit. Thus production and consumption were confined in a single area and interaction with other units—if it existed at all—was very weak. During this period long-range trade ceased almost entirely; the only exceptions were Venice in Southern Europe and the Flanders area in Northwestern Europe, both of which kept precarious trade links with Constantinople.

The high degree of self-sufficiency of the demesnes did not allow the survival of the lowest level of urban hierarchy, the market towns, and the total disruption of long-range trade curtailed the preservation of cities on the middle level of the urban hierarchy. At the same time, the breakdown of central government led to the sharp decline of capital cities, therefore eliminating the upper level of the urban hierarchy. The early medieval period was thus characterized by an almost total absence of urban systems.

The reemergence of towns and the subsequent reshaping of urban systems was closely related to the revival of trade and to the gradual recrystallization of political units (Pirenne, 1946). Even in the early phase of the medieval period, few itinerant traders wandered over the continent; they would bring some luxury items to the seasonal or annual fairs that were held outside the walls surrounding the castles or in the vicinity of ecclesiastical towns. But the volume of this trade was tiny and did not affect significantly the self-sufficient pattern of the local economy.

Trading activities received a dramatic impetus during the 11th century. The revival of trade can be related to the growth of population, increased food supplies, owing to the application of various technological innovations, and proliferation of rural handicrafts. The increased supply of commodities and handicrafts and greater demands by the expanding population provided a strong stimulus to overcome the security problems inherent in transportation and the payment of multiple excise tolls. The number of itinerant merchants increased, some even began to establish permanent trading posts, usually outside the walls of the small urban settlements, thus creating the faubourgs that, in later periods, became the major instrument of urban growth. The growing scale and complexity of trade forced the merchant class to become literate and proficient in carrying out transactions on long-distance trade routes, controlled from a fixed point of settlement (Rörig, 1967:43).

The urban center, with its merchant suburbs, also began to be recognized by the feudal lords as an extra source of revenue, sometimes their only source of cash income. Feudal barons then started to encourage the growth of these centers within their fiefdoms, to ensure the safety of the roads, and to provide legal protection for commercial transactions. The growing economic importance of these centers was one of the main reasons leading to the proliferation of new towns that added almost 1,000 settlements to the map of urban Europe. The revival of trade and the spread of new urban settlements (fulfilling the role of marketing and trade centers) broke up the cellular structure of the self-contained manorial estates that were unrelated to each other and functionally isolated, which characterized Western Europe in the early Middle Ages (Van der Wee, 1975-1976).

As the towns grew in number and size, the trading activities of the local markets and production of rural handicrafts in their vicinity were systematically stimulated. In some regions, particularly in Italy, this process of carving out local market areas was implemented by means of armed conflict and fierce competition among the towns. By using force or taking advantage of proximity and convenience of access, each town proceeded to possess a *banlieue*, a belt of rural territory, under its control. Examples of this process in the 13th century are the tributary area of Metz, which included 168 villages, and Rouen, which was the urban center for an area comprising 35 villages (Mundy and Reisenberg, 1958:35). In Germany a market town was the center of an area that included an average of 30 to 40 villages (Gutkind, 1964:193). The larger towns possessed much more extensive market areas, such as Lübeck, which had 240 villages under its control.

In the later part of the Middle Ages, from about the end of the 10th century onward, different urban systems gradually evolved in Europe. As a result of the mode of economic development in general and that of trade in particular, as well as the type of political organization that crystallized in Europe and the relations between the two, these urban systems evinced some particular characteristics. Their major features were the emerging multiple and variegated lower levels, made up mainly of market towns, the concomitant development of their upper levels, mostly through international trade, and the much weaker development and continuous shifting of the middle levels.

The local marketing activities of Western and Northern Europe gave rise to a large number of market towns whose geographical distribution was extensive. Simultaneously, the upper levels of the urban hierarchy emerged through the growth of cities that were centers of the long-range trade. These were relatively large and their market areas were not limited to their near vicinity. These large cities were able to overcome the political fragmentation of Europe by serving numerous small political entities. They made up the upper level of the continent-wide system, encompassing the larger part of Europe and comprising cities organized along the long-distance trade routes between Southern and North-western Europe, through which specialized foodstuffs, rare raw materials, and expensive manufactured goods were transported and marketed. Central and Eastern Europe were integrated, through their major cities, into this general system by means of subsidiary lines of communication.

The towns along the long-distance trade routes, although they fulfilled the same economic and marketing functions, did not necessarily have the same population size and did not rise to the same rank or level in the different urban hierarchies. The size and place of any particular town within the general system were determined by the combination of economic entrepreneurship, expressed through commercial and manufacturing activities, and the political-military power that a city could muster. The Flemish towns at the northwestern terminals of the continental trade routes that, according to their population size, were at the highest level of the urban hierarchy reached this level mainly due to their economic strength, despite their lack of territories or of a large army. So also was Venice, which despite a small land base controlled a widespread maritime commerce and became a large and important city. On the other hand, cities such as Paris, London, and Naples reached a higher level in the urban hierarchy only in the later Middle Ages because of the increasing centralization of their political power in Western Europe, the growing size of their population creating an increased demand for goods and services.

Beneath the level of the largest cities that were located on the long-distance trade routes and derived their economic prosperity from them was the rest of the urban system, which consisted of the vast majority of towns. These were organized in orderly urban hierarchies based on the economic principles of the "range of goods and services"—that is, the maximum distance people were able and willing to travel to acquire commodities or services. The number of hierarchical levels in any given region was determined by the density of the rural population, the efficiency of the transportation network, the level of affluence, and the ability of the political power to secure movement and exchange in a fragmented political structure.

Beneath the upper level of the urban hierarchy, which comprised commercial-industrial cities and some capital cities, integrated into an overall European system through the long-distance trade routes, a second level of towns existed, made up mainly of local capitals and centers of regional settlement systems (Russell, 1972; Dickinson, 1964). These towns, although strongly connected to the upper-level cities through economic interaction of high-threshold commodities and services, were nevertheless much more oriented toward their own surroundings. They often also served as the political capitals of their regions.

In an important study on Swabian towns, Ammann (1963) identified three hierarchical levels in the urban economic system, with descending intensity of population size, market areas, and type of commodities and services available in each level. On the lowest level of this urban hierarchy were the small market towns, whose market area was defined by the distance traveled by the rural population visiting the market regularly on a weekly basis. These small market towns supplied basic crafts related to agricultural activities. On the next level were the larger market towns, whose market area was determined by the business transactions carried out by the towns' merchants at seasonal fairs and roving markets. On the upper level were regional centers providing a full range of commodities and services, including the financial, administrative, and juridical functions supporting and stimulating all the economic activities carried out in the region.

In a decentralized political system such as that which existed in Medieval Europe, the economic network of towns corresponded to the spatial political structure mainly in the middle and lowest levels of the urban hierarchy. This similarity existed between the economic hierarchy and the political decentralized structure, as many market areas of the middle-level regional centers corresponded to the areas that were politically controlled and defended by the local political power. Because of the fluidity of the political system, however, competition for power among the numerous political units made the urban hierarchy subject to rapid fluctuations in relation to both population size and economic functions, thus continually changing the place of towns in the different levels of the hierarchy.

The structure of Medieval and early Absolutist European urban systems can generally be summed up as tending toward rank size, but with a relatively slow and ongoing development of the middle levels of the system.

De Vries's analysis of European urban systems gives a full description of the evolution of the city-size distribution in Europe in the later Middle Ages. The rank-size distribution for the last part of the medieval period (around 1300) is plotted as a "flat-topped curve with a shallow slope" (de Vries, 1984:95). This type of distribution depicts the almost equal-sized larger cities throughout Europe. This particular distribution is interpreted as an outcome of the low level of integration among Europe's cities and regions, the spatial organization of European medieval society being made up of a large number of small, relatively isolated regional-urban systems, each highly self-sufficient. The city-size distribution for the whole of the continent remained quite stable until the year 1500, but the urban system changed significantly in the next century. The larger cities then grew faster than the smaller ones; the distribution lost its flat top and moved toward a rank-size type. This change is interpreted as having been caused by increasing economic integration and by stronger interaction among the numerous consolidating regional-urban systems.

De Vries's analysis delineates the city-size distribution not only for the continent as a whole but for subcontinental units as well as regional systems, interpreting the various city-size distributions by a combination of the general trends of political and economic development in Europe with the particular circumstances of the various regions.

In conclusion, as long as forces of concentration constituted the major activator of the urban systems, this structure prevailed. Toward the later part of the Middle Ages, and only with the growing importance of the forces of centrality—that is, the crystallization of unified kingdoms—did the small separate regional systems merge into a large unified urban system on a national base, having a strong tendency toward a rank-size structure. The gradual amalgamation of the small fragmented political units into larger ones, covering areas on a national scale, caused the underrepresented middle levels to mature, mainly through the growth of regional capitals within the national frameworks. The addition of centrality forces, which emerged strongly at the end of this period, to the concentration forces prevailing throughout most of the Middle Ages allowed to fill up all the levels of the urban hierarchy in the era of Absolutist rulers on a national base. Thus an almost ideal type of "central place" hierarchy was created in most parts of Western Europe, usually represented as a rank-size distribution of the urban systems.

D
INTERNAL STRUCTURE OF CITIES

The pattern of the internal ecological and social structure of European cities was also greatly influenced by the specific characteristics of the forces of concentration and centrality as they developed in Europe. With respect to the elements of the physical layout of cities (walls, streets, squares, public and private buildings), cities of Medieval Europe differed markedly from those of other periods and areas yet looked very much alike, as reported by many a traveler through Europe. There was no well-defined pattern in these cities and very little planning in the narrow winding streets, the heavy gates obstructing traffic, the numerous little squares serving as markets, and the uneven skyline created by churches and houses. The appearance of medieval towns was very much the outcome of their unplanned development and, first and foremost, of their economic, political, and strategic orientations. In most places the medieval town grew out of a community of merchants and craftsmen—markets and burghers' residences also serving as places of storage and production—around the existing political administrative or ecclesiastic nucleus formed by the fortress and the church. In that period of endemic feudal warfare, these rapidly expanding settlements were in constant need of defense in addition to the protection afforded by the old fortress, and thus the rings of walls were periodically enlarged to include new suburbs. Only at the end of the period of urban growth, during the Italian Renaissance in the 15th and 16th centuries, were considerations of an aesthetic order introduced. The great majestic squares and impressive public buildings of Italy, Germany, and the Netherland are not, for the most part, of Gothic but of Renaissance origin (Froelich, 1933).

Medieval towns were characterized by a strong physical separation between town and country, achieved by the walls encircling the towns. The walls expanded gradually with the growth of cities but always maintained the strong demarcation between life under their protection and life without it. The walls were built for security purposes and at the same time served as the mechanism to control access to the city, allowing inspection of incoming travelers and collection of customary tolls. The gate taxes were a major source of income of the medieval towns, thus giving the walls a major economic importance. Building and maintaining the walls required tremendous financial outlay and technical efforts, and therefore they had to be kept to the rule of a minimal circumference at any given state of the town's development.

The gates in the walls regulated traffic in and out of the city; accordingly, areas in front of the gates were ideal locations for large, extended markets. Around these areas and along the road leading to the town were inns and workshops to cater to the needs of travelers. These activities outside the walls generated the faubourgs—areas of urban dwellers who, although living outside the walls, relied on the town for their economic activities and found refuge there in an emergency. The process through which the faubourgs were gradually included within the expanding walls and new ones established outside the new urban perimeter was the essence of the urban growth of medieval towns. Expansion of the city walls to include the faubourgs reflects the operation of concentration forces by means of which the economic power of the faubourg's markets made it attractive for the town authorities to include them within the urban perimeter. The cycle of emergence of a faubourg, expansion of the walls, and demolition of the previous ones occurred in many cities more than once as the number of urban dwellers of the faubourgs increased significantly and their economic activities reached a high level. This pattern of growth through the gradual expansion of the cities' ever-widening walls shaped the basic plan of European towns until the modern period (Burke, 1956).

The street pattern of medieval towns was defined by the constant interaction and competition between public and private space. Because of the limited size of medieval towns, restricted by the surrounding walls, competition for space was fierce, and this resulted in very dense building, narrow streets and alleys, and in the construction of multistoried houses. Competition between public space in the form of streets and private space in the form of buildings for dwelling and business use led to an irregular pattern of streets having different widths and numerous cul-de-sacs.

The internal use of land in these towns was characterized by little separation between the economic and residential functions and by a larger degree of differentiation, according to occupation and social class. The same house could serve residential, commercial, and industrial uses and, at earlier times, even public use. Similarly, there was little differentiation between occupational and residential functions. Spatial differentiation occurred according to occupational groups: Blacksmiths and innkeepers set up shop at or near the city gates; dyers and weavers of woolen cloth needed running water and were located near a spring or stream. The wealthier merchants tended to live in stately houses around the main market square, and from there it was an easy walk to the nearby town hall to attend city council sessions. Some cities had an extraterritorial ecclesiastical district around the cathedral, where the clerical population resided. Poorer artisans and laborers lived in cheaper suburbs or in rented tenements. On the whole, however, there was a fair sprinkling of the different sectors of the urban population all over town. The one great exception to the rule of locational nondifferentiation was constituted by alien, ethnic, or national groups—mainly Jews, who, from a very early period, were segregated in ghettos. The medieval town population, however, was also xenophobic toward other groups, such as foreign merchants. In Venice, for instance, the authorities erected a veritable and harshly regimented semighetto for German traders, the Fondaco dei Tedeschi, and similar institutions for strangers existed in almost all medieval towns.

This spatial organization was preserved in most cities over centuries, until the Industrial Revolution brought decisive changes. During the Renaissance, as mentioned, many architectural innovations were introduced, but the rebuilding of central parts of towns on geometric and classical lines—such as the marketplace with its public buildings—did not on the whole alter the spatial organization of the population.

E
THE TOWN CENTER AND ITS COMPONENTS

A major element of the medieval town was its center, usually including the town church, town hall, guild buildings, public buildings, and inns. The open square itself was used as the town's major market. Often the town center was adjacent to the castle, around which the medieval town evolved. The uniqueness of the European city is evident in the almost universal inclusion—even if in varying degrees of importance—in the town center of these four elements and their relations to street layout. It was this fact that epitomized the special impact of the combination of concentration and centrality forces on the construction of European cities.

All medieval towns had a certain area, or several, that were used as a market. It should be stressed that "the existence of these specialized spaces dedicated to trade should not blind us to a basic fact: the entire medieval city was a market. Trade and production went on in all parts of the city, in open spaces and closed spaces, public spaces and private spaces" (Saalman, 1968). As a result, the narrow, irregular lanes, mainly those leading from the gates to the center, were linear extensions of the marketplace as well as communication routes. Street frontage was a valuable commercial asset, especially near the gates and the marketplace. To expand this frontage, narrow passageways were formed off the main lanes, providing access to new, minor inner courtyards where commercial and craft activities were carried out. Forces of concentration thus caused the structure of the medieval town to become more elaborate and expand functionally within its limited confines.

Because marketing was the basic function of the medieval town, it was accommodated in two basic patterns, in the first of which the market occupied a square at the town center or at the town gates. In planned medieval towns that had a regular gridiron structure, the market square was a void within the grid, bounded by streets on all four sides. The surrounding buildings were usually of uniform height and connected at ground level by arcades. Whereas the market square was used for commercial activity carried out mainly from stalls, the arcade accommodated more permanent structures for marketing and handicraft activities. The second land use pattern of market activities, prevalent mainly in unplanned towns, was found in the widening of the main street and in its offshoots. No two layouts were identical, and each had its own distinct spatial character, but markets existed in all medieval towns and formed a focal point for commercial activity at that period (Zucker, 1959).

At the same time, the church was usually the largest and most conspicuous building in the medieval town. Its spires dominated the urban skyline, towering over the walls. In many towns the church stood apart from the main body of the built-up area. The open space in front of it, the *parvis*, was intended to communicate a sense of distance and respect to those approaching the place of worship. "It was on the parvis that the faithful gathered before and after the services, here they listened to occasional outdoor sermons and here processions passed" (Zucker, 1959). As it was a place of congregation for townpeople and for those from out of town, stalls of various kinds were quite often set up there. It should be emphasized that the parvis was not intended to compete with the marketplace, but, rather, merged with it, mainly in those medieval towns in which the cathedral or the church was located in the town center, which had the shape of a large, open plaza.

During the early Middle Ages, when urban life was almost extinct, some vestiges or urban reality and urban ideals were preserved in the large cloisters and episcopal centers. In fact, the monastery of the early Middle Ages became a new sort of town or the kernel of a town. The renowned plan of Saint-Gall of the 9th century was actually that of a town with all the physical and functional attributes of an urban center. Although the impulse for urban revival often originated from outside the city due to the growth of commerce, a medieval town was born in several instances out of the expansion of a monastery. The cloisters often had the same facets of organization that characterized urban life of later medieval and modern periods: punctuality, restraint, order, time regulation of the various activities, and intense interaction among members of the community (Mumford, 1961).

From the urban monastic model, the medieval town preserved certain features that affected the symbolism of its internal structure. The decisive influence of this religious kernel, was expressed in the symbolic orientation of the building of towns on earth according to Heavenly Jerusalem (Heitz, 1963; Müller, 1961). From the beginning of the 12th century, the vision of the town as a microcosm, an earthly replica of the Holy City, became widespread (Le Goff, 1972). From this source emerged the form of some medieval towns, set within a circle of walls and divided into four quarters by the intersecting four main streets. The four quarters plan could be found in such towns as medieval London and Bern, as well as Fribourg and Nuremberg in 12th-century Germany. In the 13th century the quadripartite town became common, reflecting in spatial terms the new spirit of civic organization. This particular shape was conducive to the emergence of a strong and dominating town center, located at the intersection of the main streets. At this central point an open square was carved out to become the major meeting place of the town. The law court, the pillory, the fountain, and the market cross were usually found in and around this square. Thus the judical and economic functions of the town, which were the two basic components of its existence, were located at the center of the medieval town and were accorded spiritual legitimacy and blessing (Le Goff, 1972:76).

Another crucial element in the construction of the town center, acquiring great importance with the growing urban autonomy of medieval towns, assumed physical representation in the form of splendid and large public buildings. In the early stages of the medieval period, the town council held its sporadic meetings in private abodes but later started to build its own seat, which eventually became a major public building in medieval towns. A fine example of it is the Bargello in Florence, the oldest major public building in that town, constructed in the second part of the 13th century for the *Podesta* (the knight administrating justice) and his retinue.

The guild houses formed another type of important public building, usually located in or near the center of town. The guilds constituted the effective economic political organization of the merchants and artisans, keeping records and archives, and their members met regularly. Many affairs were transacted on guild premises, which became the focal point in the economic networks of the medieval period. A typical guild house had a lower floor that included offices and archives and an upper floor on which meetings of the guild councils were held (*sala magna*). Toward the end of the medieval period, the guild halls of Italian, German, and Flemish towns grew in size and splendor and became the major public buildings.

This combination in the town center of the church, the marketplace, the guild halls, and the castle reflected in a vivid and concise manner the joint operation of the centrality and concentration forces in shaping the structure of the European medieval town. Clear variations can be observed in the layout of the town center and in the types of buildings surrounding it. In some places it was the church that dominated the town center, whereas in others the castle or the town hall. In still other towns the marketplace was the major feature and other functions were less conspicuous. In most medieval towns, however, the four basic components were present, usually defining the town center, a physical expression of the delicate balance of forces that participated in shaping the medieval town.

F
SOCIAL ORGANIZATION OF CITIES

The uniqueness of European cities was most evident in their social organization in general and in the types of autonomy that had developed within them in particular. It was not just by chance that the medieval city was one of the most favored subjects of research of the predominantly legal-minded historians of the second half of the 19th century and the political sociologists of the first half of the 20th century: Many of them saw in it the early blooming of a society dedicated to the common good, unfettered by the whims of the aristocracy and yet not inimical to the peaceful pursuit of personal wealth. It was, in short, the image of a bourgeois city-state. Modern historical research has preserved little of the ideal picture intact, yet cities of Medieval Europe still stand out as a unique form of societal organization within the panorama of urban phenomena in human history.

This uniqueness was manifest in the combination of corporate urban autonomy (defined in specific urban terms) with a predominance in the city of groups having autonomous access to the city centers and to some degree also to the centers of the society. It was also evident in the development of a special type of institutional creativity, mainly manifest in the urban mode of governance and autonomy, which culminated in a distinct, probably unique, pattern of sociopolitical identity.

A few major points will help to highlight this uniqueness. The Medieval European city differed from the rest of medieval society in its major economic activity: manufacture and exchange. At the same time, however, it shared with the society surrounding it many basic religious and also, but only to some extent, broad political orientations, although the medieval town did harbor the potentialities of surpassing the political premises of the surrounding medieval feudal society. Indeed, as the famous saying "Stadtluft macht frei" ("city air makes free") attests to, the construction of many cities or the moves into them entailed some breaking away from the limiting bases of the feudal order.

Until late medieval times, however, most cities remained within the framework of that order. That is, they were ruled by ecclesiastic or secular lords, and only in their internal institutions did important deviations develop, bearing great potentialities for the future in periods of a declining feudal order or in places (such as Italy or Flanders) in which it was relatively weak.

On the whole, these potentialities bore fruit later but existed earlier, rooted in the fact that in the medieval town proper no elite or collectivity enjoyed an inherent or immanent status of predominance, autonomy, or access to the center of urban life. Towns could thus claim their natural place in the overall institutional arrangements of the European countries and yet at the same time constitute a new type of institutional nucleus within them. This nucleus was manifest mainly in the fact that in medieval towns, the major groups were those formed along occupational lines: the associations of merchants and guilds of economically independent artisans. In almost every case these groups possessed the character of a religious brotherhood dedicated to the cult of a patron saint, remembrance of dead members, and concern for the social welfare of the latter's families. When they coincided with professional groupings, or in earlier phases when the town's militia still assumed viable military functions, neighborhood organizations indeed played a role. In cases in which the clergy belonged to closed groups, such as the chapters of the cathedral or some religious orders, it was less integrated in the city. But there were strong and permanent connections linking the clergy to the city, especially to the town's leading families. Students and teachers at the universities, which were located almost exclusively in the towns, were members of professional groups similar to the guilds, but they enjoyed the privileges of ecclesiastical status. This meant that, like the clerics, they were only slightly integrated into the fabric of urban life. Superimposed on all these groups were the factions of the ruling aristocracy—or town patriciate, as it came to be called—that were divided on political issues and found allies in the different occupational groups.

All these groups did not, of course, exist only in Europe, but what was unique there was, first, the relative weakness (although certainly not absence) of the groups generated by centrality forces, of officials, and, to a lesser extent, of religious groups. Second, and probably more important, was the fact that the potentiality of growing autonomous access to the city's central institutions developed in European cities among all these groups. This potentiality was connected to no small degree to the existence of such a tendency among all the (free) strata of the broader society and their coming together in a common bond—the *conjuratio*—so often stressed by Weber.

Thus in many European cities merchant and artisan associations were gradually integrated into the towns' political and social life, either by slow evolution or as a result of open strife, such as the 13th- and 14th-century revolts of artisans who struggled to obtain participation in town government. They became part of the institutional framework, braving the opposition of higher groups as well as of those groups related to the Church, which, until the Reformation, had successfully obstructed the grant of townsmen's duties and privileges to the members of those associations.

To some degree, this was also true of the numerically strong but politically and socially powerless lower classes or layers. Many of them, however, found a place in alternative, extrainstitutional groups that existed then, not necessarily only at the lowest levels of society. Envisaged broadly, however, the plurality of groupings in the medieval town, each with potentially free access to the center of urban life, was open enough to allow the gradual incorporation and institutionalization of almost any form of marginal or pressure groups, except for the ethnically or religiously alien ones, such as the Jews. In other words, a large measure of social mobility existed, accentuated by economic ups and downs in an essentially commercial way of life, but almost always also connected with the possibility of political expression. Needless to say, many differences existed between the various European cities from this point of view, but only in a few of them—Venice and Nuremberg were outstanding examples—was the social body too rigid to allow personal social ascent to or near the ruling group from the lower levels. But even in those towns, an individual family could rise, within the span of two or three generations, to the level of rich merchants, positioned just beneath the politically powerful aristocracy.

G
URBAN INSTITUTIONS AND AUTONOMY

The European city's heterogeneous social structure and the tendency of most social groups to gain some autonomous access to the centers of urban life have greatly influenced the nature of the city's institutions, autonomy, and conflicts, as well as the pattern of urban identity that has evolved in it. This combination gave rise to the unique nature of the European city and to its impact on the general social and political order.

The municipal institutions were centered on the city council in most cases, usually ruled by some 20 to 50 prominent citizens representing different groups and interests and epitomizing the nature and limits of internal urban autonomy. The city council started as an organ of self-government of the merchant community and as its court of law vis-à-vis the feudal lord and his court. In a gradual process, paralleling the incorporation of additional population groups, the council expanded its jurisdiction, encroaching upon such sovereign rights of the lord as taxation, raising tolls and customs duties, policing the market, jurisdiction, and defense. In most cases, by the 13th century the town council had established itself as the sole representative of the town population vis-à-vis the outside world and as the constitutional authority of the population. It fulfilled all the legal, judicial, fiscal, and economic functions and provided the services expected in the Middle Ages excluding education, which was a task assumed by the clergy. Toward the end of the Middle Ages, however, and especially following the Reformation, education as well as the provision of religious services became the responsibility of the urban authorities.

During and even before the 14th century, the social and economic development and incorporation of new groups into the central urban institutions slowed and finally stopped in many medieval cities. The council type of government was maintained, but an oligarchy of some two dozen families evolved in almost every town that reserved the exclusive right to fill the town council seats. Thus election to the council was replaced by nomination and co-optation of new families, as the old patriciate gradually faded out. It was against this closure, among other things, that the artisans' revolts were directed during the late 13th and 14th centuries—mostly without success, as the new ruling groups promptly turned into new oligarchies. Following the great wars of the 13th and 14th centuries, as a result of which Italy acquired wide territories, a further development occurred in some of the large Italian cities, which transformed them into entities not dissimilar to the regular feudal principalities. In these huge cities, with their almost unsolvable political and social contradictions, accentuated by continuous interference by foreign powers, the institution of a temporary one-man rule (the *Podesta* system) led, quite naturally, to a hereditary one-man rule and thence to the so-called Renaissance state. In other countries the typically medieval political organization was also abolished slowly by the force of royal centralization, by political and financial weakness, and by the limitations of an increasingly parochial view of the political institutions.

Yet this new political order also generated some of the specific political orientations that developed in these cities and went far beyond the premises of the medieval political order. These orientations were also engendered by almost continuous feedback relations with strong tendencies toward external political autonomy.

H
EXTERNAL POLITICAL AUTONOMY

In many European cities this tendency toward internal autonomy was indeed combined with specific attitudes toward external political autonomy. It was this combination that usually generated the particular type of urban identity and conflicts that developed in many European cities and that in many ways constituted a central aspect of the uniqueness of European cities.

With respect to the medieval towns' political autonomy, a differentiation must be made between periods, regions, and towns. In the 10th and 11th centuries—the early phases of medieval urbanization—almost all of the towns were ruled by an ecclesiastic or secular lord. The struggles of the late 11th and 12th centuries succeeded in abrogating this dependence almost everywhere. The great exchange and manufacture centers of Italy, the Netherlands, and Germany, as well as most of the large towns in these regions, managed under politically favorable circumstances and at considerable expense to remain virtually independent during the Middle Ages and parts of the early modern period. On the other hand, in countries of increasing centralization, such as England, France, and Spain, most towns were under strong royal supervision. Many of the smaller towns founded as centers of administration by kings and nobles between the 12th and 14th centuries depended on their lords from their very beginning.

This situation of oscillation between autonomy and dependence, with the numerous gradual shadings between them, was first and foremost political and administrative. Given the size of larger cities and the technological and political limitations on transport of food supplies and raw materials, no real economic independence could exist. For most of the Middle Ages, the cities' religious and cultural autonomy was also strongly hindered by the near monopoly of the Church in these fields. These conditions were mirrored in the self-consciousness and self-view of the medieval burgher: The burgher formed part and parcel of the broader religious and cultural entity constituted by Christian Europe.

And yet with all these limitations, a strong tendency toward external urban autonomy developed in Europe. This tendency was rooted in the combination of four basic factors: (1) the existence of producer cities; (2) the prevalence of autonomous strata within the society and the cities—that is, of strata with autonomous access to the center, generating their own autonomous elites; (3) the special type of geopolitical decentralization that evolved in Europe—that is, a factual decentralization, yet one based on the multiplicity of mutually impinging centers with strong common orientations; and (4) the relatively strong tradition of legal autonomy prevailing in Europe and that led to the creation of special types of *conjuratio* of freemen, which often constituted the nuclei of autonomous city organizations. Similarly, the regional location of cities—as was the case in Venice—could also easily influence the development of this autonomy in the direction of a city-state-like identity. Furthermore, as demonstrated by A. Pizzorno (1973), different regional power relations between upper landed and urban classes could greatly influence the degree of such political distinctiveness of cities. This feature was weakest where close interrelations existed between these different ruling groups and strongest, often accompanied by intensive conflicts, where these upper groups were relatively distinct.

I
CULTURAL ROLE OF EUROPEAN CITIES

The full impact of this type of urban autonomy can be understood only when considered against the background of the social and cultural creativity and urban identity that developed in many European cities. This creativity was manifest in two different dimensions. The first was in the field of cultural creativity proper; it was not very different in principle from that of other civilizations, although some specific characteristics did develop. The second (and, in combination with the first, the most specific to European civilization) was in the realm of civic and political ideology and institution building.

We have already noted the fact that many cultural functions, aimed at benefiting the town population, were not initially assumed by the urban authorities but by the Church. The Church oscillated between two different orientations to city life and, often, a strong ambivalence to it. On the one hand, the ecclesiastic institutions—especially the monasteries—often served as repositories of the memory and image of city life and even sometimes constituted kernels of urban development, as did the bishopric palaces and castles. On the other hand, the Church often viewed the town, in theory and for a long time in practice, as an essentially negative form of life, conflicting in almost all its aspects with religious teachings. This attitude was strongly influenced by biblical texts and philosophical writings of late antiquity and found concrete bases in what was defined as the improper gains implicit in trade, the essential urban occupation. The relatively high social mobility that existed in towns and constant changes in fortunes were regarded disapprovingly by the Church, both ideologically and out of fear of losing its dominance over the cities. Indeed, the towns had often rebelled against their ecclesiastic lords, a fact that weighed heavily against the towns not only politically but also in philosophical-religious terms.

The dynamics of urban life were nevertheless stronger than the Church's reservations, and the city rapidly became the major center of cultural creativity. A few major forces predominated in this process—the quick concentration of wealth in the hands of the *nouveaux-riches*, the fast turnover in the upper echelons of society—leading to a conspicuous increase in consumption including massive encouragement and patronage of art. With its tradition of literacy, the merchant class produced a vast amount of written works, some of which could well be classified as art. Finally, being located in the towns, the universities, from their very beginnings in the 11th and 12th centuries, were instrumental in promoting a continuous influx of intellectual elements into the cities. When urban-based heresies posed a real threat in the 13th century, a strong ecclesiastic presence by the mendicant orders became manifest in the towns, adding to the diversity and heterogeneity of urban cultural life.

The concentration of cultural institutions together with constant contacts with other European (if not overseas) countries engendered by international trade and by the wanderings of artisans and scholars, made the medieval city a center of cultural innovation. This urban culture was closely related to the general cultural orientations of the time, to their largely religious character, despite the beginning of the secular cultural creativity mentioned earlier. By the end of the Middle Ages, the men of the Italian Renaissance and their followers in other countries took decisive steps toward cultural thought and expression and, by so doing, closed the door behind the Middle Ages, heralding a new cultural era. The European cities stood out among those of other civilizations as a unique case because they were centers of great cultural creativity—a feature found in many, if not all civilizations (in China, for instance)—and because it exceeded here the basic premises of cultural creativity of the broader society.

J
URBAN IDENTITY

This special type of cultural creativity, however, can be understood only in relation to the distinct institutional creativity of European cities, the specific type of social autonomy, and the development in Europe of a distinct urban identity, beyond the one based on local attachment. Such an identity was made up of the following elements: the distinction between city and country, a special type of sociopolitical consciousness and symbolism, and sometimes, the extension of this symbolism to the centers of society.

Strong emphasis on the distinction between city and countryside thus emerged in most European centers; but although it varied among the cities, it predominated in numerous European towns, unlike in the city-states of antiquity. Such an identity developed, as it did perhaps even more in some of the city-states of classical antiquity, around a special type of political urban consciousness that may have been transposed to the centers of the respective societies. The distinctiveness of such an identity lay in the emphasis placed on the relative autonomy of the civic order, the common civic good, and the participation of active citizens in governance of the city or polity.

The tendency toward the development of a special civic consciousness, of a specific culture of civility, became especially articulate whenever cities evolved due to some combination of forces of centrality and concentration (predominantly the latter). This tendency was articulated above all due to the attempts of active urban elites, rooted in the forces of concentration, to permeate the city's or polity's center and imbue it with their own orientations. The combination of such forces gave rise to new types of orientations, political activities, and values, such as freedom and openness. Such values were focused on the city and their articulators regarded it and city life as the major arena for implementing their principles and developing a specific civic identity and cultural consciousness of civility in these cities in general and in the city-states in particular.

Many of these elements of political symbolism that were strong in antiquity were later transposed to the European cities. They became interwoven there with specific elements of the European political tradition, facilitating the development of the new civic culture and consciousness, and culture of civility, in conjunction with the different dimensions of urban autonomy. This was also connected to the specific corporate identity based, in extreme cases, on the *conjuratio* or on other types of organization of the urban community.

K
URBAN AUTONOMY, IDENTITY, AND CONFLICTS

It was through the combination of external political autonomy, development of internal autonomy, and specific urban identity that the particular types of social and political conflicts and protest movements that evolved in the European cities can be explained. Many of these conflicts—such as the various local uprisings and rebellions—did not differ greatly from those occurring elsewhere. Besides these, there were additional types of struggle, briefly alluded to earlier, which are of special importance to our analysis. One such type, which occurred in the medieval cities of Flanders and Italy and, to a lesser degree, in other European cities, was caused by the attempts of various lower- or lower-middle urban groups—often revolutionary or semirevolutionary—to reconstruct the corporate structure and governance of the cities. Another type of conflict, that could best be observed in some of the Italian city-states of the Renaissance, and later in a more complicated form in most Western societies, was characterized by citizens' tendency to orient themselves toward the center, demanding greater participation in their respective societies but on their own terms. At the same time, however, they articulated their distance from the more passive elements of the periphery in the specific types of their social organization.

The setting within which these conflicts developed was under continuous tension caused by the inclination of many urban elites and subelites to enlarge their participation in the regulation of urban affairs as well as their self-regulation in the major spheres of their activities. The outcome of such tensions depended on the balance of power among these forces, a balance that necessarily changed over time.

It was indeed the continuous confrontation of such urban groups, nourishing tendencies toward self-regulation in the control of cities against the central (external and/or internal) authorities, that gave rise to the very dynamic type of social conflicts and struggles unique to European society. These could also culminate in the attempts of urban or mostly urban-based elites and strata to change the centers of society and the premises of its governance according to principles derived from their own urban experience. This experience was couched in terms of a special civil order, very often resembling that of the classical city-states of antiquity. These movements of protest and conflict constituted perhaps the most distinctive characteristic of European cities.

PART III

ANALYSIS AND CONCLUSIONS

CHAPTER 12

Distinctiveness of Cities and Urban Systems

A
THE BASIC VARIABLES

1 Introduction

In the preceding chapters we presented brief analyses, based on reexamination of the basic secondary literature, of some major characteristics of cities and urban systems in several civilizations and societies, as they developed in the broader context of configurations of forces of concentration and centrality.

We first examined how different combinations of concentration forces of economic, technological, and geopolitical factors generated different levels of resources. Then we described how the different forms of centrality manifest in coalitions of elites predominated in a society that carried different cultural orientations and, acting in various political-ecological settings, exercised different modes of control over the flow of resources in a given society. Through this exercise the forces of centrality shaped not only the structure of institutional and economic markets but also access to them and the flow of resources among them, the structure of centers and center-periphery relations, the structure of boundaries of the major collectivities (particularly of the political ones), and also the nature of information flows and of reference orientations within the respective societies.

Interaction between the forces of concentration and centrality shaped the spatial concentration of resources and information and consequently the major aspects of urban life, the structure of cities and urban systems, and cities as centers of distinct patterns of social and cultural creativity. We have observed that these institutional forces shaped major aspects of cities and urban systems by structuring the different modules of spatial concentration of resources, the spatial organization of the flow of such resources among modules of concentration and their functional interrelations, and the internal and spatial social organization of such modules.

The case studies have substantiated three points previously specified. The first was that contrary to the assumptions made in many of the classical studies examined in Chapter 1, the combination of the different aspects of institutional structure, and hence also that of cities and urban systems, tended to crystallize in more variations than were assumed in the classical approaches. The second point was that our analysis has confirmed the premise that although each of the concentration and centrality forces generated specific tendencies with respect to the structure of urban systems and cities, it was only through the interaction between the forces that a concrete urban system crystallized. The third point was that this analysis has also supported the premise that in different societies or at different periods of development of any society, far-reaching changes occurred in the strength and constellations of the various forces that engendered changes in the major aspects of cities and urban systems.

We have indicated here more extensively than in Chapter 2 how the institutional components—that is, the levels and types of technology and economic development, modes of control over the flow of resources exercised by different elites in the different geopolitical settings—combined in the types of societies analyzed.

2 Technologies and Economic Structures

From the standpoint of the levels of technology and economic structure, most of the case studies (with the exception of Southeast Asia) were relatively advanced agrarian societies, based mainly on sedentary settlement, some types of intensive agriculture, and few internal and international contacts. It might be worthwhile to review briefly the economic base and types of technology that developed in some of these societies.

Southeast Asia

The economy of Southeast Asia has been characterized by a predominance of wet-rice cultivation. The humid climate and soil fertility resulted in a highly productive agriculture. Irrigation and drainage technology was localized and the hydraulic system was maintained by villagers. There was little differentiation in agricultural production, and thus exchange and trade could not reach a significant level (Coe, 1961). The peninsular character, combined with the rugged topography, limited accessibility to parts of the region. Low specialization and differentiation of agricultural crops, the abundance of locally produced foodstuffs, and the poorly developed transportation system all contributed to the cellular structure of the numerous self-contained village units forming the region.

Colonial Latin America

A different type of economy prevailed in colonial Latin America based on mineral extraction and plantation agriculture. This space economy was characterized by three major components: (1) a series of mining cores in Mexico and Peru; (2) the development of agricultural and ranching areas, peripheral to the mining cores, intended mainly to supply the mining areas with foodstuffs; and (3) a delivery system designed to funnel silver and gold to Spain (Stein and Stein, 1970:28). Exploitation of the mines required labor and mining technology. The first was provided mainly by drafting Indians (Mitas); the second was imported from Europe and consisted mostly of iron and steel tools, pit props, and of mercury used in amalgamating silver from crude ore.

Agricultural development in Spanish Latin America was aimed at ensuring the success of the Spanish mining enterprise. Colonial agriculture tore apart the preconquest agrarian structures and replaced them, as from the 16th century, with the *hacienda* as the instrument to supply the mining economy and to allow the Spanish conquerors to reestablish in America the major status symbol of Southern Spain—the landed estate with a largely immobile labor force. The agricultural methods applied in the *hacienda* were labor intensive and resulted in low productivity.

The extensive agriculture of Spanish Latin America characterized by "slash and burn" was a far cry from the intensive, sedentary agriculture of the pre-Columbian civilizations that resulted in abundant and dependable yields. In the semiarid areas of Meso-America and the Central Andes, intensive agriculture had flourished without benefit of plow, wheel, or draft animals mainly through the application of sophisticated irrigation technology and superb organization of manpower. This highly productive, labor-intensive agriculture yielded an abundance of maize—the Indian staple food—as well as ancillary crops of beans, squash, tomatoes, and chili peppers. In the valley of Mexico water for irrigation was captured from melting mountain snow by applying skillful techniques and enlarging the huge natural basin of interlocking lakes. In the Central Andes rivers were harnessed through an intricate canal system; mountain slopes were elaborately terraced and water was diverted there through well-kept canals. The conquerors allowed this irrigation system to disintegrate both in the central valley of Mexico and in the Central Andes, thus substituting the previous high-yield agriculture with the extensive *hacienda* farming of the Spanish colonial period.

Chinese Empire

The Chinese economy was based on wet-rice cultivation, practiced to the highest level of intensity and able to support a dense population over large areas. The agricultural civilization spread from the original core area along the middle course of the Huang-Ho River to the south and east. By the end of the 14th century every patch of land in the Eighteen Provinces making up China proper, south of the Great Wall and east of Tibet, was under cultivation. This huge territory, encompassing many regions with diverse geographical characteristics, was dominated by the two core areas of the Huang-Ho and Yangtze basins linked by the Grand Canal, which possessed the most developed irrigation and drainage systems, flood control, and canal transportation facilities. Long-standing investments in irrigation and transportation technologies made these core areas the most productive and affluent regions of the empire, thus assuring the empire's hegemony and protection from the development of multiple polities.

Despite great variations in the physical features of China's many regions, the rice cultivation landscape was quite uniform and based on the use of tremendous manpower and sophisticated water management techniques for irrigation and drainage. This intensive use of land for rice growing resulted in a scarcity of wood for fuel and timber and fodder for cattle raising.

Industrial technology made an early start in China, and by the 14th century methods of production of iron and textiles were highly developed. At that period industry was much more advanced in China than in Europe, the quality and complexity of the goods produced attesting to a high degree of technological ingenuity and entrepreneurship organization.

Islam

The Islamic civilization presents a very different picture. The basic economic feature of the Islamic world in the Near East and North Africa, mainly from the 7th to the 11th century, was the existence of economies of scale, uniting in one faith, one culture, and one language a diversity of population spread in low densities over a huge territory extending from Spain to Asia (Jones, 1981:175). Situated at the edge of deserts, both in the Near East and in North Africa, Islamic civilization was characterized by a semiarid agriculture that depended heavily on winter rains and was often plagued by draughts. A small part of the rural economy—on the narrow river basins along the Nile and in Mesopotamia—was intensive, relying on complicated irrigation techniques to produce two to three crops annually, which was unusual for that period. The Arab Agricultural Revolution, which diffused Indian crops as far west as Spain, was enlarged by widespread cultural contacts and travel (Watson, 1974:18). Most of the rural economy, however, was extensive, with a strong element of oasis cultivation, and its resource base was quite poor, especially in timber. Indeed, the European Christian powers, having spotted the scarcity of the Muslim Empire's resources, prohibited the export of iron and timber to them (Strayer, 1974:403). Trade was mostly conducted over long distances and concentrated on luxury items, but it could not compensate for the paucity of mineral resources.

India

The economy of the Indian subcontinent was characterized by a refined mosaic of tens of thousands of villages, almost indestructible, which survived natural catastrophes and changes of regimes and rulers. Especially when compared with China, Indian agricultural areas had low productivity despite the potential for double crops in a good rainy year (Jones, 1981:199). The low average yield was due mainly to the low level of the irrigation technology. Only a small part of the cultivated land was irrigated, the simple "Persian wheel" being the only lifting device used. At the beginning of the colonial period, most of the irrigation projects, some of which were ancient, had fallen into decay.

Above the basic layer of the villages, a mesh of political units recurred persistently for long periods. Various conquerors had left intact the cellular structure of "village India," which provided them with a reliable major tax base. Most of the villages were self-contained and had few contacts with the nearest small garrison town. The villages' and even small towns' relative isolation stemmed from the extraordinarily scanty communication facilities, both intra-regional and interregional, probably due to the tremendous size of the country, difficult terrain, and the rarity of navigable rivers. India was thus split into a large number of nearly separate markets, with interaction and competition being almost prohibited by the excessive cost of transportation over land.

Japan

Until the modern period, Japan's economy was based on intensive agriculture, mainly rice cultivation, common to many regimes in South and East Asia and subject to monsoon seasons. The abundance of water and its most efficient use by intricate systems of irrigation overcame the scarcity of cultivable land. The mountainous nature of the islands of Japan leaves arable only about one-sixth of the land, subdivided into tiny patches along river beds and estuaries, isolated inland bays, and the shores of the inner sea and lakes. Scarcity of land and heavy inputs of irrigation technology and manpower necessitated the development of intensive agriculture that would produce the highest possible yields.

The need to establish and maintain sophisticated irrigation and drainage systems required a high level of social organization to allocate and coordinate the agricultural work. Because rivers in Japan flow for only a short distance before reaching the sea, and given the patchy distribution of arable land, these complicated irrigation systems were necessarily limited to a local scale. Social organization thus focused on a framework of family and village, encouraging consensus decision making in small groups.

The unique physical nature of Japan, consisting of a number of islands, rendered the country less prone to external influences and more self-contained than continental societies of similar size. At the same time the surrounding sea, together with the inner sea, allowed for easy communication among the islands and along the coastline and also provided a highly important economic contribution—fish, which supplied the population's main animal protein.

The traditional Japanese agricultural economy succeeded in overcoming the physical limitations imposed by isolation and meager resources through hard work and the application of sophisticated technologies, and also owing to a particular social organization that enabled highly efficient use of land and sea resources.

Medieval Europe

The Western European economy during the early Middle Ages can be characterized as one of agrarian subsistence. The pronounced decline in Mediterranean trade due to Islamic military expansion, the total breakdown of the secular administrative order following the disintegration of the Roman Empire, and frequent invasions of eastern nomadic tribes all contributed to a radical reshaping of the Western European economy, reducing its scale and complexity and linking it to local land resources. The main economic activity was thence agricultural, on a subsistence level, based on a cellular structure of small isolated domains having negligible economic interaction. The small scope of economic transactions was the outcome both of low agricultural productivity and insurmountable communications difficulties as, owing to the deterioration of the road network, travel and transport became extremely hazardous. Tolls had to be paid upon crossing the numerous boundaries resulting from the cellular political fragmentation. "Over a journey of a hundred miles, a travelling merchant might fall under a dozen sovereignties, each with different rules, regulations, laws, weights, measures and money" (Heilbroner, 1968:50). The agricultural products were consumed locally; foodstuffs were virtually not marketed and rarely transported more than a few miles. People who were not engaged in the production of foodstuffs had to live or stay temporarily in the households of the producers or in their immediate vicinity. The settlement pattern of this subsistance agricultural economy comprised a multiplicity of small villages among which were located castles, monasteries, burgs, and ecclesiastical towns, of which only the latter, usually a survival of Roman towns, offered any resemblance to urban settlements. It should be emphasized that the limited population residing in towns or in other forms of settlements had to live on the land, much the same as the villagers.

Until the 11th century the greater part of the land was unoccupied and uncultivated; there were areas of dense forests, and the hilly slopes and low-lying valleys were untilled. Beginning in the 9th century, with the growth of the population and its slow but steady progress, land reclamation started on a large scale and agriculture and settlement expanded. During these developments, leadership was assumed by the various monastic orders (Mundy and Reisenberg, 1958); feudal lords also encouraged the clearing of their forests and the drainage of their swamps. As large tracts of wasteland were rendered arable, the economic base of Western Europe significantly improved (Pirenne, 1946:60-70).

The expansion of settlement on idle land and increased productivity of the agricultural sector would not have occurred without the advent of several technological innovations. In this respect it is worthwhile to quote E. L. Jones: "The improvement of the means and context of production by gradual steps from ancient times is a puzzling feature of the European record. . . . Europe became a continent where the fundamental scientific work went regularly forward" (1981:46). Among other technological and organizational innovations that affected European economy in the later Middle Ages, the moldboard plow, reintroduced in the 9th to 10th century, made it possible to turn over the heavy soils of Northern Europe. Closely related to the adoption of the moldboard plow was the substitution of the three-field rotation of land use for the two-field system. This increased agricultural productivity, spread the work more evenly over the year, and, by diversifying crops, encouraged greater stability of the food supply. The new rotation system created the possibility of using the spring plantation to produce crops of animal fodder, which, coupled with some technical innovations, led to the use of horses and carts, thereby revolutionizing the transport technologies of the period (White, 1962:72).

Increased agricultural productivity and improvement of transport technologies led to a growing food surplus that fed a large population not exclusively engaged in agriculture, and brought about the emergence and flourishing of rural handicrafts. In numerous rural areas throughout Western Europe, such local handicrafts as weaving woollen cloth, wine production and copper and brass work soon far exceeded local demand. The products of these rural handicraft activities joined the luxury items transported along the long-range trade routes for distribution in other parts of Europe. The revival of trade and the subsequent growth of towns in the later Middle Ages were thus strongly related to the growing productivity of the agricultural base and the emergence and spread of rural handicrafts.

3 Structure of Elites and Modes of Control

Within the societies analyzed in the case studies, these technologies and economic structures combined in diverse patterns with the other institutional factors mentioned—that is, the various structures of the elites exercising different modes of control. These economic and technological bases also merged with ecological formations, compact and centralized societies or relatively open or decentralized ones, small societies with a strong orientation toward external markets, and relatively large societies possessing stronger internal markets.

Rather weak compact regimes, in which relatively low levels of technological development and economy prevailed, emerged mainly in Southeast Asia and in some Near Eastern polities. Strong compact regimes developed in Latin America, mostly combined with extractive economies, in some Islamic regimes and mainly in the Russian, Byzantine, and Chinese empires. Complex feudal and semifeudal decentralized structures evolved in Europe, Japan, and India. In some patrimonial Sultanic Islamic regimes, however, a lower degree of compactness (but usually no full decentralization) combined with less developed agrarian and commercial systems.

The combination of technology, economy, and political-ecological formations merged in turn with different structures and coalitions of elites, articulating various types of cultural orientations and exercising different modes of control over the flow of resources. The most crucial distinction existed between the embedded and autonomous elites, a distinction closely related to their major cultural orientations. Strong autonomous political and cultural elites were found in the ancient Greek city-states (not included in our case studies), as well as in the Chinese civilization and most Christian and Islamic societies. Important variations occurred in the latter cases, however, among which the Latin American and a few Sultanic Islamic societies constituted partial exceptions. Such elites conceived that strong tension existed between the transcendental and the mundane orders and, at the same time, showed some strong elements of a this-worldly conception of salvation—that is, that at least a partial overcoming of this tension can be achieved through the reconstruction of the mundane world.

Conversely, the more embedded elites, or a mixture of embedded and autonomous ones, were found in "pagan" Southeast Asia, in the ancient Near East, or in pre-Colombian Meso-America (the latter not included in our case studies). A relatively strong emphasis was usually placed on a conception of homology or parallelism between the transcendental and the mundane orders.

Within the realm of Hindu or Buddhist civilizations—where the strong emphasis placed on the tension between the transcendental and mundane orders was related to a strong other-worldly conception of salvation—elites emerged that were autonomous in the religious sphere but relatively embedded in the political one. Such a pattern of elites and cultural orientations also evolved, though in a more limited way, in imperial Latin America and during many Sultanic periods of Islamic history, but there the other-worldly orientations were never as dominant as in the realm of Hindu and Buddhist civilizations.

From the viewpoint of the structure of elites and their cultural orientations, Japan constituted a rather special case, characterized by highly embedded but regionally differentiated elites. These elites tended, however, to break out of their embedment in narrow ascriptive settings owing to their strong orientation toward broader societal frameworks and mainly toward the center. It was this characteristic, together with strong commmitment orientations, that connected them to the social and economic orders.

The degree of organizational distinction, specialization, and relative segregation of the different elites—mainly cultural, political, and administrative—cut across the distinction between embedded and autonomous elites. In most cases of fully embedded elites—such as in Southeast Asia—a relatively small degree of such specialization and differentiation occurred, although there, as in other such societies, specialized priestly groups and organizations tended to emerge. Where more autonomous elites existed, the degree of their differentiation depended on the nature of the prevalent cultural orientations, mainly the conception of salvation. In civilizations where pure this-worldly conceptions were prevalent (as in China, the Greek city-states, and the Roman Empire), relatively monolithic elites tended to develop with little differentiation between their political and cultural orientations. In civilizations where other-worldly orientations predominated, (such as Hindu and Buddhist ones), multiple elites evolved inclined to segregation in their own realm. In the Christian and Islamic civilizations, however, where this- and other-worldly conceptions of salvation were combined, multiple specialized elites tended to emerge, segregated in varying degrees but potentially impinging on one another.

These different types of elites, articulating different cultural orientations, tended to exercise diverse modes of control over the flow of resources and information and to structure different center-periphery patterns. The more embedded elites thus reinforced tendencies toward narrow markets and toward a limited flow of resources among them and to adopt relatively homogeneous and undifferentiated conceptions of social structures and of reference orientations. They also tended to generate relatively small symbolic differences between center and periphery and a low degree of permeation of the periphery by the center.

Conversely, the more autonomous elites tended to foster wider markets, to regulate both the access to such markets and the flow of resources among them, and to devise varied and differentiated symbols and conceptions of the social order and reference orientations. They also tended to generate symbolically autonomous centers that attempted to permeate the periphery according to the premises of the center. Among these autonomous elites and the modes of control they exercised many variations emerged that were indeed closely related to their specific cultural orientations.

The structure of the coalitions of elites and modes of control they exercised were also greatly influenced by the relative autonomy of the major strata and by the nature of their access—autonomous as against mediated—to respective centers of society and to major attributes of the social and cosmic orders. This autonomy was most developed in Medieval Europe, less so in the Byzantine Empire, and, to an even lesser degree, in the realm of Islamic civilizations.

The modes of control, combining in a variety of ways with the various kinds of technology and political-ecological settings, coalesced differently in the imperial, patrimonial, and decentralized regimes. There were also great variations within each of these patterns.

The institutional forces that coalesced in these regimes generated the distinct types of cities and urban systems, as well as the social and cultural urban creativity or the cities as loci of such creativity. Changes in the momentum and direction of these forces also influenced the degree to which cities became activators of processes creating new levels of resource, activity, and information generation. The fact that spatial concentration of resources and population in specific spatial modules rendered them distinct from their surroundings, allowed the emergence of such creativity and shaped its nature. It was this distinctiveness that created problems in the internal organization and structuring of such concentrations of population and resources, as well as in their relations with the other sectors of the society.

The modes and degrees of intensity of the cities' social and cultural creativity were expressed in ways that distinguished the social structure, conflicts and protests, and cultural activities that evolved in cities and urban systems from those of the rural sector or from the major societal centers usually found in cities.

The very concentration of population and resources in cities created, as we indicated in Chapter 2, a marked distinctiveness from both center and periphery. The major dimensions of social and cultural creativity of cities pointed to the ways in which cities interrelated with both center and periphery of their society. These distinctions among the urban modes of social and cultural creativity became evident when constructing the social structure of cities and urban systems. The distinctiveness of this urban social structure distinguished the cities from the societal center and from the broader rural periphery, and also strongly influenced the ecological structure of cities.

The development of distinctive social organization and ecological structure has raised several crucial questions with respect to the relations of these specific organizations with their surroundings, especially the center and its rural periphery, and to the potential conflicts generated between them. The combination of such developments and conflicts gave rise to another dimension of the distinctiveness of cities—namely, their autonomy in relations with the other sectors. Such autonomy applied primarily to the extent of regulation of internal and external affairs and to the political control held by the center over the periphery of society, as well as to confrontations between such regulation and the tendency of urban groups toward self-control and active participation in such regulation.

Emerging from this continuous interaction or confrontation between the distinct social organization and ecological structure of the cities and their tendencies toward regulation and self-regulation were three additional dimensions of the social and cultural creativity specific to urban settings: (1) definition of the urban identity prevalent in a society, the conception and self-conception of cities, of its active groups in relation to other sectors of society; (2) conceptions of the city and the moral evaluation of cities prevalent in a society; and (3) the nature of social conflicts and movements of protest that evolved within the cities.

The distinctiveness of cities found expression not only in their internal structure or organization but also in the interrelations among the different types of cities and urban systems.

The nature of such distinctiveness and creativity, however, varied greatly among the different societies or sectors thereof, according to the nature of the different aspects of urban life, shaped by institutional forces. These variations occurred in accordance with the relative strength, autonomy, and diversification of the forces of concentration and centrality and their interrelations as they developed in the respective societies. These social forces influenced the concrete modes of construction of various modules of spatial concentration of population and of resources, the spatial organization of the flow of such resources among such modules of concentration and the functional interrelations between them, and different aspects of the social organization of such modules of concentration.

We shall now analyze systematically the manner in which the interaction of institutional patterns, shaped by the forces of concentration and centrality, influenced major aspects of the structure of cities and urban systems, as well as major types of distinctive urban, social, and cultural creativity.

B
DISTINCTIVENESS OF URBAN SOCIAL STRUCTURE

We will now analyze the constellations of the forces of concentration and centrality and the conditions under which distinct types of urban economic and social structures and of political organization emerged. We will also determine what distinguished them from other elements of the social structure, from the rural, usually peripheral, sectors and from the centers of their respective societies.

The first condition necessary to the formation of a distinct urban social organization is demographic density leading to the concentration of a relatively large population within a small, confined space. Every urban place, every spatial concentration of population was to some degree distinct from its surroundings and, accordingly, other conditions being equal, the denser the concentration, the greater the possibility that diverse types of specifically urban social organization would develop.

Such distinctiveness may have been of different kinds, and the best starting point for their analysis is the distinction made by Redfield and Singer (1954) between orthogenetic and heterogenetic cities, mentioned in Chapter 2. One such distinction, best exemplified in heterogenetic cities, was evident in the development of a mode of differentiation, based on economic and occupational diversification, and in the types of productive and mercantile activities. These sectors were distinct from the urban ones rendering different services, mainly political-administrative and religious functions that epitomized the organization of centrality. Another distinctiveness of heterogenetic cities was manifest in the spatial concentration of forces of centrality that fostered a quantitative differentiation from the social structure of the periphery. Within the framework of this concentration, diversification of the forces of centrality may also have taken place, as well as concomitant development of the social structure and organization that to some degree were distinct from those prevalent in predominant centers in this society.

Each of these types of distinctiveness of urban societies may have evolved with different degrees of intensity and variation, and overlapping may have occurred among them, leading beyond Redfield and Singer's original distinction between orthogenetic and heterogenetic cities. Each type of distinctiveness was the focus of potential tensions, conflicts, and protest movements in cities and in the relations among different sectors of the urban population and other sectors of societies and among their respective centers and peripheries.

The evolution of these types of distinctiveness of the urban social structure was influenced by several factors: (1) the relative importance of forces of concentration as against forces of centrality in the construction of any city; (2) the specific structure of these forces—whether the major elites were autonomous or embedded, monolithic or heterogeneous—the concomitant cultural orientations they articulated, and the modes of control they exercised; (3) the degree of autonomous organization of the major strata and their free access to the center; (4) the interrelations between forces of centrality and concentration; and (5) the size of the society and its international standing—whether a relatively small society oriented to external markets or a relatively large one with wide internal markets. These general arguments will now be elaborated.

The heterogenetic city, characterized by a particular occupational specialization and class structure—usually a producers' city—was generated mainly by forces of concentration or by their combination with forces of centrality. As demonstrated in the case studies, the emergence of forces of concentration gave rise to types of a social—especially occupational—structure distinct from the periphery, in terms of greater economic diversification, mainly of artisan and commercial activities. A strong tendency toward more specialized and distinct social organization also evolved, based on a relatively high degree of occupational and economic specialization. The social structure of such cities differed from that of the respective centers of society because specific social carriers of centrality within them—political, administrative, and religious elites—were of secondary importance and did not play an important role in city organization or in control over urban land. Following the distinction between producers' and consumers' cities, already stressed by Weber and many other 19th-century economic historians, these were mostly producers' and merchants' cities as opposed to orthogenetic cities, which were mainly consumers' cities.

The differentiation of such cities from the centers of their respective societies was evidenced by the fact that their most active groups were distinct from the central ruling elites. Their distinctiveness from the periphery could be seen in the emergence within them of social and economic organizations—merchant and artisan organizations, associations, or guilds, and special class and elite structure. This situation could be found in other, especially rural, sectors only in embryonic form, but usually tended to be identified with the city. Accordingly, the milieu of such cities or urban quarters was characterized by mobility and relative openness of the fluctuating population and by the features often derided by moralists of various civilizations.

This distinctiveness of the social structure of cities often tended to be defined as specifically urban. Examples occurred in the Byzantine Empire, in some Islamic cities, and especially in Europe insofar as the social strata were characterized by status autonomy and autonomous access to the center and could generate relatively autonomous elites more or less independent from those of the center.

Conversely, the emergence of an orthogenetic urban society was generated largely by the forces of centrality mobilizing or attracting broader strata to the cities. The distinctiveness of the cities, mostly the so-called orthogenetic ones, generated by forces of centrality, thus differed from that of the heterogenetic cities. The sharp distinctiveness of the cities created by the centrality forces from the periphery developed in terms of a quantitatively different occupational and economic structure. These cities contained a greater concentration of those social categories that were the carriers of centrality—the political, administrative, military, and religious elites—and that, of course, also existed and functioned in the periphery but, obviously, in a minor way.

The articulators of the forces of centrality congregated more densely in such cities and rendered them distinct by epitomizing the major premises of the ruling classes of society as well as its centrality. Such a concentration of elite functions characterized the ecological distinctiveness of the forces of centrality. These cities thus emphasized their distance from the periphery by symbolizing the major premises of the society, but this distance was based on a certain basic continuity or similarity of the principles of social organization in both center and periphery.

Such cities did not usually foster the development of strong, distinctive types of economic production but tended to remain overwhelmingly consumer centers. They were imbued with the principles of upholding the social, moral, and cosmic order of society by legitimizing the political power instituted by religious sanction. These cities were characterized by a relatively strong emphasis on kinship or tribal ties and tended to emphasize strongly the basic differences between core and periphery as evidence of the center's efficient control over the entire space. It was only in quarters of the economically distinct but socially marginal groups that a "looser" atmosphere tended to develop.

The two types of urban distinctiveness are, of course, ideal ones. In concrete cases they intermingled, although one type would usually predominate. Great variations tended to emerge in their structure, however, and in their concomitant distance from center and periphery alike. Cities differed by the degree of distinctiveness of their social structure, organization, and cultural creativity. Such differences were influenced by several variables: the extent of autonomy and heterogeneity of the central elites, in relation to the cultural orientations they articulated, to the modes of control they exercised over the flow of resources, and to the types of center-periphery relations they generated. Such differences were also influenced by the relations that existed between the ruling elites and the broader strata of the population, by the autonomy and cohesion of such strata, and by the extent to which relatively autonomous elites emerged.

The distinctiveness of the orthogenetic urban centers from both center and periphery tended to weaken when the forces of centrality shaping these cities were supported by relatively embedded and homogeneous elites, and tended to become stronger when such forces were composed of autonomous and heterogeneous elites. The more embedded and homogeneous the major elites were, with little transcendental or strong other-worldly orientations, the more relatively restricted modes of control were exercised and less distinction made between center and periphery and permeation of the latter by the former. This was the case in Southeast Asia, Latin America, India, and the more Sultanic or patrimonial Islamic societies. Beyond the religious centers, special types of urban social organization emerged, distinct, to a minor degree, from both the center and the periphery of their respective societies. Even in areas where the population density was relatively high, urban groups were rarely different—and were so perceived—from those of the other sectors of society. In such cases the upper urban groups did not differ much from the ruling central elites and from the upper rural groups, and the lower urban groups were not very distinct from the rural ones.

Conversely, the more autonomous the ruling central elites were, the more they carried strong transcendental orientations with a strong this-worldly emphasis, and exercised a differentiated mode of control, the greater became the distinction between center and periphery. This was the case in all imperial and imperial-feudal societies, such as the Chinese, Russian, Byzantine, Abbassid, and Ottoman empires, as well as in feudal Europe and Japan. In these societies the social elements and types of social organization and cultural creativity promoted by the carriers of centrality generated not only quantitative but also qualitative differences in the social structure of cities from those prevalent in the periphery of society.

The extent to which such distinct forms of urban life developed within the urban centers constructed by autonomous elites varied greatly according to the homogeneity or heterogeneity of the controlling elites and the status autonomy of the major strata, all connected to the major cultural orientations prevalent in that society. The more monolithic elites tended, as was the case in China, to block the emergence of autonomous secondary elites. This also discouraged the evolution of new, specifically urban types, of social organization supported by the articulators of centrality.

Conversely, the more pluralistic their structure and the greater the number and diversity of such elites, with relatively independent bases of resources, the greater was the probability that at least some of them would generate distinct forms of social organization and institutional creativity in these cities. This situation, found in Byzantium as well as the Russian or Islamic empires and mainly in Europe, was usually connected with the prevalence of different mixtures of this- and other-worldly orientations. Many of the forms of social organization and creativity that evolved in these cities were different from those articulated by the official ruling elite of the centers, even if these elites were mostly located in the capital cities. In such cases the variety and distinctiveness of the central autonomous elites, usually concentrated in the cities, lent a specifically urban flavor to their activities and distinguished at least some of them from the more central ruling elites as well as from the upper groups of the periphery, even if they were not defined in specific urban terms.

Significantly, such tendencies could be found not only in the imperial and imperial-feudal systems but also in the more decentralized systems evolving within the framework of other-worldly Great Civilizations, whose centers and frameworks were supported by relatively autonomous cultural-religious elites. In India, for instance, it was within some specifically religious urban settings, epitomizing the distinct religious centers, that certain forms (albeit embryonic) of urban social organizations tended to develop. Moreover, the activities of such relatively autonomous central elites tended to reinforce—as occurred, in different degrees, in imperial and imperial-feudal societies in Europe and Japan—the tendencies of the forces of concentration to generate new types of social organization and class structure concentrated in the city and identified with it. The diversity of such a social and occupational structure tended to be greater wherever such forces of concentration were stronger and more autonomous and when these tendencies were reinforced by relatively autonomous central elites.

The distinctiveness of the urban structure, in relation to both the center and the periphery, was even more pronounced insofar as strata emerged having some autonomous access to the center, as occurred in Europe and to a smaller degree in the Byzantine and Abbassid empires. In such cases—and Europe is indeed the most outstanding one—a peculiar relation existed between the cities and the centers, as well as among most peripheries of their societies. Many groups, especially the lower ones, attempted to create new types of urban organization and government in which they could participate autonomously. Together with the upper groups of these cities, they often tended to orient themselves toward the center of their society, demanding greater participation in the center and attempting to shape it according to their own premises. At the same time they expressed, within the specific types of their social organization, their distance from the more passive elements of the periphery.

In Europe this situation was reinforced by three factors: (1) the emergence of numerous urban centers, stimulated by forces of concentration or their combination with heterogeneous, pluralistic centrality forces, led to growing economic specialization; (2) the strong emphasis placed on the autonomous access of the major strata to the centers of society, facilitated the growth of secondary elites and distinct class and status formations; and (3) the emergence of heterogeneous and specifically urban types of social organization was enabled by the decentralization of the political ecological structure.

In Japan and India such decentralization allowed for some development of a special urban social structure although, given the absence there of the factors mentioned above, the distinctiveness of the urban structure, generated by the forces of centrality, was smaller than in Europe. Cities could also attain distinctiveness when the forces that constructed them were moved by conquest—as in most Islamic societies—and were ethnically alien from the periphery, or in cases of enclaves of foreign merchants in many patrimonial regimes, alien from center and periphery alike.

The nature of the distinctiveness of urban structures also depended greatly on the size of the society in which they developed, on its standing in its respective national and international systems, and on the concomitant structure of the markets to which its activities were oriented. In small societies fostering a strong orientation to external markets, as was the case in most city-states of antiquity and in some of Medieval Europe's countries, the economic differences between city and countryside were smaller and greater symbiosis developed between the urban patriciate and the rich landowners. The cities then became fuller embodiments of their respective forces of centrality against the more peripheral groups.

The predominance of internal markets within different societies or regions encouraged the emergence of specific urban groups and organizations, but distinctiveness from the center depended on the other factors mentioned earlier.

The preceding analysis indicates that the emergence of each type of distinctiveness of the urban social structure and of the potential tensions and conflicts it engendered was influenced, above all, by the strength, autonomy, and diversification of the respective forces of centrality and concentration. It further points out that the strongest distinctiveness of the urban social structure existed when cities were constructed by a combination of concentration and centrality forces, either when they were supported by multiple and autonomous elites and by autonomous social classes and strata attempting to penetrate the centers, or when such cities developed in situations of relative political decentralization within common civilizational frameworks or within special types of political-ecological niches or enclaves.

C
INTERNAL STRUCTURE OF CITIES

The preceding study of the distinctiveness of urban social structures leads to the analysis of the spatial expression of such organization and distinctiveness—that is, the internal ecological structure of cities. Such a structure can best be distinguished according to three major dimensions that have been at the forefront of sociological and ecological research: (1) the nature of the internal ecological differentiation of cities, of the spatial social organization of urban areas; (2) the nature of city centers; and (3) the physical distinctiveness of cities from their surroundings.

Four aspects can be discerned in the internal differentiation of the urban ecological structure: (a) whether the city comprises the complete array of urban land uses—residential and nonresidential—or whether it is a unifunctional city (made up mainly of temples, military garrisons, and the like); (b) the degree of spatial differentiation between residential and nonresidential functions; (c) the spatial differentiation among various nonresidential functions; and (d) the principles of segregation of population in the residential areas of the city according to ethnic origin, cultural background, religious affiliation, socio-economic status, and so forth.

Of special importance in the second dimension relating to the city center were, first, the presence or absence of a well-defined center, serving and attracting population, and the degree to which it differed physically and architecturally from other parts of the city; second, the major functions of the center—political, administrative, religious, and so forth; and, third, the degree to which these functions overlapped spatially in a single area of the city center or were dispersed throughout the city.

The third dimension of the distinctiveness of the urban ecological structure found expression in the amount of space left open for public use, as well as in the level of connectivity of the street pattern.

An additional dimension was the role played by planning the construction of the city—that is, whether the form of the city was predetermined or occurred through incremental growth, by the gradual addition of built-up blocks. In preconceived cities, the plan reflected aesthetic values, security measures, social hierarchy, cosmological representation on earth, or a combination of some of these elements. The extent to which the city plan symbolized the cosmic order is of special interest—an aspect of the construction of cities that has been elaborated upon by Paul Wheatley in connection with Chinese, Japanese, and Southeast Asian cities.

Finally, the distinctiveness of the urban ecological structure is evident in the degree to which the city area was sharply defined, whether enclosed by walls or moats, or allowed to spread unhindered and merge gradually with the rural areas. This difference usually influenced the density of buildings and population in the city.

Needless to say, much variability existed among cities in the elements of their ecological structure, due to historical circumstances, features of the terrain, regional location, ethnic composition of the population, and so forth. Beyond these variations, some major contours of the basic elements can be explained by the operation of the constellations of the two major social forces, those of concentration and centrality. Regarding the distinctiveness of the urban ecological structure, the explanation will be related to the strength, diversity, and autonomy of these forces, whose impact on the structuring of the internal organization of cities was effective mainly through the modes of control over the use of urban land, in which the distinct modes of structuring the urban ecological space found their fullest expression.

The forces of centrality articulated by the religious, political, and administrative elites tended to allocate land to serve their respective functions, to the detriment of occupational and residential uses, the latter being subsumed in the former. Such subsumption was evident in the tendency to ensure close contiguity of functional and residential quarters and a predominance of political and ritual as against occupational considerations, the lower classes being forced to the outskirts of the city.

The forces of centrality also tended to structure the control of urban space so as to differentiate between the areas devoted to religious and political functions and those allocated to economic, commercial, and manufacturing functions. Commerce and manufacture tended to be relegated to the city periphery or to the edge of areas allocated to the former functions. In those peripheral quarters, the economic functions were allowed some ecological segregation. The forces of centrality were also operative in constructing distinct city centers, whose nature and diversification depended on these forces' structure and on the major orientations they articulated. The degree of internal differentiation of the spatial organization of the various functions of centrality was related primarily to the social organization of the elites, mainly to their autonomy and diversification, their cultural orientations, and to the extent of the major strata's free access to the respective centers of the society and the city.

Some important differences arose, stemming from prevalent cultural orientations. In societies such as those in Southeast Asia and Meso-America, in which no tension existed between the transcendental and the mundane orders and which were characterized by the predominance of embedded elites, a strong concentration of religious and political-administrative functions tended to evolve in a single, relatively undifferentiated compound. In the Hindu and Buddhist societies, in which strong tensions existed between the transcendental and the mundane, a perception of such tension arose from the other-worldly orientation of autonomous religious elites, the religious and political functions were usually spatially segregated. They were both, however, still located in one broad, central compound, separate from other parts of the city. In purely religious centers of such societies, the religious function, of course, predominated, but in political centers the religious and political functions tended to intermingle. In some peripheral compounds, often of a lower order of holiness and purity, attempts were made to replicate such patterns on a smaller scale.

In Latin America, where embedded political elites with relatively strong other-worldly orientations were guided by the distant imperial center, cities were highly planned. Their centers evinced little internal differentiation, and usually the religious, political, and administrative functions were concentrated within a single central space.

Autonomous elites, exercising differentiated modes of control, inspired more differentiated patterns of land control and spatial organization. Among the societies under study, some important differences can be discerned in the homogeneity or heterogeneity of the prevalent elites, their cultural orientations, and their consequent attitude toward the organization of space. In Chinese cities, where tension existed between the transcendental and mundane orders, together with a strong this-worldly orientation articulated by a relatively monolithic elite, spatial concentration of the central functions existed in most cities, especially in the capitals. Compared with Southeast Asian cities, Chinese spatial concentration was characterized by greater internal differentiation between ritual and administrative functions, and by a kind of organized gradation of the economic and residential quarters according to their hierarchical standing in the basic social structure.

In Christian and Islamic civilizations, where multiple autonomous elites articulated this- and other-worldly orientations and exercised different modes of control (usually related to qualitative differences between center and periphery), a tendency evolved toward some dispersion of the central functions. Many differences arose among societies based on the elites' structure and orientations, especially the degree of interweaving or separation between their this- and other-worldly orientations, as well as on the relations among the elites. Such centers tended to assume a more unitary urban character in cities where political and military functions predominated. The more closely the major elites were interwoven—as were their this- and other-worldly conceptions of salvation—the stronger would be their tendency to combine the various functions into one major (if highly differentiated) center, which formed an integral part of the town fabric. Latin American cities of the Spanish Empire provide a good example of such centers, the major plaza containing the cathedral, the governor's palace, and the municipality. This spatial concentration with clear differentiation between the major functions existed even earlier in many medieval towns of Western Europe. A similar pattern was also found in the Byzantine Empire, especially in Constantinople.

Conversely, the more segregated were the elites—as in Russia and in later Islamic societies—the more dispersed were the central functions, some of which were completely detached from the main town fabric. This was sometimes accentuated by a physical separation, as in the walled compound of the *kreml* or the royal citadels of the late Islamic capitals. The difference between Istanbul, the most imperial of Islamic capitals, and Constantinople in this respect is very enlightening.

Modes of spatial structuring of the central functions were also closely related to the degree to which cities represented the cosmic order spatially, both in the location of religious structures and places of worship—a universal urban feature—and the degree to which representation of the cosmic order constituted a basic component of the city's spatial organization. This was related to a central dimension of a society's cultural or cosmic orientations—namely, the perception of a deep schism between the transcendental and mundane orders and the doubt about the possibility of bridging it in the mundane sphere. Such a schism was weakest, and the stress resulting from attempts to bridge these orders was equally weak, among the so-called pagan societies or in those in which the tension between the transcendental and mundane orders was couched in metaphysical but not in religious terms. Conversely, the recognition of such a schism was highest in monotheistic religions, in which God was conceived as the Creator of the Universe and the task of bridging between the transcendental and mundane orders seemed, if not impossible, at least very difficult to achieve. This formulation of space prevailed in Southeast Asia, China, Japan, and in urban concentrations such as those of the Yoruba in Africa and in pre-Colombian cities in Meso-America.

As has been stated, the Southeast Asian, Chinese, and Japanese cities reflected the cosmic order in their spatial organization. Angkor Thom, Ch'ang-an Bradabur, Kyoto, and Madurai are striking examples of cities that were planned as physical representations of the cosmic order on earth, their sites being at the intersection between the cosmic and the earthly planes. Physically symbolizing these intricate religious concepts were the street plan and the architecture; the number, size, and shape of buildings, gates, and pathways; the height and shape of towers and roofs; the location of open spaces; and the town's geographic orientation and water bodies.

In cities within the Islamic and Christian realms, special sites were allocated to places of worship and religious institutions that often dominated the urban structure and skyline and influenced the internal ecological pattern. Islamic cities were often created by political-religious impetus; the location of the central mosque and citadel shaped the cities' ecological pattern. Moreover, as Olec Grabar persuasively demonstrated, the pattern and architecture of buildings in such cities strongly emphasized the basic concept of the religious and political domination of Islam. Similarly, the famous grid pattern of Latin American cities represented the basic sociopolitical conception prevalent in Spanish colonial regimes.

The tendency to control land use and the consequent structuring of the internal urban space generated by forces of concentration operated in a different manner than those generated by forces of centrality, that tended to give rise to greater spatial differentiation of social and economic functions as well as between functional and residential quarters, which were rather limited in traditional societies. In most cities the areas devoted to trades and crafts mingled with the residential ones. This rule also applied to those cities in which spatial differentiation existed among economic activities, such as quarters of the goldsmiths, money changers, tanners, and so forth. These functional differentiations usually reflected the ethnic segregation of the various groups, each of which specialized in a specific economic activity.

Far-reaching differences could develop, however, in the degree of differentiation, both within the functional and between the functional and the residential spaces, depending on the extent of the city's economic development and differentiation: The greater such development and differentiation, the greater the distinction between these functions. This differentiation also depended to some degree on the autonomy of the major urban strata and elites and on the extent of specialization. The more autonomously specialized and heterogeneous the elites were (as was the case in Western Europe, the Byzantine Empire, and somewhat in Russia and Japan of the Tokugawa period), the greater the spatial differentiation of such functions. Conversely, the more embedded the elites were (as was the case in Latin America, most Islamic cities, and India), the smaller the spatial differentiation of the social and economic functions. The more developed and autonomous were the different forces of concentration, the more they would attempt to permeate the center, as occurred in the Byzantine and early Islamic empires and in Western Europe. A unique pattern of spatial differentiation and distinctiveness evolved in these societies, especially in Western Europe, characterized by the combination of a strong tendency toward spatial differentiation of the different economic functions and residential quarters and a parallel tendency toward the construction of a distinct center. The economic and internal civic functions and the more specific political and religious functions of centrality tended to combine either in a single major center or in closely connected ones. Insofar as cities were politically autonomous, as in the Italian city-states, the political and civic functions tended to be located in the same urban space. In capital cities of empires or absolutist monarchies, the spatial location of these two functions tended to be more distinct.

An important aspect of the ecological distinction of cities was that the allocation of public spaces was made in accordance with the street pattern, primarily the degree of connectivity of the streets so as not to create a predominance of cul-de-sacs. Preservation of public spaces and the connectivity of the road network were enforced by the authority of the center—imperial or patrimonial—as well as by the corporate identity of the elites and broader strata, and by their mutual relations. The stronger and more distinct the centers and the common identity and corporateness of the urban strata and elites, the larger and more defined the public spaces of a city and better the connectivity of its road network. In both imperial and patrimonial regimes—such as Spanish Latin America, in which the center was strong and preserved its dominance—a well-defined and structured system of public spaces and a high degree of street connectivity tended to evolve, at least in the central parts of the city. Conversely, when the center weakened, as in many Islamic polities, especially when such a weakening was combined with a gradual decrease of the major elites' distinctiveness and autonomy and with their growing fusion and embedment in the respective strata of the society, there was less structuring of public spaces. At the same time, individuals and private groups tended to usurp and include public spaces in their own properties or encroach upon such spaces, public avenues, and roads, thus causing less connectivity of the streets and creating numerous cul-de-sacs.

Different constellations of concentration and centrality forces shaped other aspects of the internal ecological structure of cities—namely, their separation from the countryside, often expressed by the erection of city walls. Such ecological distinctiveness of the city from the countryside depended greatly on the degree of differentiation of the city's social structure, which was influenced largely by the combination of economic development and structure of elites and of center-periphery relations. This separation was more pronounced in cases of sharp distinctiveness of the cities' social structure in imperial and imperial-feudal regimes and much weaker in patrimonial polities. The separation of cities, expressed by the building of walls and moats around them, was strongly affected by external and internal security considerations. The more open to invasions or plagued by internal strife an area was, the more a city tended to separate itself from the countryside by walls that secured it from external threats and completely controlled movement into it.

It can therefore be concluded that the modes of distinct ecological structure of centers in many ways followed the pattern of their social and occupational distinctiveness and were similarly shaped by the strength, diversity, and autonomy of their forces of concentration and centrality and by their interrelations. The general tendencies toward structuring the cities' internal ecological structure were indeed shaped by these forces, yet the urban concrete contour, the distribution of quarters, and the level of population density were influenced by the balance between demographic forces and economic opportunities. Needless to say, all these tendencies could greatly vary from one city to another within the same society and could change in the course of the same city's history.

D
URBAN AUTONOMY—INTERNAL AUTONOMY OF CITIES—PATTERNS OF REGULATION AND SELF-REGULATION, AND SOCIAL CONFLICTS

The distinctiveness of the cities' social organization and ecological structure gave rise to crucial problems in the cities' relations with their surroundings and with other sectors of the society, mainly with the centers of society and its periphery. This interaction generated one of the most socially and politically visible aspects of the distinctiveness of cities: their autonomy in relation to other sectors. Such autonomy and the tensions it engendered emerged from the conflicts between tendencies, which were inherent in the crystallization of the different aspects of the cities' social and ecological distinctiveness. The concentration of a large population and especially the emergence of a varied occupational and class structure gave rise to problems of social control and regulation that were of special importance to the centers of these societies. These problems may have generated tendencies toward regulation by the urban groups, both of their own internal affairs and of specific urban issues, those of the city as a whole. Tensions may have arisen between the tendencies of the center and those of the different urban groups, exacerbated by the urban groups' inclination to participate not only in the regulation of their affairs by the center but also to take part in the center itself. These tensions constituted one of the major foci of protest and conflict in the cities, and these tendencies, as well as the conflicts erupting among them, were influenced by the various constellations of concentration and centrality forces. The interrelations between these constellations, operating in different political-ecological settings, shaped the various patterns of regulation or self-regulation and the tension between them.

The tendency toward active self-regulation of specifically urban affairs was developed especially among the urban groups formed by the forces of concentration, which were the more differentiated economic and occupational groups, but much less by the urban groups generated by the forces of centrality, which were the primary political and religious elites. The degree to which such tendencies toward self-regulation developed fully depended on the strength and autonomy of these groups. The more they were embedded in broader ascriptive elites—as in Indian cities for instance, where the social groups were organized on a caste system—the smaller was their specifically urban type of self-regulation and the weaker their corporate urban identity. In general, self-regulation was relatively weak in most of the cities in patrimonial regimes and was limited to neighborhood, occupational, or ethnic subgroups. It usually was not very effective and rarely structured according to specific criteria of the urban setting. The more autonomous such groups were—above all in imperial and imperial-feudal regimes—the stronger was their tendency to corporate organization, to some degree of self-regulation, and to the generation of some corporate urban identity, and so were, potentially, the tensions between these tendencies and rulers' attempts to regulate urban affairs.

The rulers' regulation of urban affairs can be characterized by its scope which may have ranged from external affairs to aspects of the municipal or specific economic and professional-commercial activities of urban groups, guilds, corporations, or of individual merchants and craftsmen. Such regulation may also be characterized by the degree of the controllers' penetration into the internal affairs of the major urban groups.

The various case studies in this book indicated that the scope of regulation of the urban structure by central authorities depended on the compactness of the boundaries of the political regime. They also indicated that the intensity of such regulation depended on the mode of control exercised by elites carrying different cultural orientations, as well as on the relations existing among the ruling elites and the broader social strata.

The scope of regulation of urban affairs by a central government was more encompassing in compact than in decentralized systems and was inlfuenced by the degree of distinction of the center from the periphery. Perhaps more important, it was influenced by the internal strength of the center. Wherever the center was less distinctive and had less autonomy, as in most patrimonial regimes, in which the major elites were embedded and exercised a more extractive mode of control, the scope of such control was smaller and confined mostly to external affairs, tax collection, and keeping the peace. Even in such cases the degree of the center's internal strength influenced the scope of such regulation. The control was certainly much stronger, encompassing different aspects of social and economic life of the urban groups in Spanish Latin America, where the centers were stronger than in Southeast Asian, Indian, and Islamic societies, whose centers were usually much weaker.

The intensity of such regulation—that is, the degree of penetration of the centers into urban groups' and of cities' internal affairs—depended on the autonomy and structure of the society's ruling elites. The more autonomous and heterogeneous they were, the greater was the intensity of their regulation. The more diversified were the modes of control exercised by the elites, the more distinctive were the centers developed by them. The greater the distinction between center and periphery, the stronger the center's tendency to regulate different aspects of the urban structure and to supervise many, if not all, the urban groups' activities. Accordingly, in Southeast Asia, India, the Sultanic Islamic countries, and (to a smaller degree) Latin America, the tendency toward overall intensive regulation of urban groups by the central authorities also tended to be weaker and limited to extracting taxes, maintaining peace and order, and upholding the major symbols of the polity.

Conversely, in those societies in which autonomous elites exercised different modes of control and generated distinctive centers—that is, in imperial and imperial-feudal decentralized centers and, to a smaller degree, in the religious centers of decentralized other-worldly civilizations, as in India—these elites tended to exercise more intensive regulation of the urban groups. The intensity varied according to the structure of the central ruling elites, their orientations, and the modes of control they exercised. In the case of more monolithic elites, carrying this-worldly orientations, regulation was relatively weak and the more or less distinct urban groups that emerged, whose internal structure did not entirely differ from that of the periphery—as was the case in China—enjoyed comparatively wide autonomy in the regulation of their internal affairs, subject, of course, to overall regulation by the center. In contrast, the intensity of such regulation was greater, and the smaller the degree of autonomy of the urban groups, the more heterogeneous these elites were. But these tendencies could be counteracted by those of other urban elites and groups toward self-regulation and toward participation in such regulation.

The extent of participation by different sectors of the urban society in such regulation of urban affairs can be explained by the combination of rulers' tendencies to regulate urban affairs and the trends toward self-regulation of the different urban groups. The scope of this participation was influenced by the structure and orientation of major elites and their relations with the major strata, as well as by the degree of the latter's autonomy and the extent of their autonomous access to the center.

The more monolithic these elites were, the smaller was the urban groups' official participation in such regulation. The more pluralistic was the structure of these elites, articulating the centrality of their society, the more numerous were the central subelites and the more autonomous the major urban groups generated by the forces of concentration. That is, the higher the degree in which they could construct their own criteria of membership and generate elites with autonomous access to the centers, the greater their tendency to attempt to participate in the central regulation of urban affairs through their own corporate activities and through their representative elites, acting in the respective centers. It was also in such cases that a relatively strong tendency emerged toward crystallization of corporate urban organization and identity.

Such cases occurred to some degree in the Byzantine and Russian empires, in the Islamic kingdoms of the Abbassid period, and in Western and Central Europe, where tension developed among the tendencies of many elites and subelites to enlarge their participation in this regulation and to develop self-regulation in the major activity spheres and the activities of elites at the center to control such regulation. This tension gave rise to manifold conflicts and movements of protest, whose concrete outcome depended on the balance of strength between the various forces, a balance necessarily changing over time in any society.

The nature of the confrontation among the different tendencies—that is, those of the external and internal central authorities to regulate the cities, and of the different urban groups toward self-regulation and participation in the central groups' regulation of urban affairs—certainly influenced the nature of the social conflicts that emerged in cities.

Patterns of Urban Social Conflicts

Social conflicts, struggles, and protests were endemic in cities, occurring among the central ruling groups, between lower and upper urban groups (some types of class conflicts), and between the lower and upper groups and the central authorities. The distinctiveness of urban life was manifest in those conflicts specific to the urban setting and in relation to other sectors of society in general and to the center in particular.

The possibility of eruption of such distinct conflicts and protests was rooted in the very nature of the cities' relations to the centers and peripheries of their respective societies. Hence, the nature of such conflicts was influenced by the same conditions that shaped the distinctiveness and autonomy of cities.

The conflicts that occurred in cities can be distinguished according to the probability of their eruption, to their position in the city's social structure, and to the ideologies generating their occurrence. The very existence of a greater concentration of population, of the many heterogeneous groups within it, and of the prevalence of high social density rendered cities even more prone to the eruption of conflicts. In the case studies numerous illustrations of the endemic nature of these conflicts and struggles were given. Their contours differed in the various types of cities in close relation to the dimensions of their urban social structure's distinctiveness. The institutional frameworks of such conflicts, struggles, and protests and the symbols through which they were articulated—especially class symbols—depended mainly on the autonomy of the elites and strata and on the degree of their autonomous access to the center.

Such conflicts and protests derived from changes in the balance of power among the different groups and social elements and on the intensity of the control wielded by the ruling elites. The results of such conflicts varied greatly, however, depending on the modes of control exercised by the different types of elites. The least distinctive types of urban conflicts were the popular rebellions of urban elements—demanding, for instance, the distribution of food—that were usually generated by lower sectors of the forces of concentration. These did not express distinct political demands or some form of broader social or political consciousness. Such a consciousness, often couched in terms of class conflict, struggle, and protest, tended to emerge in cases in which more distinctive urban conflicts erupted, and were focused on the different dimensions of urban autonomy—the structuring of power within the cities—on the participation of various (especially lower) groups in urban government or on the relation of such power to that wielded by the centers of society. These specifically urban types of conflict and protest were influenced by the same conditions that gave rise to the different dimensions of urban autonomy—namely, the relative strength, diversity and autonomy of the forces of concentration and centrality and their interrelations.

Popular rebellions and uprisings were thus relatively predominant in cities in patrimonial and patrimonial-like societies, such as Southeast Asia, Latin America, or Sultanic Islam. These cities were characterized by the predominance of embedded elites, little distinction between center and periphery, and by a relatively low level of social heterogeneity.

E
URBAN AUTONOMY—
EXTERNAL POLITICAL AUTONOMY

Just as different types of distinctiveness of urban social structure combine in various ways with the dimensions of regulation and self-regulation of urban groups, so do all of them combine in different constellations in the external autonomy of cities and urban communities. This autonomy is manifest in the degree of the cities' political independence with respect to the overall external affairs and their relations to broad political units. External political autonomy may be of several types, three of which were prevalent in history. One type of independence existed in cities described as sovereign or semisovereign. This sovereignty was most developed in ancient city-states and, to a smaller degree, in some European towns, especially Italian, Flemish, and Northern German Hanseatic. Such cities usually differed by some internal characteristics from the other types of sovereign entities incorporated in international systems. A second type of political independence of cities, sometimes resembling the first type, could be found in some parts of Medieval and early modern Europe: the relative autonomy of a city regarded as a distinct corporate political unit but not wholly sovereign. Such units, defined mostly in terms of their urban nature, enjoyed some freedom or autonomy vis-à-vis the prevailing sovereign power whose sovereignty they acknowledged but from which they derived or usurped their own distinct corporate identity and their internal and, at least partially, external identity. The third type, found in Latin America, sometimes in India, and in some Islamic countries, was characterized by de facto usurpation of central power by various magnates or potentates located in the cities. This group and its activities, however, were not necessarily perceived or defined as specifically urban.

The most important conditions encouraging the emergence of such political autonomy were geopolitical, mainly some kind of decentralization of the internal political or international systems, within which the cities tended to develop. Those three different types of political autonomy, however, emerged in various kinds of decentralization. The crucial difference was found among the conditions giving rise to the first type of independence (city-state) as opposed to the other two types, although some admixture or overlapping may have taken place, as occurred in Europe—for instance, in Venice or other Italian city-states or in India. The city-state type tended to arise and persist in special political ecological niches of heterogeneous international systems and in the merging of these systems in which strong political powers interacted but did not necessarily constitute part of a political or even civilizational framework. The ability of such cities to survive as independent units often depended on the strength of their respective hegemonic international centers, most of which ultimately lost their sovereignty and were incorporated into other components of their international system, mainly patrimonial kingdoms or empires.

The other two types of the cities' political independence arose and persisted within relatively more unified internal political or civilizational frameworks. Such independence prevailed in situations of either decay or decline of compact centralized systems, or in relatively decentralized ones constituting parts of somewhat unified civilizational frameworks and characterized by noncoalescing boundaries of the different collectivities and cross-cutting markets.

The first situation—found in Southeast Asia, the Islamic countries, Latin America, and seldom in late Byzantium or China, in periods of decline of the central authority—usually occurred as a result of a weakened central autonomy of either patrimonial or imperial states and of subsequent contraction of their markets. It usually led to usurpation by different urban groups within their municipal or regional confines of the regulation exercised by central authorities of both the internal and external functions. The different types of center-periphery relations, prevalent in patrimonial and imperial regimes, greatly influenced some of the concomitants of this usurpation. In patrimonial systems, characterized by little distinction between center and periphery—as occurred in Latin America or Sultanic Islam—local or regional regimes centered in cities developed. They were supported by a combination of rural landlords and urban notables, sometimes in cooperation with military potentates, who attempted to establish some mini-patrimonies of their own. In such situations the external, mostly trading enclaves existing in many such systems may also have become more independent. In imperial systems, in addition to these manifestations, such usurpation could sometimes be connected with the emergence of a specific urban political identity, similar to that found in city-states. This was especially the case in the late Roman and Byzantine empires, where the tendency was close to the heritage of city-states or to attempts to reestablish imperial centers.

The second situation, found in Europe and, to a lesser extent, in India and Japan, developed within a decentralized political setting that possessed a relatively unified civilizational framework. This type was characterized not only by the weakening of central power but also by the crystallization of multiple centers bound together in a common civilizational framework. In such situations, a certain corporate distinctiveness and organizational autonomy may have easily developed in cities, varying from place to place according to the power relations of the respective political actors. This distinguished the cities from the other political settings in the same broad frameworks.

The nature of such autonomy and political organization—especially of the degree to which its political corporateness was defined in specific urban terms—greatly depended on the structure of the predominant elites, the modes of control they exercised, the cultural orientations they articulated, and the types of center-periphery relations they generated. The comparison between India and Europe—and, to some degree, Japan—which possessed decentralized political systems within a common civilizational framework, emphasizes the importance of the structure of elites to the development of these different dimensions of urban autonomy and identity.

In Europe strong emphasis was placed on combined this- and other-worldly orientations to salvation, articulated by multiple, autonomous, and mutually impinging elites. Relatively strong stress was also placed on the autonomous access of different strata to the major attributes of salvation, to the centers of power, and to the social order. A strong distinction existed between center and periphery, combined with a multiplicity of primary and secondary centers. All these led to a tendency toward a corporate urban political autonomy, defined in distinct urban terms, combining sociopolitical and socioeconomic components, and a perception of the city as a distinct corporate unit enjoying some freedom or autonomy vis-à-vis the prevailing sovereign power, whose sovereignty was, however, acknowledged.

Conversely, in India strong other-worldly orientations prevailed, carried by elites embedded in broad ascriptive groups. Hence, neither the type of urban identity related to the political autonomy of cities found in feudal Europe emerged there, nor that which developed in the ancient polis. In Japan the situation was rather mixed. In some periods, especially the pre- and early Tokugawa periods, some autonomy did evolve because a smaller number of elites existed who were much less autonomous and were guided more by commitment to the center than by a clear perception of tension between the transcendental and the mundane orders. Hence, in periods of relative decentralization, some strong autonomous urban corporate structures emerged in Japan but did not develop a strong corporate political and social identity.

The definition of external urban autonomy, in specifically urban political terms, thus tended to be strongest in situations of relative decentralization in a common civilizational framework; Europe is the best case in point. There the ruling elites were autonomous and heterogeneous and the major strata were relatively autonomous and had access to the center of the society in which existed a multiplicity of centers. But even in Europe, where these conditions were most prevalent, they did not always lead to a full-fledged crystallization of all the dimensions of autonomy. Local conditions there were often of crucial importance, especially the political situation and the balance of power among social groups, as well as the placement of the respective cities in special international niches. Insofar as a strong political unification developed—as under absolutist rulers and, in such countries as England, even before that period—a certain weakening of the major dimensions of the second type of urban autonomy occurred, but the element of urban self-identity persisted. Here also the international placement of cities—perhaps best illustrated in the case of Venice—could also have influenced the development of such an autonomy in the direction of a city-state identity. Similarly, different power relations between upper landed and urban classes in their respective regions could have greatly influenced the degree of political distinctiveness of the cities, which was lowest when a close interrelation existed among the ruling groups and highest, often combined with intense conflicts, when the upper groups were relatively distinct.

F
URBAN IDENTITY

The various modes and degrees of cities' distinctiveness usually manifest themselves in elusive but very crucial aspects of city life—patterns of urban identity and moral evaluation of cities—or are closely related to them. These two dimensions of city life have been researched much less systematically than the structural dimensions we discussed, and it is therefore more difficult to present a systematic and well-founded case. The following analysis assumes the form of conjectural hypotheses on which further research could be based rather than of fully documented conclusions.

Needless to say, some urban identity, bound with local patriotism, evolved in conjunction with population concentration and the attainment of some distinctiveness from other sectors of society. The nature of that specific urban identity, however, was contingent on that of the urban society's distinctiveness, both from the center (political and religious) and from major groups of the periphery. Such an identity, however, may have been of a different order or may have comprised certain components, denoting a different type of "distance" of the urban setting from the center or from the periphery, each of which was related closely to the aspects of urban life analyzed above. This identity thus may have been couched in terms of local attachment or patriotism or through distinct dimensions of internal and political autonomy. Beyond these components, such an identity may have developed in a specific direction, the city being conceived not only as the area where broader societal forces found more articulated expression but also as the carrier of a particular political culture, of a distinct type of sociopolitical creativity.

These components of urban identity naturally developed under the same conditions that gave rise to the different aspects of urban social structure. The degree to which this identity was couched in terms of a distinct political type but not of distinction between the urban and rural sectors was most prominent in city-states, especially those of antiquity. This was closely related to the fact that cities constituted autonomous political entities. These cities functioned as special enclaves in international economic and political systems within which they were oriented toward external economic and political markets that usually minimized the distinction between city and countryside and between upper urban and rural classes. Hence the internal social structure of such city-states tended to differ from that of kingdoms or empires, but not necessarily from their own hinterland. Indeed, one of their outstanding characteristics was their tendency to embrace within their political confines combined urban and rural elements. The same population also tended to fulfill both rural and urban functions, the difference being more apparent in the setting within which these functions were performed and less so among the different social groups.

The self-identity of these cities as sociopolitical communities was hence, as indicated by M. J. Finley (1973), couched less as an urban than as a rural setting and more in terms of a distinct sociopolitical form, separate from the other types of political systems with which it interacted in a common international system. This type of collective identity was reinforced in cases in which the prevalent type of cultural orientations, structure of elites, and consequent center-periphery relations differed from those existing in other parts of the international system. Significantly, once the city-states lost their political independence and became incorporated into wider, culturally nonalien political units, this type of urban identity veered toward local attachment or patriotism, or assumed some local or historical distinctiveness within the framework of a new common political and civilizational setting.

Conversely, the tendency toward a specific urban identity, expressed mostly in terms of distinction from the rural setting, may have been greater when specific economic and possibly political and religious activities developed in cities (especially heterogenetic ones) as a social and class structure distinct from that of the periphery, constructed by differentiated occupational groups and classes. In contrast, the urban identity that developed in the more orthogenetic cities emphasized the articulation of central values and symbols of society. The degree to which such types of urban identity were connected with the political orientations specific to the urban setting depended, however, on the type of elites, their orientations, and the center-periphery relations they engendered within their respective societies. Wherever the elites were more heterogeneous, a more distinct type of urban identity tended to develop.

The emergence of a differentiated urban identity, which included distinct political components together with an emphasis on differentiation from the rural setting, was thus weaker in most patrimonial societies. Such was the case in Latin America, Southeast Asia, India, and most Islamic societies, in which embedded elites predominated and little distinction existed between center and periphery. Common to most of these patrimonial cities was the fact that unless they constituted enclaves inhabited by foreigners, their social structure was similar to those of the broader strata of their societies and their centers. The elements of urban identity evolving within these cities hence were expressed as local identity and attachment to place, quite detached from the central powers and not as a distinct urban community or entity. In this case the major exceptions were the external urban trading enclaves found in many patrimonial societies (mostly in Southeast Asia), which were distinct not only economically but also ethnically and religiously from both the centers and the periphery. It is obvious that a strong, distinct identity developed here, but the specific urban element was probably secondary in structuring this identity.

Conversely, the emergence of a specific urban political identity, as occurred in the Byzantine and Abbassid empires and in Western Europe, tended to be greater when the center was distinct from the periphery and a relative plurality of centers and elites existed, as well as autonomous strata. In Europe this tendency was also enhanced by political decentralization within a common civilizational framework and cross-cutting markets.

Under these broader social conditions a distinct type of collective political identity may have emerged—beyond the types of urban identity closely related to the aspects of internal organization and political autonomy of cities already analyzed—probably constituting the fullest manifestation of a distinctive urban institutional creativity. Such an identity crystallized around a special type of political urban consciousness that may even have been transposed to the centers of the respective societies. This identity differed in the emphasis placed on the civic order's autonomy, the common civic good, and citizen participation in governance of the city or polity.

This type of specific urban identity evolved in the classical city-states of antiquity and in those of Medieval Renaissance Europe. As previously indicated, this identity arose under specific conditions in which differentiated modes of control were exercised in societies characterized by a multiplicity of autonomous elites with strong transcendental orientations and strong, sometimes predominant, this-worldly elements. These conditions also gave rise to a symbolic distinction of centers, autonomous strata having independent access to them, existing either in political-ecological enclaves or in situations of political decentralization or multicentricity within broader civilizational frameworks.

The emergence of such a specific civic consciousness, and culture of civility, became especially articulate within such frameworks when cities developed due to a combination of the forces of centrality and concentration, primarily the latter. This tendency was expressed by attempts of the active urban social and cultural elites, rooted in the forces of concentration, to permeate the center of the city or polity and imbue it with their own orientations.

The combination of such forces gave rise to new types of political orientation or ideology and to such values as freedom and openness, which focused on the city. The city was regarded by urban elites as the major arena for the implementation of these principles and for the development of specific civic identity and cultural consciousness of civility. Such a civic identity evolved as a result of the political independence of the urban settlement combined with its particular social structure, political symbolism, and the structure of authority.

In the classic city-states, such political symbolism was based on several important principles. Foremost was the recognition of the moral order's autonomy as distinct from the tribal or social order, coupled with a quest for the integration of these orders through the autonomy of the individual. Another principle was recognition of the possibility that tension and conflict existed between the moral and social orders; an example is the moral protest of the autonomous individual depicted in Greek tragedy. This quest for integration was also based on the conception of complete identity of the social and political orders and on the realization of intensive participation of citizens in the body politic.

It was in the socially and culturally diversified city-states that these orientations ushered the breakdown of some of the structural and symbolic limitations of traditionalism. This was evident in the fact that the periphery was able to participate in the center's formation and that the very contents of the sociocultural order were no longer considered as given but as susceptible to change. Many of the components of political symbolism were later transposed to the European cities, where they became interwoven with the specific elements of the European political tradition, facilitating the emergence of a new civic culture and consciousness, the culture of civility, in conjunction with the different dimensions of urban autonomy. This was also connected to a specific corporate identity based, in extreme cases, on the *conjuratio* or on other similar types of organization of the urban community. Significantly, the very conditions that led to the emergence of these tendencies also exacerbated the level of social and political strife and gave rise to those types of urban conflicts previously analyzed.

G
MORAL EVALUATION OF CITIES

Analysis of the different types of urban identity, the development of civic identity or the identity of civility, and the patterns of urban social conflicts, leads us to the different kinds of moral evaluation of cities and the conditions that gave rise to them. When speaking of the moral evaluation of cities, we do not mean the rather natural feeling of distance or difference between country and city, usually related to some moral evaluation of each other. Such a feeling need not be universal and, insofar as can be ascertained, was probably rather weak in ancient city-states, in some of the urban centers of the Renaissance, in India, and in some Islamic countries, where continuity of the social structure and habitat existed between city and country.

Whatever the degree of such natural, relatively diffuse, and not fully asserted attitudes, however, they should not be confused with the more articulated moral evaluation of cities, created by distinct intellectual and political elites, expressing such attitudes in sophisticated terms while often feeding on more general popular orientations. Of course, such articulated evaluations may have been both negative and positive and quite often were confronted controversially in the same historical context. They varied in different civilizations and at different periods of their development in their degree of articulation as well as in the importance of the positive evaluation. What were, then, the conditions that gave rise to a relatively highly articulated evaluation of cities and to either a positive or a negative articulation? We can only offer preliminary conjectures that we hope may stimulate further research.

It seems that the conditions generating the types of moral evaluation were indeed similar to those shaping the different dimensions of urban distinction and identity and their various constellations. A highly articulated moral evaluation of the city, whether positive or negative, was mainly the concern of autonomous elites who attempted to impose their own orientations on the construction of the central moral orders of a society and on its own respective centers, which they regarded as closely related to cities or located in them. It was therefore natural that a positive or negative moral evaluation of cities, and their frequent confrontation, tended to be articulated most often in situations of competition or contention among such elites.

This was why the tendency toward such an articulation was relatively weak in patrimonial societies. The predominant elites were embedded there and, accordingly, little distinction existed between center and periphery and also little continuity or similarity between the social structure of the city and countryside. Even in those societies the usually negative attitude toward the city could be fed by concentrating foreign elements in the cities—conquerors or foreign merchants—or whenever an overwhelming economic concentration occurred that could not be controlled by the center.

Contrasting with this muted evaluation of cities, a more articulated positive or negative evaluation emerged in societies that were characterized by a plurality of autonomous, mostly political, religious, and intellectual elites, in which a concomitant strong symbolic but also organizational distinction existed between center and periphery. Different groups in such societies tended to uphold either a positive or a negative evaluation of cities, and these different views could clash in political or literary controversies. The specific relations between cities and the social and sociopolitical orders of their societies, or of society in general, as perceived by these elites constituted the major parameters of such a negative or positive evaluation.

A negative evaluation of cities was usually based on the claim that city life undermines and corrupts communal solidarity and morality, on the fundamental precepts of the moral and political order and of the ruling elites, and on their adherence to these precepts. This negative view endured from the time of Cato (2nd century B.C.) until the romantic negative evaluation of modern society and cities. It tended to be articulated mainly by those elites representing the forces of centrality or communal solidarity, who saw that their position and orientations or access to the center were weakened or threatened in two ways by impinging new social forces. This threat could have arisen when, within a society and particularly around its cities, direct clashes occurred between the forces of centrality and concentration and among the elites articulating and organizing these forces. Such clashes could also occur in various interconnected or separate situations when the forces of concentration expanded quantitatively, when heterogenetic cities or sectors so extended their scope that they could no longer be regulated by the respective (political or religious) centers. These cities then claimed for themselves wide negative autonomy (that is, noninterference by the central elites) while being seen, especially by the secondary elites, as undermining the structure of the countryside.

The first type of such clashes tended to intensify, in Europe and the Byzantine and Roman empires, when the forces of concentration impinged on the center, claiming a share in its construction according to the specifically urban values they represented. In such cases the negative evaluation of cities was articulated by the primary or secondary elites representing the traditional values of centrality or communal solidarity and expressing the negation of the "open" urban life-styles as against the traditional charismatic values of the broader moral and sociopolitical orders of the society, of its center, and, possibly, also of the periphery.

The other type of negative moral evaluation of the city was expressed by traditional, religious, and sociopolitical secondary elites, mainly when imperial regimes solidified or declined and such elites felt too removed from the center to be able to influence it. This pattern of criticism occurred in all civilizations; its fullest illustration can be found among the preachings of the Hebrew prophets and in the writings of Ibn Khaldun, who regarded the cities as corrupters of the ruling elite. They considered that city life led such elites to forego their adherence to pristine values and to precepts of the moral order of society. The people articulating this negative evaluation of cities usually advocated renouncing the luxuries of city life, adopting the simplicity of the tribal, rural, or sectarian life and returning to the moral order of society.

A positive evaluation of cities tended to evolve, in contrast to the negative one, extolling the city as the locus of special virtues, cultural and political creativity, the moral order of society, or as a manifestation of the cosmic order. Such a positive evaluation of cities was usually articulated by those elites who were either incumbents of the respective centers or were able to construct them in accordance with their own broader orientations and premises. There may have been two types of such elites and their concomitant positive evaluation of cities. The first included those groups that were the carriers of forces of centrality representing a broader traditional or charismatic order, not specific to the cities but located within them, and who tended to eulogize the virtues of the city as the epitome of such charismatic forces of centrality. The second type comprised the elites representing and extolling the specific urban life-styles derived from the development of forces of concentration and attempting to permeate these elements into the central premises of society. These elites praised the cities as loci of distinct political or sociopolitical culture or identity as well as of freedom, openness, and civility, which distinguished them from other kinds of social and political structure of regimes. In the ancient and medieval city-states and primarily in modern Europe, such lauding of the virtues of the city's civil and moral orders may have been transposed to the broader society and portrayed as the epitome of a new type of sociopolitical and moral order that, in its basic premises, was closely connected to the specificity of city and city life.

The moral evaluation of cities in modern times, relatively fully documented and researched, ranged from extolling the city and the civic virtues in the Renaissance and Enlightenment periods, to the more ambivalent, if not negative, view that evolved in the wake of the Industrial Revolution, especially within the Romantic Movement. These views, which greatly influenced the basic perception of cities in modern social science, constitute perhaps the most articulate manifestation of the similarity between the positive and the negative evaluation of cities articulated by the different elites.

H
URBAN SYSTEMS AND HIERARCHIES

The preceding analysis focused on the major dimensions of the social, political, and cultural distinctiveness of cities and on the understanding of cities as centers of distinct social and cultural creativity. The picture would not have been complete, however, without studying the major aspects of the interactions among cities that form urban systems and hierarchies, and without indicating how the distinctiveness of cities and the eventual emergence of special types of social and cultural creativity became manifest in the structure of urban systems. We also attempted to show that a more intensive interaction could have developed among different cities, beyond their embedment in their respective environmental settings, and to proceed beyond the major distinctions between the systems of primacy and those of rank-size, which have been elaborated in previous studies.

We will now examine the major aspects of urban hierarchies analyzed in our case studies: the number of hierarchical levels in those systems; the order and sequence in which new hierarchical levels were added; the quantitative and functional relationships among levels of the urban hierarchy; the average size or density of urban places in each hierarchical level; the major functions of such hierarchies; the extent to which the hierarchies were uni- or multicentric; and the extent of self-sufficiency of the different levels.

We will distinguish between the ecological contiguity of different urban concentrations within a seemingly common societal or civilizational framework and the functional interrelations among the levels, based on a continuous flow of resources among them. The basic form of "mere" ecological contiguity can be identified in the early stages of urbanization in the inland civilizations as well as in the completely decentralized systems, as were found in the early periods of Medieval Europe; in most other cases it is possible to talk of functional interrelations among such levels.

The case studies generally indicated that various aspects of urban systems were shaped by the interrelations among the levels and types of economic structure and the communication and transport technology—as they molded the structure of markets and the flow of resources among them—as well as those existing among the modes of control exercised by the various coalitions of elites embracing different cultural orientations and operating in different political-ecological settings. These were the forces that shaped the nature of the urban hierarchies through their impact on the construction of the spatial concentration of different modules of population and resources, on the spatial organization of the flow of resources among such modules, on the functional interrelations between such modules, as well as through their control over the spatial organization of the society and the economy.

The case studies indicated that the internal momentum of the forces of concentration of demographic and economic activities tended to generate mainly the lower and the middle levels of the urban hierarchies, and that the interrelations between those adjacent lower and middle levels greatly depended on the relative importance of the internal or external markets as well as on the level of the available technology of transportation and communication.

The higher the level of production and consumption within the agricultural sector of the traditional preindustrial societies, which could serve as a basis for wide internal markets, the stronger was the tendency toward the emergence of several closely related lower and middle levels in the urban hierarchy (as occurred, for instance, in China). Moreover, the greater the volume of combined commercial and agrarian transactions in a society, the stronger was the tendency toward a relatively high density of settlements in the respective levels of the hierarchy. In cases in which consumer demand in the agricultural sectors and the development of commercial internal markets were weak (as in the Byzantine Empire) and the externally oriented trade was stronger, the more pronounced was the tendency for a smaller number of lower and middle levels in the urban hierarchy to exist. The lower was the density of each level, the greater would be the autonomy of the respective levels, especially the lower ones.

The number of levels and the degree of functional interrelations among them also depended on the development of transport and communication technology in a given society. As these technologies developed, the number of levels increased and the functional interrelations between such levels strengthened. It should be emphasized that the nature of the functional interconnections among the various levels of an urban hierarchy was shaped not only by the economic and technological development but by the demographic resources as well as their potential flows.

Such crucial aspects of these interrelations as the number of settlements in the respective levels and the different types of functional interrelations between the levels of a hierarchy were influenced by the pattern of the forces of centrality, mainly by the different modes of control over access to the major markets and to the flow of resources between markets as exercised by various coalitions of elites articulating different cultural orientations and molding the center-periphery relations of their societies. The most important distinction among the modes of control that influenced these aspects of the urban systems was that between the control exercised by coalitions of relatively autonomous elites and the one wielded by relatively embedded elites.

The differentiated control over access to major markets was thus connected with the control over the convertibility of resources between relatively open markets and with the distinctiveness and symbolic autonomy of centers. It was connected as well with the concomitant permeation of the periphery by the center, as practiced by relatively autonomous elites, most of all in imperial and imperial-feudal systems. These modes of control tended to create multiple institutional linkages among the different regional and institutional markets and to generate a great number of levels of the urban system, a high level of functional interrelation between them, and a smaller degree of autonomy of each level.

Conversely, a more restrictive mode of control, minimizing the flow of resources among markets and limiting access to them, was connected with low symbolic autonomy of the centers and a low level of penetration of the periphery by the center, as exercised by relatively embedded elites. This situation, which occurred mainly in patrimonial systems, generated tendencies toward a small number of levels in an urban system and restricted functional interrelations among these levels, toward the predominance of administrative-extractive functions as opposed to generative-economic and political-functional relations among the different levels, and thus toward a concomitant disjunction between the levels and their relative autonomy.

Other aspects of the structure of the urban systems, such as the nature of the various levels' major functions, the degree to which any urban system was uni- or multicentric, and the closure of systems (defined as the degree of external orientation manifest in the amount and intensity of interaction with areas outside the system), depended mainly on the composition and orientations of the ruling elites, on the direction of the control they exercised over the convertibility of resources in major institutional markets, and on several political-ecological variables. These variables were, principally, the relative compactness of the political system, coalescence of the boundaries of the major collectivities, and the emphasis laid on internal as against external markets.

The nature of the major functions of an urban hierarchy—whether it was mostly economic, religious, political, or administrative—was influenced, in the context of any given society or social sector, by the strength of the specific type of forces of centrality or concentration and of the elites and groups organizing and articulating these forces. The degree of filling all levels of an urban hierarchy depended on compactness of the boundaries and coalescence between the boundaries of different collectivities. The greater the compactness, the more likely the center was to build up and control all levels of the hierarchy, whereas decentralization and openness of boundaries weakened the center's relative control of a hierarchy's different levels and left vacant some of the hierarchical levels. The number of levels in an urban system was also influenced by the areal extent of the respective political community and by the concomitant emphasis on internal or external markets. Thus, on the whole small states tended to allow a smaller number of such levels to evolve.

The effects of the relative predominance of monolithic elites on the structure of urban systems in patrimonial regimes emphasize the importance of the various modes of control exercised by the different elites, especially political-ecological ones, in structuring the various aspects of urban systems. Even when a monolithic elite predominated in a patrimonial system, as in Southeast Asia, the tendency toward primacy of the urban system was undermined by the greater social and, often, geographic distance of the periphery from the center, the weakness of functional links among the different levels of the system, and by the low level of coalescence among the boundaries of the different collectivities.

Insofar as multiple elites developed within such societies, as was the case to some degree in Latin America, India, and the patrimonial Islamic countries, a multiplicity of urban systems—economic (often outside-oriented), political, and religious—tended to evolve. In these cases the different levels impinged on each other to a lesser extent than in imperial systems. Conversely, a relatively close functional interrelation between adjacent levels of urban systems tended to emerge within imperial systems, pushing them in the direction of a rank-size pattern. Within such systems a relatively close functional interrelation among the different levels of an urban system also evolved, as well as a small degree of autonomy of most levels, except for the extreme peripheral ones.

Comparative analysis of the factors shaping urban systems in decentralized regimes reveals the importance of political-ecological factors in comparison with the effects of modes of control exercised by different elites. In general, decentralization affected the structure of urban systems in the direction of a growing separation between levels and weakness among the higher levels, and a strong ecological disjunction tended to grow among the various levels. This led to primacy in many relatively isolated regions, to restricted functional interrelations among them, and to the relative autonomy of each region. A closer look at the material indicates that the full impact of decentralization on the structure of urban systems varied according to the combination of decentralization features with the different modes of control exercised by the various elites. In such cases, insofar as some greater distinctiveness of the religious centers evolved—as occurred in India and, to a lesser degree, in Islamic countries—a tendency emerged toward the crystallization of a somewhat greater number of levels in the urban system and a closer interrelation among the different levels of temple networks and pilgrimage centers. But even this tendency was rather muted, and it could not be sustained constantly in the absence of political and economic interrelations among such concentrations of pilgrimage and temple cities.

More crucial in this respect is the comparison between imperial-feudal systems (mainly in Europe but also in Japan) and other decentralized societies or civilizations, such as Indian or several Islamic societies and, in extreme cases, decentralized ones in purely patrimonial settings such as in Southeast Asia or in some Near Eastern civilizations of antiquity. This comparison indicates that the modes of control exercised by the various elites influenced the degree of functional interrelations between adjacent levels of the hierarchy and the tendency to transform the ecological contiguity into spatial concentrations with more systematically interrelated links among the levels of the urban system. The importance of the various modes of control and types of center-periphery relations in decentralized regimes is best illustrated by the differences between the feudal systems—as they evolved in Europe and Japan—and the other types of decentralized systems as occurred in India, in some Islamic regimes, and in those patrimonial regimes that developed in Southeast Asia or the ancient Near East.

Center-periphery relations that developed in Medieval European and Japanese societies contrasted with those of other decentralized systems, characterized by strong distinctiveness of the multiple centers and their continuous penetration of their respective peripheries. These relations were maintained by autonomous elites with either strong orientations toward the tension between the transcendental and the mundane orders or—as was the case in Japan—a strong commitment to the centers. In Europe and Japan these modes of control gave rise to a tendency toward strong and continually expanding functional interrelations, at least between adjacent levels of the system, as well as to a gradual transformation of the ecological contiguity into more systematically interrelated links among the different levels of the urban system. In these cases, specific urban systems emerged that were characterized by a greater number of levels than in situations of primacy, but a smaller number than in compact imperial systems. Such developments in the urban system were then connected with the emergence of high (or at least continually expanding) agricultural production, consumer demand, and concomitant commercial activities.

In comparison with the full-fledged imperial case, the highest control levels, those of the capital city, were absent in imperial-feudal systems, as also were often the lower levels. Insofar as lower levels did evolve, however, they tended to be dissociated from the middle levels and to retain a relatively high degree of autonomy. In feudal or imperial-feudal systems, a high degree of functional interdependence developed between adjacent levels of an urban hierarchy, especially the middle ones, but not necessarily among all of them and differences in the extent of development of functional relations may then have arisen among levels, even adjacent ones. The degree to which interrelations, as opposed to mere ecological contiguity, evolved among such levels depended mainly on the combination of the level of economic production and the effectiveness of political control of the territory.

In early (9th-10th centuries) as opposed to late (14th-15th centuries) Medieval Europe, when the production level as well as the political control of broad regions were relatively low, few functional interrelations existed among different regional clusters and among the few levels of each regional unit. Each level tended to be connected with its own immediate hinterland or with outside economic international centers. The momentum generated by economic development and closer political control later transformed the ecologically contiguous levels into more functionally interrelated and denser levels of the urban system.

Throughout Japanese history there was a high degree of interrelation among regional levels. This persisted even through periods of decentralization and tended to increase with economic development and more centralized political control, mainly in the Tokugawa period.

In the decentralized systems of Indian or Islamic societies, relative disjunction occurred among the different levels, or sets of levels, of the urban hierarchy. Significantly, however, those urban networks—especially the religious ones found in India and, to a smaller degree, in other Buddhist countries and in Islam—engendered by the autonomous elites tended toward greater functional interrelations in the religious spheres among the respective levels of their systems, especially when they combined with political or economic centers that could influence the flow of resources among them.

A similar picture emerged with respect to the development of monolithic, unifunctional as opposed to multiple urban hierarchies. The stronger the compactness of the political regime and the coalescence between the boundaries of the political and other collectivities, the greater was the tendency toward a monolithic urban hierarchy, whereas a decentralized political system and cross-cutting markets generated a greater diversity of functional hierarchies. Insofar as political decentralization was combined with autonomous civilizational frameworks carried by distinct elites—as was the case in India, Islamic countries, and Europe—it led to the development of relatively autonomous religious or cultural urban networks that cut across narrower local hierarchies. The predominance of one type of ruling elite, which partially evolved in Japan and in many semipatrimonial states, gave rise to multiple local urban hierarchies in a decentralized system. Only when the independent merchant forces were able to shed the control of the political elites did some rudiments of autonomous economic urban hierarchies of national scope tend to develop.

The relative importance of the modes of control exercised by elites can also be revealed by analyzing the different degrees of closure of the urban system—that is, the degree to which the urban system has an external orientation—manifest in the amount and intensity of interaction with areas outside the system. Such closure seemed to be related to the degree of compactness of the political system and of interweaving among the different internal institutional markets and coalescence of boundaries of the major collectivities. The stronger these forces were, the greater was the tendency toward closure of the urban system. The imperial systems, which were usually compact, with a high degree of coalescence between the boundaries of their major collectivities and with close interrelations among their major institutional markets, were thus distinguished by a high level of closure of their respective urban systems, as long as they were able to maintain the compactness of their political system. Such closure can be observed in the particular type of urban system whose different urban hierarchies tended to coalesce, mediated by the central political power.

The urban hierarchies that developed in these societies may also have had a strong autonomous relation to some external markets. In patrimonial societies the structuring of the relations between internal and external markets differed entirely from that in imperial systems. In such patrimonial societies, the economic activities were oriented toward external markets and based more on the extraction and use of the country's natural resources than on internal trade and were mediated by a primate center or centers that were themselves oriented toward external, often different, markets. The smallest degree of closure of urban systems was found in decentralized regimes, whether in Medieval and early modern Europe, in India or in Japan (at different periods of its history), and, to some degree, in Islamic countries. A shift occurred in the boundaries of different system components as well as a continuous reshuffling of the flow of resources among them and many subsystems with changing boundaries.

It was not just the decentralization that was of crucial importance in those regimes, but its combination with the degree of coalescence of the major institutional markets. In this respect the distinction between Islamic societies and India on one hand and Europe and Japan on the other is of great interest. Nonimperial Islamic societies were characterized by relative compactness and slight coalescence of the boundaries of their political and religious communities and by their predominance over the economic ones. Hence relatively closed regional urban systems tended to evolve, with two partial exceptions: centers of external trade and religious centers. India's societies paralleled those of Islam and were characterized by a much higher degree of decentralization and blurring of political boundaries. However, some interweaving and de facto coalescence occurred between the boundaries of the different collectivities; accordingly, with the exception of the religious centers and those of external trade, relatively closed systems developed (although they were more limited than in other urban systems).

Conversely, feudal Europe, characterized by much less coalescence of the boundaries of the major collectivities, gave rise to what was probably the smallest degree of closure of the urban system, which began to change in the period of Absolutism. The lack of compactness of the political system weakened its closure and rounding off in both Europe and Japan. Due to economic conditions and to the structure of elites, this also led to fluidity and openness of the boundaries of the urban networks and to frequent changes and shifts in the location of centers.

An entirely different type of open system tended to develop in societies, especially small ones, characterized by a strong orientation toward external markets. This openness of the urban system was connected in Europe with a growing tendency toward the evolution of a rank-size system and increased interaction among cities of different sizes and functions. When combined with the dynamics of internal structure of cities, this interaction greatly facilitated the emergence of specific types of urban social and cultural creativity, characteristic of Europe, and increased the extent to which such cities became autonomous foci for the generation of social and cultural forces.

In the preceding chapters we analyzed the general tendencies of the different components of cities and urban systems and the conditions leading to them. We showed that urban systems were shaped by combinations of economic and demographic forces and by modes of control exercised by different elites, by center-periphery relations, and by various political-ecological variables. These forces joined together in diverse constellations in different civilizations and shaped the contours of urban systems in different periods and regions. We will now analyze several such concrete constellations, as illustrated in the case studies.

CHAPTER 13

Patterns of Cities and Urban Systems and the Institutional Creativity of Cities

On the basis of the case studies presented in Part II, we have analyzed the ways in which the forces of concentration and centrality, their strength, variability, and autonomy, the concomitant levels and types of resources, and the modes of control over them have shaped the spatial concentration of people, resources, and information. We have also described the major aspects of the structure of cities and of urban systems that led different cities to become centers of distinct patterns of social and cultural creativity.

The combination of economic, technological, and geopolitical factors generated different levels of resources, while coalitions of major elites, having different orientations and operating in various sociopolitical settings, exercised diverse modes of control over the flow of resources in a society. These factors jointly influenced the institutional framework that shaped the components of the urban structure.

Only the general tendencies of the different constellations of these social forces or the directions in which they influenced the development of such components of urban life have been studied. It was emphasized that only through the interaction of several of these forces, coalescing in specific historical settings, did the concrete patterns of cities and urban systems crystallize. The concrete constellations of social forces can be discerned from the combination of the levels of economic development, the control over the flow of resources and information, the degree of compactness or centralization of the political system, and the coalescence of the boundaries of the major collectivities. Such combinations coalesced differently in various types of regimes—among which we have analyzed several patrimonial, imperial, and decentralized ones—as well as in city-states (mentioned only tangentially), with numerous variations in each of these regimes.

We will now analyze the impact of these concrete constellations on the development of different patterns of cities and urban systems.

A
PATRIMONIAL REGIMES

We will start by studying the societal constellations in which the relatively embedded elites constituted the key element, exercising extractive or restrictive modes of control, combined with a compactness of political boundaries but cross-cut by different levels of economic development. Such a combination was found in patrimonial regimes—represented in our case studies by Southeast Asia, Latin America, and many Sultanic Islamic countries—as well as in numerous Near Eastern and Meso-American societies. These systems were characterized by a relatively low level of economic development, weak internal markets, strong extractive policies, and orientation to external markets, and a low degree of coalescence between the boundaries of the collectivities and the civilizational frameworks, insofar as these emerged within some of the great post-Axial Age civilizations. Their predominant coalitions comprised non-autonomous political and religious elites that, as in most so-called pagan societies, were also embedded or, as in the post-Axial Age civilizations, were autonomous in the religious but not in the political field.

These elites exercised an extractive mode of control over the flow of resources. They generated relatively narrow institutional markets and a restricted flow of resources among them. These societies were accordingly characterized by weak internal markets and strong disjunction between internal and external markets. Trade was oriented outwards, regulated mostly by the rulers or by external groups who impinged rather weakly on the narrow internal markets, increasing only marginally the internal flow of resources. Consequently these societies were characterized by a weak propensity to adopt technological innovations that could have intensified their bases of production.

The ruling elites of patrimonial regimes devised ways to regulate their relations with both the external groups and external trade. They eagerly absorbed and coopted social groups, structurally different from the ones predominating in their societies. These groups, usually living in economic or religious enclaves, were either specialized economic groups that could contribute to the extraction and accumulation of resources or religious orders with transcendental orientations and could add luster and prestige to the centers. This co-optation, however, seemed attractive only when the enclaves were external to the structural core of the society or so long as they did not impinge on the society's internal structural arrangements, especially on the relations between center and periphery. This explains the strong tendency of patrimonial regimes to permit ethnically alien groups, which could more easily be segregated than indigenous elements, to engage in differentiated economic or religious activities.

Most patrimonial societies were also characterized by a relative lack of structural, as opposed to ecologic and symbolic, distinctiveness of their center from the periphery and mostly by an adaptive attitude of the center toward the periphery. Weaker countrywide class consciousness and symbolic articulation of the major types of collectivities also tended to develop in these societies.

Patrimonial regimes varied in the type of their economy, the degree of their dependence on external markets, the strength of their center and the efficiency of their control over the periphery. Above all, these regimes varied in the structure of their elites, according to the cultural orientations they carried. Here the major difference was found among the regimes forming part of the great post-Axial Age civilizations—Hindu and Buddhist—as well as those in Latin America and Islam and those belonging to pagan civilizations. Whereas in the latter case all of the major types of elites (cultural and political alike) were embedded in basic ascriptive frameworks, in the former the religious elites—and to a smaller degree the political ones (as in the Latin American Spanish Empire)—were mainly autonomous. The cultural elites who carried models of cultural and social order were those that generated the Great Traditions of their civilizations and special broader civilizational frameworks, based on a strong perception of the tension between the transcendental and the mundane orders, the like of which was not found among other pagan patrimonial regimes. Concomitantly, these elites created centers that were distinct from their own periphery in the religious sphere, as well as special networks linking these centers with the periphery.

These combinations of societal forces, characteristic of patrimonial regimes, generated a pattern of internal social organization and of the occupational structure of cities within them that differed in a limited way from those in the periphery or in the centers of society. Insofar as a strong distinction of these cities evolved, it was mostly due to a sharper articulation, within a relatively condensed space, of the society's basic premises and symbols of its specific form of centrality, and of a concomitant denser concentration of the upper political, administrative, and religious elites. The internal ecological structure of those cities was characterized by a special urban space allocated to religious, political, and administrative central functions, subsuming the residential under the religious or administrative duties, and by a strong tendency to group all these functions in one basic ecological framework. In nonmonotheistic civilizations these tendencies were often constructed in such a way as to represent the cosmic order in their organization of space.

The internal corporate self-regulation of these cities tended to be very limited both in respect of their own affairs and the affairs of their respective groups, whose regulation by the centers of their societies had wide scope but not much intensity. External political autonomy was very limited too, and the urban identity that emerged among such elites was mostly of the local patriotism type with few specific social or political components. In these cities the social conflicts were either engendered by the lower classes or were rebellions of local magnates against the center. The rather muted evaluation of cities evolved without a strong confrontation between the negative and positive appraisal of city life. Needless to say, cities generated by forces of concentration also emerged within such regimes, but they were either foreign enclaves, as in many Indian or Southeast Asian societies, or were of secondary rank and relatively disconnected from the main urban systems.

Variations in the structure of urban systems evolved according to their level of internal economic development. An economy of low level, as found in Southeast Asia, tended to generate an urban system of weak primacy in which, beyond the primate city, only few lower levels of urban concentrations existed. These urban settlements were widely dispersed and had limited functional interrelations. As though by default, these lower levels enjoyed a high degree of autonomy. In extreme cases, in which low level of technology and economic development, as well as very weak political centers were found—in Southeast Asia, India, or the ancient Near East—a central primate city may not have evolved but merely a series of ecologically contiguous cities, each of these seeming to be of a primate nature, but with a restricted hinterland that had few interrelations with such a city.

Insofar as a higher level of agricultural productivity developed, it usually generated a more structured primacy pattern, characterized by closer interrelations among adjacent levels of the center and the periphery, the former absorbing resources from the periphery and channeling them into its own export-oriented outlets. In cases in which more autonomous social and political forces emerged, generating greater internal markets and consumer demand, as in Latin America and China, the primacy system tended to be transformed into one evincing some characteristics of rank-size order.

Differences in the level and type of economic development also affected the internal features of cities that developed within these patrimonial regimes. The higher the level of their economic development, the more heterogenetic the cities were and the stronger was their impact on other cities. In addition, the more acute the level of the social conflicts within the cities, the sharper became their moral evaluation.

The far-reaching differences in the nature of the cultural orientations prevalent in the various patrimonial societies, in the structure of elites and in the modes of control they exercised, also affected some crucial aspects of cities and urban systems. The main difference existed between the more embedded elites found in the so-called pagan Southeast Asia and the ancient Near East (or in pre-Colombian Meso-America, not analyzed in the case studies) and in those that were autonomous in the religious sphere, owing to their strong other-worldly orientation, but embedded in the political sphere. The latter elites could be found in the realm of Hindu and Buddhist civilizations but also in imperial Latin America and in many Islamic polities where other-worldly orientations were never as dominant as in the realm of Hindu and Buddhist civilizations. More distinctive urban centers developed, at least in the religious sphere in the latter cases, and were associated with a more articulated urban identity and with stronger confrontation among the different evaluations of the center. As regards more embedded elites, as in Southeast Asia, a strong ecological disjunction occurred among the different levels of the urban systems which were organized in regional sectors having restricted functional interrelations among the different regions as well as among more distant levels.

B
IMPERIAL AND DECENTRALIZED REGIMES

Another major pole of constellations of the different forces of concentration and centrality developed in imperial systems and is represented in our case studies by the Chinese, Byzantine, Russian, and, to a smaller degree, the Abbassid and Ottoman empires. These imperial systems were characterized by (a) the combination of highly compact political boundaries and centralization, (b) relative coalescence of the major collectivities' boundaries, (c) relatively developed economic systems, (d) a preponderance of internal markets, and (e) highly autonomous elites.

In imperial and imperial-feudal societies most of the elites tended to define themselves in autonomous terms and had autonomous resource bases and a potentially autonomous access to the center and to each other. This was true of the articulators of the cultural and social order—that is, the cultural and religious groups. This was also true of the political elites and, to a smaller degree, of the representatives of different collectivities and the economic elites. Moreover, a multiplicity of secondary elites evolved in these societies, such as various sectarian groups in the religious sphere or various social and political groups and movements. These elites impinged on the center and on the periphery and shaped the movements of protest and political struggle. Each of the "primary" and "secondary" elites were able to constitute the starting point of movements of protest and political struggle, evincing a high level of organizational and symbolic articulation. As with all autonomous elites, these generated the emergence of wide institutional frameworks cutting across the major ascriptive communities and exercising diversified but intensive control over access to the markets, as well as over the flow and conversion of resources among these markets, and adopting varied social orientations and reference symbols.

These elites also generated specific types of center-periphery relations. The major characteristics of these elites were a high level of symbolic and ecological distinctiveness from their respective centers and the continuous attempts of the centers not only to extract resources from the periphery but also to permeate and reconstruct it according to their own premises. The political, religious, and cultural centers thus constituted the foci and loci of the various Great Traditions, distinct from the local traditions that developed in these societies. The permeation of the periphery by the centers was manifest in the latter's promotion of widespread channels of communication and in their attempts to break through the ascriptive ties of the periphery.

Closely connected to this type of center-periphery relations was the high level of articulation of symbols of countrywide social hierarchies, some political consciousness of the upper strata, and high ideological symbolization and mutual orientation among the major religious, political, and even ethnic and national collectivities. Although each collectivity tended to generate a relatively high degree of autonomy, they also constituted mutual referents for each other; for example, in the Byzantine state being a good "Hellene" was identified with citizenship and vice versa. In imperial and imperial-feudal societies, this high degree of symbolic articulation and distinctiveness of the major institutional aspects was firmly linked to certain types of cultural orientations that were articulated by these elites.

Most of these empires developed in close relation to some of the Great Civilizations or Great Traditions in human history, such as the Chinese civilization, which crystallized around a blend of Confucianism, Taoism, and Buddhism, and the various forms of Christian and Islamic civilizations. Most of these shared with the other civilizations that emerged in the first millennium B.C.—the so-called Axial Age—the conception of a basic tension or chasm between the transcendental and the mundane orders and the strong emphasis placed on the necessity of bridging the gap between them. They differed from other civilizations—especially Hindu and Buddhist—in sharing the concept that some kind of this-worldly occupation could act as a bridge spanning the transcendental-cosmic and the mundane world—or, to use Weber's terminology, as a focus of salvation.

The imperial systems examined in the case studies varied in the nature of their economic structure, in the degree of coalescence between the boundaries of the civilizational or religious, political, and national frameworks and collectivities, and in the emphasis placed on this-worldly as opposed to some blend of this- and other-worldly conceptions of salvation. They also differed by their elites' monolithic or pluralistic structure and the degree of development of autonomous strata in them.

This combination of components generated a greater variety and much more intensive dynamics of cities in imperial systems, as well as greater distinctiveness of cities and urban systems and more varied types of urban social structure and cultural creativity. A greater variety of cities emerged as well as closer interrelations among them. The problem of control over cities and urban hierarchies thus became more acute, as did the potential impingement of the diverse heterogenetic cities on their respective centers of control. The cities constructed by forces of centrality in the imperial regimes were characterized by stronger distinction of their social and occupational composition from the periphery, albeit in the direction of articulation of the society's major symbols and premises. Their ecological structure was more differentiated than that of the "patrimonial" cities. On the whole, however, they were still structured around the central religious, political, and administrative frameworks, and they evinced a strong tendency to encourage the evolution of several centers within the cities and to unite functional and residential quarters.

Heterogenetic cities in imperial regimes tended to be rather stable and diversified, depending mainly on their place in their respective level of the urban hierarchies. All, however, were characterized by a social and occupational structure differing from those of the periphery and of the centers of the society. Moreover, the strong economic and occupational groups that emerged within the orthogenetic cities in general, and capital cities in particular, led to greater ecological differentiation among the political, religious, and economic functions. This led to the creation of very distinct and visible occupational districts with varying degrees of differentiation between the residential and the economic quarters. Such cities were generally imbued with a strong and vibrant urban flavor.

The societal centers tended to exercise intensive regulation over the affairs of the cities and urban groups. Insofar as the prevalent major strata and the economic and occupational groups were relatively autonomous, a tendency toward more corporate internal regulation of these areas may have evolved, as well as possible participation by urban groups in the central authorities' regulation of urban affairs. Within these cities, not only did sporadic outbursts and rebellions tend to occur, but lower or lower-middle urban groups also attempted, often in a revolutionary or semirevolutionary manner, to reconstruct the corporate urban structure and governance. A more articulated moral evaluation of cities also emerged, as well as sharper confrontations between negative and positive evaluations of the moral order of the cities.

The constellations of economic and social forces found in imperial societies, and the high level of consumer demand and growth of internal markets in the agricultural sector, combined with intensive commercial urban activities, with flexible distinctive autonomous centers and with a strong permeation of the periphery by the center, to generate specific types of urban systems. These types were characterized by close functional interrelation between adjacent levels of the urban hierarchy with little autonomy of most levels, except for the most peripheral ones. Most urban systems in imperial regimes, mainly of an administrative but also of an economic nature, tended toward rank-size distributions. Within this basic framework, variations arose among the different imperial systems, according to the predominant type of economic structure, the degree of compactness of the political systems, and the structure of the elites, their cultural orientations, and the autonomy of the major strata. These variations were reflected in the structure of urban systems and the different dimensions of internal organization and distinction of cities as well as, ultimately, in the major modes of social and cultural creativity that evolved within them.

With respect to the types of economies in imperial regimes, the major differences occurred between the economies based on internal markets of large volume and those based on external markets. The emergence of such internal markets, based on extensive production and consumption in the urban and rural sectors alike (existing in China and, to a smaller degree in the Byzantine Empire), strengthened the tendencies toward the development of a rank-size distribution and close functional interrelations among the different levels of the urban hierarchy. Conversely, when consumer demand was low in the agricultural sectors—as occurred in the Roman Empire, based on a slave economy that necessarily inhibited the emergence of an internal agricultural market, or at certain periods in the Byzantine and Russian empires—the multiple functional relations among the various levels of the hierarchy weakened, leading to a rupture between the different levels. The more robust the internal markets were, the stronger was the tendency to create heterogenetic cities having incipient leanings toward autonomy and a stronger impact on the upper levels of the urban hierarchies. These economic factors also increased the likelihood of intensive social conflicts erupting within the cities.

Imperial systems also varied according to the degree of compactness of their boundaries, which influenced the effectiveness and scope of the center's regulation of urban affairs. The more compact the system was, the more effective such a regulation tended to be. The degree of compactness also influenced the extent of development of a full-scale urban system characterized by a strong tendency toward a rank-size order but controlled by the capital city. The crucial importance of an effectively controlled compactness in shaping urban hierarchies was best illustrated in times of decline of imperial power, as occurred in the late T'ang period in China and in the Byzantine Empire. Then the weakening of the effective control by the center resulted in rupturing the relations among the different levels of the urban hierarchies, weakening the links among the different regional subsystems, and narrowing their scope.

Imperial regimes differed by the structure of their ruling elites, by the cultural orientations they articulated, by the modes of control they exercised, and by the relative autonomy of their major strata. The most crucial differences existed among the monolithic elites, usually evincing strong this-worldly orientations, and the more heterogeneous ones, usually carrying some combination of this- and other-worldly orientations. The latter patterns could also be discerned in the degree to which heterogeneous elites were segregated or interwoven. Both the monolithic and segregated elites tended to exercise relatively restricted modes of control. Whereas segregated elites inclined to exert more intensive control than did monolithic ones, the more heterogeneous and closely interwoven elites' control was more flexible, though often also very intensive.

These differences shaped the structure of urban hierarchies, especially the degree to which they tended to coalesce and were directed or controlled by a single center and, conversely, the degree to which these hierarchies were autonomous or, in technical terms, the degree to which the overall urban hierarchy of a certain society was uni- or multicentered. In general, the more fused or monolithic the controlling elite, the greater the tendency would be to develop a monolithic urban hierarchy, whereas a multiplicity of elites tended to generate greater diversity of functional hierarchies or subhierarchies.

The homogeneity or heterogeneity of elites, their relations to the broader strata and the structure of these strata influenced the degree of coalescence and the effectiveness of central control, as well as the relative predominance of the urban hierarchies mostly having a rank-size distribution type. Thus we find that a strong combination of this-worldly cultural and political orientations was evinced in China by a monolithic political-cultural elite having solidary but distant relations with broader strata, especially with the free peasantry. This elite controlled the comparatively well-developed economic sphere, but without denying it some autonomy and momentum, thus allowing for close interweaving and overlapping between the political and economic urban hierarchies. In periods of decentralization more dissociation between these hierarchies tended to occur.

In the Byzantine and, to a lesser degree, in the Ottoman and the Abbassid empires—in which this- and other-worldly conceptions and a mixture of proselyting, cultural and religious, military, political orientations were prevalent and adopted by multiple autonomous elites and in which continuous tension existed between the aristocracy and the free peasantry as well as between the religious and the political elites—several autonomous urban systems tended to emerge, each fulfilling different functions: administrative, economic, religious, or military. Most of these systems, however, tended toward a rank-size type and inclined to cut across one another. The relation between these hierarchies and the intensity of their tendency to develop into a rank-size order varied throughout Byzantine and Ottoman history, depending on the strength of the different elites and the level of the economy's development.

Combined this- and other-worldly orientations were prevalent in Imperial Russia, together with strong military and power orientations of different, hierarchically organized, segregated elites, among whom the political one predominated. At a time when only slightly autonomous strata and a bonded peasantry prevailed, mercantile systems could only evolve slowly. The combination of all these factors gave rise to a multiplicity of segmented urban hierarchies, among which the political-administrative one predominated, with variations in the strength of such hierarchies depending on the extent of dominance of the political vis-à-vis the other elites.

In the realm of Islamic civilizations, the situation was more complex. Given the elites' basic orientations and structure, multiple urban hierarchies tended to emerge, especially distinct religious, political, and economic ones. Their degree of overlap depended on whether an imperial, patrimonial, or semitribal regime was prevalent and on the relative strength of the center.

In the period of Absolutism in Western Europe, strong tendencies evolved toward a unicentered urban system of primacy type controlled from the capitals by the respective rulers. Yet because of the basic cultural and structured pluralism prevalent in European societies, which was connected to the multiplicity of urban systems that evolved in the feudal and early modern periods, the societies being characterized by multiple elites, the rulers could not easily abolish the different systems and radically change the urban system from a rank-size to a primacy pattern.

In some imperial systems the modes of control exercised by the elites possessing diverse cultural orientations generated important differences in the various dimensions of the cities' internal structure, in their relations to the centers and peripheries of their respective societies, in their distinctiveness, and in the modes of creativity that emerged within them. The greater the heterogeneity of such elites and the more flexible the mode of control exercised by the ruling elites—as was the case in the Byzantine and Russian empires, as opposed to the more restricted control exerted in China by the monolithic elite—the more intensive was the regulation of urban affairs. At the same time, especially insofar as the major strata and particularly the dominant urban strata were autonomous and had potentially autonomous access to the center—as occurred in the Byzantine Empire, to a limited degree, and was fully developed in feudal and Absolutist Europe—the more did they tend toward self-regulation and possible participation in the center's control of urban and societywide affairs. In those cases the conflicts erupting in such cities were intensive and center-oriented, and strong political dimensions of urban identity emerged in large sectors of the urban population together with an articulate moral evaluation of cities and a strong confrontation between the positive and negative aspects of this evaluation.

The instances of more flexible control exercised by heterogeneous elites may have given the forces of concentration that emerged in these centers not only a wider scope for the construction of cities but also the possibility to impinge more directly on the society's centers, whether the political or cultural ones. In such a situation, the two modes of creativity specific to cities tended to become interconnected: those generated by forces of centrality and characterized by full articulation of the society's major central symbols, and those of the forces of concentration, characterized by the development of specific types of production and of economic and class structures. This interconnection enlarged the scope of activities of the urban forces and enabled them to impinge on the centers of their societies. Consequently, the nature and basic premises of the forces of centrality were transformed, a transformation that led beyond and blurred the distinction between heterogenetic and orthogenetic cities.

In decentralized systems, both the imperial-feudal ones, as in Europe and Japan, and the basic decentralized ones, as in India or Southeast Asia, were characterized by (a) noncoalescence of the boundaries of major political and civilizational collectivities, (b) the lack of stability and compactness of the boundaries of political collectivities, and (c) a tendency to cross-cut the major institutional markets. They differed, however, in the structure of elites and modes of control exercised by them and in the concomitant structure of centers and center-periphery relations, as well as in their respective economies.

The feudal or imperial-feudal regimes (especially in Europe but also, though to a more limited extent, in Japan) were characterized by relatively autonomous and multiple elites exercising flexible modes of control. These regimes were also characterized by a strong distinctiveness of their often multiple and non-coalescent centers, by a high degree of permeation of the periphery by the center, and by the prevalence of cultural orientations similar to those identified in imperial systems. In Japan, however, it was not the perception of tension between the transcendental and mundane orders but the strong element of commitment to the center that constituted the predominant cultural orientation; hence, the autonomy of the elites was much smaller.

C
THE INSTITUTIONAL CREATIVITY OF CITIES

The preceding discussion brings us back to the starting point of our analysis: namely, that the conditions generating the distinctiveness of cities as centers of social and cultural creativity, manifest in the ways in which social structure, organization, and cultural activities developed in cities and in the organization of urban systems, were distinct from those existing either in the rural sector or in major societal centers. These centers were usually located in the cities, and the two were very often almost identical.

The concentration of population and resources created their social and ecological distinctiveness from both the centers and the peripheries of their respective societies. The ways in which the cities interrelated to both the centers and the periphery of their societies indicated the major dimensions of the institutional, social, and cultural creativity emerging in them. Such distinctiveness of the urban modes of institutional creativity became manifest when reviewing the major dimensions of the social structure of cities and urban systems described in the case studies. This distinctiveness was also evident in the urban social structure and in the degree to which the distinct types of social organization, ecological structure, autonomy, and social conflicts developed in cities.

In the case studies we analyzed the basic social forces that gave rise to different modes of distinctiveness of cities as centers of institutional creativity, and the extent to which they not only constituted the spatial epiphenomena of the basic social forces but became the generators of new social forces. These different modes of distinctiveness and creativity were fostered by the strength, diversity, and autonomy of the forces of concentration and centrality that operated in the various societies or social sectors at different periods of their development.

The strength of these forces has thus affected the degree of development of the different types of cities, approaching Redfield and Singer's definition of heterogenetic and orthogenetic cities. The diversity and autonomy of each of these forces led to the different dimensions of distinctiveness in the directions taken by their development: full symbolization of the forces of centrality in the orthogenetic cities and diversity of economic and occupational specialization of heterogenetic cities, as well as the resulting modes of social and cultural creativity.

The greater the strength, autonomy, and diversity of the forces of centrality, the more diversified were the orthogenetic cities generated by them. The more distinctive these cities were from both the periphery and the societal center, the more diversified were the patterns of general and cultural creativity, and those of the evolving elite's social organization. The greater the strength and variability of the forces of concentration, the greater was the diversity of the occupational economies and ecological structures, as well as of the modes of productivity that developed mainly in the heterogenetic cities, or in sectors thereof. Of great importance in this respect was the combined diversity of the occupational and economic structure, the strong autonomy of the strata to which these groups belonged and of their access to the center.

The forces of concentration did not merely exist in a kind of ecological symbiosis with those of centrality, but they attempted to construct those cities as distinct communities or corporate entities. These forces often developed antagonism toward centrality and even attempted to permeate the respective centers of their societies. It was in such cases that the conflicts and protests that erupted in cities were couched in broad societal terms and symbols and occasionally but not exclusively in terms of class symbols. Thus, they often impinged not only on the structure of cities and governance, but also on the structure and symbols of the centers of their respective societies. Cities and urban systems then became the generators of new social forces. The special distinctiveness of the cities and their institutional, social, and cultural creativity found expression in the creation of great works of art and architecture and provided special ecological sites within which cultural activities took place. Urban autonomy and identity could indeed be transformed into specific, distinct, institutional, social, and cultural formations, fostering the development of new types of civic order that could spread, even beyond the cities, to the very centers of society.

Although nuclei of such possibilities could already be discerned in some imperial regimes, especially in Roman and Byzantine systems, their full development occurred only in the ancient city-states and in early and late Medieval Europe. These developments depended on certain geopolitical conditions, either political decentralization within the framework of a common civilizational setting, or occupation of special niches within relatively heterogeneous international systems. In both cases the geopolitical factors were not the ones that were crucial in these developments but rather their conjunction with the specific characteristics of their format—namely, the autonomy and heterogeneity of autonomous elites who possessed multiple cultural orientations and exercised intensive modes of control. In the absence of these social and cultural conditions, the geopolitical factors could not, as in India or Japan, have generated such a far-reaching transformation of the urban autonomy. Given the existence of these structural and social forces, however, the geopolitical factors could, indeed, serve as crucial catalysts of such a transformation.

Ultimately, it was the combination of all these forces that gave rise to the special mode of creativity of cities we have mentioned. It was in those cities that a higher level of distinctiveness of the social structure evolved and that strong ingredients of centrality were found, in addition to a combination of intensive regulation and attempted self-regulation and the confrontation between them. These factors led to greater self-regulation and participation in the center's intensive regulation of urban affairs. The conflicts and struggles often occurring were aimed at the societal center in the primate city and toward its possible reconstruction. These cities were characterized by a very intensive self-identity that upheld them as carriers of a specific sociopolitical order. It is in connection with the development of those cities that their articulate moral evaluation emerged, as well as a sharp confrontation between the positive and negative evaluation of urban life.

This specific creativity focused on the development of civility, a special civic consciousness that was originally rooted in the movements of protest and change of these societies. This tendency was implemented primarily by the efforts of the active urban elites, mainly stemming from attempts of the forces of concentration to permeate the center of the polity and imbue it with their own orientations. Thus new types of orientations, political concepts, values, and symbols, such as freedom and openness, were created that focused on the city or were epitomized by it. As mentioned, these orientations developed fully in the classical city-states, were later transposed to Medieval European cities, becoming interwoven with the specific elements of the European political tradition, and facilitated, in conjunction with the different dimensions of urban autonomy, the emergence of a civic culture and consciousness—the culture of civility.

Significantly, the very conditions that gave rise to these tendencies also exacerbated social and political conflicts in the cities and generated urban strife and protests. These conflicts attest to the emergence in these cities of a mode of institutional creativity that went beyond the distinction between heterogenetic and orthogenetic cities and transformed them by combining some elements of both these urban types. This creativity was connected to intensive conflicts but also to a distinct and unique transformation of the relations between, on one hand, the cities and, on the other, the centers and peripheries of their respective societies. It was only under these conditions that the forces of concentration were transformed into a full-fledged component of centrality, which, in turn, changed the centers of the societies.

It is therefore not surprising that these were the two cases singled out in the literature as the prototypes of pure, ideal types of cities and of their distinct creativity, at least in the premodern world. But although there can be no doubt that there is indeed validity in such an approach, our analysis has indicated that it would be wrong to evaluate all cities and their creativity only according to these frameworks. The conditions generating these distinct types of creativity could have changed even in these cases, as occurred both in the ancient world and in Europe. Beyond that, the fact remains that in other civilizations, with different combinations forces of concentration and centrality, other modes of creativity emerged in cities. In most, if not all, such civilizations, once cities were constructed they generated intensive patterns of different modes of creativity.

Epilogue

In this book we have taken a long journey, and it might now be worthwhile to look back and see where the road on which we have traveled has led us.

Our basic approach to the study of the urban phenomena—the nature of cities and urban systems—has been a comparative civilizational one. The major assumption of this approach has been that the best way to understand the specificity of the urban phenomena, at least in historical premodern societies, is to analyze them in their civilizational context in the context of the ideological premises and institutional frameworks of these civilizations.

This approach has been, in many ways, a sequel of some of the major studies of cities that developed in classical sociology and comparative civilization analysis conducted by Durkheim, Tonnies, Childe, and, above all, Weber. Like these studies, the approach presented here also attempts to combine the analysis of historical societies with a comparative macrosocietal perspective. The main emphasis of the comparative civilizational approach developed here has been greatly influenced by the Weberian approach, especially in his historical and comparative works, as well as by some of Marx's basic insights in the analysis of cities.

But while continuing the tradition of these studies, especially that of Weber, our approach has attempted to go beyond them in several ways. First, it took into account the numerous theoretical developments and controversies in social science that have taken place since the "classical" period. Second, the analysis presented here was based on the rich and manifold researches of many aspects of the urban phenomena that have been undertaken in history, sociology and anthropology since that period. Third, it brought together the sociological and anthropological approaches, on one hand, and the geographical approach on the other. Fourth, it applied the same range of analytical variables to the analysis of various aspects of the urban phenomena, mainly the internal social composition and organization of cities, the spatial organization of cities and urban hierarchies, and the different aspects of autonomy and cultural creativity of cities.

In terms of basic analytical assumptions we have attempted to combine the analysis of technological development, of the social division of labor and social differentiation, and the structure of resources generated by them with the study of power relations and conflicts and cultural orientations and traditions. Contrary to the strong trend in recent research in the social sciences to dissociate between studies of culture and of social structure—a trend connected with the declining predominance of the structural-functional school in studies of modernization of the 1950s—we have attempted to combine these dimensions of social action and use them as basic analytical tools in the comparative civilizational approach.

Our analysis has endeavored to specify the ways in which these dimensions of social interaction are brought to the fore through specific social actors and social processes. Among these actors and processes, we have stressed the importance of socioeconomic groups and of the resources they generate, the structure of the major elites and modes of control of both symbolic and material resources they exercise, as well as the opposition to such control.

The emphasis placed on civilizational premises and frameworks implied that in order to understand fully the crystallization and dynamics of a political regime or an economic formation, the civilizational dimension should also be considered. In other words, relatively similar regimes (for instance, agrarian empires or various types of decentralized polities) tend to generate different dynamics. These dynamics have a decisive impact on shaping many aspects of the urban phenomena, according to the civilization within which they evolve. At the same time, within the same civilization, different types of such regimes may also emerge, thus attesting to the importance of analyzing the cultural, structural, and political-ecological variables that combine in shaping institutional formations in general and different aspects of the urban phenomena in particular.

We have shown that the strong emphasis on cultural dimensions does not necessarily imply a "value-consensus" view of society. It not only accepts conflict or change as inherent in the constitution of any society but also indicates, in opposition to Marxist scholars, that such conflicts—especially the more dynamic ones and with far-reaching transformational possibilities—cannot be understood without taking into account the cultural dimension of the civilizational premises.

In analyzing the social actors, forces, and processes that, in various civilizational contexts, influence the development of the aspects of the urban phenomena, we have subsumed these forces under the two headings: "concentration" and "centrality." Needless to say, this distinction is an analytical one and not a description of concrete social phenomena.

These analytical terms were chosen (a choice triggered many years ago by the fact that, in Hebrew, they stem from the same root) because they seemed to indicate two distinct modes of crystallization of different social forces, as they bear on some of the most crucial aspects of the urban phenomena. They also denote two major features of the urban phenomena—namely, both constitute an area of spatial concentration of population and economic activities, as well as a center of regulation and control in the political, religious, and economic realms. The very distinction between them indicates that these two aspects of urban life differ greatly in their institutional implications, that they generate different dynamics that do not always operate in the same mode but that may also produce tension between them.

Under the term "concentration" we have designated that mode of institutional organization and dynamics evolving out of the relatively autonomous impact of demographic, technological, and socioeconomic forces and the concomitant production of resources. Under "centrality" we have designated that mode of institutional organization and dynamics generated by the combination of politial and cultural control, the control of production and flow of resources, as well as of the symbolic universe of a society.

As has been demonstrated abundantly throughout our analysis, these two types of broad forces can be subdivided in various ways, each having distinct autonomous tendencies. But the subsumption of such different types of social action under the terms "concentration" and "centrality" indicates that these two broad types are indeed of the utmost importance for understanding the impact of the social actors and processes on shaping the different aspects of the urban phenomena.

The distinction between concentration and centrality brings us to another distinction pervading our work, that between center and periphery. It is clear that these terms are also applied here in an analytical way. In the context of the analysis of cities and urban phenomena, it is of crucial importance to differentiate between the center-periphery and the urban-rural distinction. The latter has been of special importance, even being most controversial, in the historical and sociological research of urbanization and community studies.

The gist of our analysis is that not all cities constitute centers in an analytical sense and that most of the heterogenetic cities—to use the Redfield-Singer nomenclature—do not usually constitute such centers. Similarly, certain centers, such as some religious ones in the realm of Indian or Islamic civilizations, may barely be regarded as cities in terms of spatial concentration of population.

Moreover, even in those centers in which the element of centrality is crucial—in the various types of orthogenetic cities—one has to distinguish between the purely urban aspects of social organizations and those of centrality, although close relations exist between them. On one hand, the very concentration of population in cities and the dynamics of urban life generate specific problems, which are not related to those of centrality as a distinct analytical aspect of social organization and which create within the cities many sectors that must be regarded as periphery. On the other hand, the very forces of centrality may generate within the orthogenetic cities patterns of creativity that are not just their reflection or manifestation but that also work against the premises of the predominant modes of centrality. It is, above all, the possibility of the transformation of forces of concentration into those of centrality—a possibility that seems to be actualized only in cities—that is of special interest.

This brings us to the central problem of our analysis—namely, whether cities constitute only a reflection or spatial manifestation of various social forces (as has been claimed by contemporary Marxists, going to some degree against

Marx's own insights), or whether they also evince some specific, autonomous characteristics. Our approach, which stresses the importance of the civilizational context and different social forces as being crucial for understanding the diverse aspects of the urban phenomena, seemingly veers heavily in the direction of such a view of cities as epiphenomena, as reflections of broader or "basic" social forces. Yet, our analysis, with its strong emphasis on the modes of social organization in cities, on cities as arenas or loci of distinct, specific types of cultural creativity, implies a view of cities not merely as constituting spatial reflections of social forces, be they social classes or forces of production, political, and cultural formations. Our approach assumes—and we have attempted to illustrate this point throughout our analysis—that the very crystallization of cities generates the possibility of new modes of social and cultural creativity. Here, the central fact, stressed by Paul Wheatley, is that spatial concentration in cities generates new forms of production, control, and diffusion of information—technical, instrumental information as well as the basic symbolic universe of the mental maps of societies—and modes of control of resources that transcend the inherent "autonomous" dynamics of such forces and creates the potentiality of new types of social and cultural creativity.

Such new modes of creativity may develop in multiple directions in different cities according to the basic premises or parameters of their respective civilizational contexts. They all contain, however, the seeds of basic societal processes of centrality and concentration that may transcend the different ideological premises and institutional frameworks of civilizations and regimes. The fact that in most societies a distinct symbolism of cities develops attests to the recognition of this possibility in the tradition of self-reflexivity of human societies.

We give only two illustrations from our analysis. The first, based on some imperial regimes, such as those in China and particularly in the Byzantine Empire, where the very concentration of different types of social activities and economic forces in the capital and provincial cities and the concomitant constant confrontation between multiple autonomous elites and social strata, have generated the potentiality of basic changes affecting the fate of these regimes. These basic changes would be rather difficult to envisage without such confrontation in specifically urban frameworks. The second example, based on the city-states of antiquity (which have not been analyzed in any detail in this book) and in Medieval European towns, indicates that the crystallization of new civilizational premises was indeed engendered by the distinct combination of the forces of concentration and centrality within the spatial framework of the city, with its specific institutional and symbolic dimensions.

Another central facet of our approach has been the attempt to explain different aspects of urban phenomena, which have hitherto been studied separately, by means of the same explanatory variables, even though different combinations of such variables are required for the interpretation of the various phenomena. In this approach we have been able to link, in a systematic way, such different aspects of the urban phenomena as the geographical and social or cultural ones. We have also shown that such a combination, beyond its inherent interest, is also of great importance for understanding the generation of different modes of institutional and cultural creativity and of potential transformation.

The emphasis on cities as arenas of autonomous social and cultural creativity brings us back to another important problem in the comparative analysis of cities. This problem is whether it is possible to compare cities across different societies and civilizations, whether it is possible to talk of the city as a universal phenomenon having some common meaning or characteristics across different civilizations, or whether the cities of any civilization constitute a unique phenomenon, fundamentally different from those of other civilizations.

Such a basically historicist view is quite close to, although not necessarily derived from, that which sees cities only as a reflection of other social forces. As already indicated, we have strongly rejected this view and have attempted to show not only that cities can be understood best in their respective civilizational contexts but that the contexts themselves can be analyzed comparatively. But can the same be said about the cities themselves? Our review of attempts to find universal traits of cities or urban systems may have indicated that beyond some minimal common characteristics—especially the fact of some greater, denser concentration of population—the cities of diverse civilizations are so distinct and different from one another that they cannot be compared in any meaningful way.

We have endeavored to show throughout our analysis that even if many of the existing approaches to the study of traits of cities are unacceptable, so also is the view that cities of different civilizations are not comparable. We have attempted to show—especially upon concluding our analysis in Chapters 12 and 13—that although such a comparison cannot be made in terms of one variable or trait singled out in the myriad studies of cities, or in terms of the dimensions of the urban phenomena in our own analysis, they can yet be compared in terms of the constellations of such variables or dimensions and the different combinations of the dynamics of forces of centrality and concentration, as they generate different modes of social and cultural creativity of cities. It is indeed around this focus that, it seems to us, the most meaningful comparisons of cities can be made, although this does not necessarily preclude the comparison of at least some of the different dimensions of city life.

This claim of ours is substantiated by a very important fact, a fact that has largely been taken for granted, to which no great attention was paid and with which we started our analysis: In most known civilizations a distinction arose between city and noncity, as well as an ambivalent fascination with cities. The "contents" of such a fascination, as also the concrete designation of cities, varied greatly among the different civilizations, and systematic analysis of such variations in the connotation of cities is still pending. But even a preliminary review indicates that such an ambivalent fascination focused on those aspects of the city which denoted the specific modes of social and cultural creativity developing within them. Although the concrete expression of such creativity varied among the cities of different civilizations as well as among different cities within the same civilization, the existence of such creativity, of the ambivalent attitude to it and of the very term "city" can be found in most civilizations (some hints of which we have given in the chapter on the moral evaluation of cities) are considered universal and can serve as foci of comparative analysis.

Our analysis has been devoted entirely to cities and urban systems in historical "premodern" civilizations. The question necessarily arises here of whether this approach is also applicable to the study of the modern urban phenomenon, as it developed after the Industrial Revolution with the emergence of capitalist (and later socialist) economic systems, with the expansion of the capitalist system, the later evolution of the so-called postindustrial societies, and following the great revolutions that transformed the basic political and ideological premises of the civilizations that originated in Europe and spread throughout the world. It is only through further research carried out on the lines of such an approach that this question can be answered, but some preliminary indications (or at least suggestions) are not out of place here.

The central assumption, that the combination of the different social actors, forces, and processes—as they shape the basic premises of civilizations and their institutional frameworks—is of special importance in influencing aspects of the urban phenomena, is also valid concerning the analysis of modern civilizations. The crucial problem arising here is that these premises, as well as the nature of the concrete social and economic forces, have greatly changed in modern as compared with premodern settings. It is true, however, that contrary to earlier assumptions about theories of modernization and about the convergence of industrial societies, these changes have not given rise to a single, relatively homogeneous, modern civilization; rather, they have led to various modern societies or civilizations, the differences between them being greatly enhanced by the specifically modern economic, political, and ideological settings and by the older civilizational premises and the continuous interaction between them. It is indeed such an interaction that influences the shape of cities and urban systems in the various modern or modernizing societies.

Yet, whatever the differences between the transformations in postindustrial societies and those in the so-called Third World, the common core of the forces of modernity—of communication and information technology in general and of the spread of new types of political ideologies in particular—has greatly changed the nature and structure of the instrumental and symbolic institutional aspects of urban-rural relations, as well as of the distinction between center and periphery.

The question of how to relate all these developments to the basic assumption of our approach and how to modify them in that process, it seems to us, should be taken up by further research.

References

CHAPTER 1

ABRAMS, P. and E. A. WRIGLEY [eds.] (1978) Towns in Societies. Cambridge: Cambridge University Press.

ABU-LUGHOD, J. (1969) "Testing the theory of social area analysis: the ecology of Cairo, Egypt." American Sociological Review 34: 189-212.

ADAMS, R. M. (1966) The Evolution of Urban Society. London: Weidenfeld and Nicolson.

AGNEW, J., J. MERCER, and D. SOPHER [eds.] (1984) The City in Cultural Context. Boston: Allen & Unwin.

BACON, E. (1967) Design of Cities. London: Thames and Hudson.

BARKER, D. (1978) "A conceptual approach to the description and analysis of an historical urban system." Regional Studies 12: 1-10.

BERRY, B.J.L. (1961) "City size distributions and economic development." Economic Development and Cultural Change 9: 573-587.

———and A. PRED (1961) Central Place Studies: A Bibliography of Theory and Applications. Philadelphia: Regional Science Research Institute.

———and P. H. REES (1969) "The factorial ecology at Calcutta." American Journal of Sociology 74: 447-491.

BINFORD, L. (1968) "Post Pleistocene adaptions," in S. R. Binford and L. Binford (eds.) New Perspectives in Archaeology. Chicago: Aldine.

BIRD, J. (1977) Centrality and Cities. London: Routledge & Kegan Paul.

BLANTON, R. (1976) "Anthropological study of cities." Annual Review of Anthropology 5: 249-264.

BLAU, P. (1964) Exchange and Power in Social Life. New York: Wiley.

BOURDIN, A. and M. HIRSHHORN [eds.] (1985) Figures de la Ville: Autour de Max Weber. Paris: Aubier.

BOURNE, L. [ed.] (1982) The Internal Structure of the City (2nd ed.). New York: Oxford University Press.

BRAIDWOOD, R. J. (1974) Prehistoric Men (8th ed.). Glenview, IL: Scott, Foresman.

———and G. WILLEY [eds.] (1962) Courses Toward Urban Life. New York: Aldine.

BURGESS, E. W. (1925) "The growth of the city," in R. E. Park, E. W. Burgess, and R. D. McKenzie (eds.) The City. Chicago: Chicago University Press.

———and D. BOGUE [eds.] (1964) Contributions to Urban Sociology. Chicago: Chicago University Press.

CARTER, H. (1956) "The urban hierarchy and historical geography." Geographical Studies 3: 85-101.

———(1972) The Study of Urban Geography. London: Arnold.
———(1983) An Introduction to Urban Historical Geography. London: Arnold.
CASTELLS, M. (1976) "Is there an urban sociology?" in C. G. Pickvance (ed.) Urban Sociology: Critical Essays. London: Tavistock.
———(1977) The Urban Question: A Marxist Approach. London: Arnold.
———and F. GODARD (1978) City, Class and Power. London: Macmillan.
CHANDLER, T. and G. FOX (1974) 3000 Years of Urban Growth. New York: Academic Press.
CHASE-DUNN, C. K. (1985) "The system of world cities, A. D. 800-1975," in M. Timberlake (ed.) Urbanization in the World Economy. Orlando, FL: Academic Press.
CHILDE, G. V. (1947) History. London: Cobbett Press.
———(1950) "The urban revolution." Town Planning Review 21 (April): 3-17.
———(1954) "Early forms of society," in C. Singer et al. (eds.) A History of Technology. Oxford: Clarendon Press.
CHRISTALLER, W. (1966) Central Places in Southern Germany (C. W. Baskin, trans.). Englewood Cliffs, NJ: Prentice-Hall.
COF, M. D. (1961) "Social typology and tropical forest civilization." Comparative Studies in Society and History 4 (January): 65-85.
COHEN, E. (1976) "Environmental orientations: a multidimensional approach to social ecology." Current Anthropology 17, 1: 49-69.
DAHRENDORF, R. (1964) Class and Class Conflict in Industrial Society. London: Routledge & Kegan Paul.
DE VRIES, J. (1984) European Urbanization 1500-1800. London: Methuen.
DICKINSON, R. E. (1951) European City: A Geographical Interpretation. London: Routledge & Kegan Paul.
DURKHEIM, E. (1964a) The Division of Labour in Society (G. Simpson, trans.). New York: Free Press.
———(1964b) The Rules of Sociological Methods. New York: Free Press.
EISENSTADT, S. N. (1973) "Traditional patrimonialism and modern neopatrimonialism." Research Papers in the Social Sciences, Vol. 1. Beverly Hills, CA: Sage.
———(1981) "The schools of sociology," in J. F. Short (ed.) The State of Sociology. Beverly Hills, CA: Sage.
———and M. CURELARU (1976) The Form of Sociology—Paradigms and Crises. New York: Wiley.
———(1977) "Macro-sociology: theory, analysis, and comparative studies." Current Sociology 25, 2: 1-112.
ELLIOT, B. and D. McCRONE (1982) The City, Patterns of Domination and Conflict. London: Macmillan.
EL SHAKS, S. (1972) "Development, primacy and systems of cities." Journal of Developing Areas 7: 11-36.
FANON, F. (1965) The Wretched of the Earth. London: MacGibbon and Kee.
FINLEY, M. I. (1977) "The ancient city: from Fustel de Coulanges to Max Weber and beyond." Comparative Studies in Society and History 19 (July): 305-327.
FIREY, W. (1947) Land Use in Central Boston. Cambridge: Harvard University Press.
FLANNERY, K. V. (1969) "Origins and ecological effects of early domestication in Iran and the Near East," in P. Ucko and G. Dimbledy (eds.) The Domestication and Exploitation of Plants and Animals. Chicago: Aldine.
———(1972) "The cultural evolution of civilizations." Annual Review of Ecology and Systematics 3: 399-426.
———[ed.] (1976) The Early Mesoamerican Village. New York: Academic Press.
FOX, R. G. (1977) Urban Anthropology: Cities in Their Cultural Settings. Englewood Cliffs, NJ: Prentice-Hall.
FRANK, G. (1969) Capitalism and Underdevelopment in Latin America. New York: Monthly Review Press.
FRIEDMANN, J. (1973) Urbanization, Planning and National Development. Beverly Hills, CA: Sage.

FUSTEL DE COULANGES, N. D. (1980) The Ancient City (trans. of La Cité Antique). Baltimore: Johns Hopkins University Press.
GALE, S. and G. OLSSON (1979) Philosophy in Geography. Reidel: Dordrecht.
GIEDION, S. (1963) Space, Time and Architecture (4th ed.). Cambridge: Harvard University Press.
GOTTMANN, J. and J. LAPONCE [eds.] (1980) "Politics and geography." International Political Science Review 1, 4.
GRAMSCI, A. (1971) Selections from the Prison Notebooks. London: Lawrence and Wishart.
GUTKIND, E. A. (1964-1972) International History of City Development, Vols. 1-8. New York: Free Press.
HANDLIN, O. and J. BURCHARD [eds.] (1963) The Historian and the City. Cambridge: MIT Press.
HARDOY, J. E. (1972) Pre-Columbian Cities. New York: Walker and Company.
HARLOE, M. [ed.] (1977) Captive Cities. London: Wiley.
HARVEY, D. (1973) Social Justice and the City. London: Edward Arnold.
———(1978) "The urban process under capitalism: a framework for analysis." International Journal of Urban and Regional Research 2: 101-131.
———(1982) The Limits to Capital. Chicago: University of Chicago Press.
HERBERT, D. T. and R. J. JOHNSTON [eds.] (1978) Social Areas in Cities: Processes, Patterns and Problems. New York: Wiley.
HINDESS, B. and P. Q. HIRST (1975) Pre-Capitalist Modes of Production. London: Routledge & Kegan Paul.
HOMANS, G. C. (1961) Social Behavior: Its Elementary Forms. New York: Harcourt Brace Jovanovich.
HOSELITZ, B. F. (1955) "Generative and parasitic cities." Economic Development and Cultural Change 3 (April): 278-294.
JEFFERSON, M. (1939) "The law of the primate city." Geographical Review 29: 226-232.
JOHNSON, G. J. (1980) "Rank-size convexity and system integration: a view from archaeology." Economic Geography 56: 234-247.
KENYON, K. (1957) Digging Up Jericho. London: Benn.
LAMBERG-KARLOVSKY, C. C. and J. A. SABLOFF (1979) Ancient Civilizations. Menlo Park, CA: Benjamin Cummings.
LAVEDAN, P. (1926) Histoire de l'Urbanisme: Antiquité et Moyen-Age. Paris: Laurens.
———(1941) Histoire de l'Urbanisme: Renaissance et Temps Modernes. Paris: Laurens.
LEFEBRE, H. (1972) La Pensée Marxiste et la Ville. Paris: Castermann.
LEVI-STRAUSS, C. (1963) Structural Anthropology. New York: Basic Books.
LEWIS, O. (1965) "Further observations on the folk-urban continuum and urbanization, with special reference to Mexico City," in P. M. Hauser and L. F. Schnore (eds.) The Study of Urbanization. New York: Wiley.
LINSKY, A. S. (1965) "Some generalizations concerning primate cities." Annals of the Association of American Geographers 55: 506-513.
LOPEZ, R. S. (1963) "The crossroads within the wall," in O. Handlin and J. Burchard (eds.) The Historian and the City. Cambridge: MIT Press.
MAINE, H. (1931) Ancient Law, Its Connection with the Early History of Society and Its Relation to Modern Ideas. London: Oxford University Press.
MARCUSE, H. (1964) One-Dimensional Man. London: Routledge & Kegan Paul.
MARX, K. (1964) Pre-Capitalist Economic Formations. New York: International Publishers.
———(1965) Selected Writings in Sociology and Social Philosophy. New York: McGraw-Hill.
McELRATH, D. C. (1962) "The social areas of Rome: a comparative analysis." American Sociological Review 27: 376-391.
———(1968) "Societal scale and social differentiation," in S. Green et al. (eds.) The New Urbanization. New York: St. Martin's Press.
McGREEVEY, W. P. (1971) "A statistical analysis of primacy and lognormality in the size distribution of Latin American cities, 1750-1960," in R. M. Morse (ed.) The Urban Development of Latin America, 1750-1920. Stanford: Stanford University Press.
MELLAART, J. (1967) Çatal Huyuk: A Neolithic Town in Anatolia. London: Thames and Hudson.

MELLOR, J. R. (1977) Urban Sociology in an Urbanized Society. London: Routledge & Kegan Paul.
MERTON, R. K. (1963) Social Theory and Social Structure (rev. ed.). New York: Free Press.
MOLOTCH, H. (1976) "The city as a growth machine: toward a political economy of place." American Journal of Sociology 82: 309-332.
MORGAN, L. (1877) Ancient Society. New York: Charles H. Kerr.
MORRIS, A.E.J. (1972) History of Urban Form. London: Godwin.
MORSE, R. M. [ed.] (1971) The Urban Development of Latin America, 1750-1920. Stanford: Stanford University Press.
———(1972) "A prolegomenon of Latin American urban history." Hispanic American Historical Review 52 (August): 359-394.
MUMFORD, L. (1938) The Culture of Cities. New York.
———(1961) The City in History. London: Secker and Warburg.
PEET, R. (1977) Radical Geography: Alternative Viewpoints in Contemporary Social Issues. London: Methuen.
PICKVANCE, C. G. [ed.] (1976) Urban Sociology: Critical Essays. London: Tavistock.
PIRENNE, H. (1946) Medieval Cities (F. D. Halsey, trans.). Princeton: Princeton University Press.
PRED. A. (1973) "The growth and development of systems of cities in advanced economies," in A. Pred and G. Tornquist, Systems of Cities and Information Flows. Land Studies in Geography, Series B, No. 38. Lund: Gleerup.
———(1977) City Systems in Advanced Economies. London: Hutchinson.
RASMUSSEN, S. E. (1969) Towns and Buildings. Cambridge: MIT Press.
REDFIELD, R. (1947) "The folk society." American Journal of Sociology 52, 4.
RICHARDSON, W. (1973) "Theory of the distribution of city sizes: review and prospects." Regional Studies 7: 239-251.
RODWIN, L. and R. M. HOLLISTER [eds.] (1984) Cities of the Mind. New York: Plenum Press.
ROKKAN, S. (1980) "Territories, centers and peripheries," in J. Gottmann (ed.) Center and Periphery: Spatial Variation in Politics. Beverly Hills, CA: Sage.
ROSENAU, H. (1972) The Ideal City. New York: Harper and Row.
ROZMAN, G. (1973) Urban Networks in Ching China and Tokugawa Japan. Princeton: Princeton University Press.
———(1976) Urban Networks in Russia, 1750-1880 and Pre-Modern Periodization. Princeton: Princeton University Press.
———(1978) "Urban networks and historical stages." Journal of Interdisciplinary History 9: 65-91.
RYKWERT, J. (1976) The Idea of a Town. London: Faber and Faber.
SANDERS, W. T. and B. PRICE (1968) MesoAmerica: The Evolution of a Civilization. New York: Random House.
SAUNDERS, P. (1981) Social Theory and the Urban Question. London: Hutchinson.
SENNET, R. [ed.] (1969) Classic Essays on the Culture of Cities. New York: Appleton-Century-Crofts.
SERVICE, E. R. (1975) Origins of the State and Civilization: The Process of Cultural Evolution. New York: Norton.
SIMMEL, G. (1950) The Sociology of Georg Simmel (K. Wolff, trans.). New York: Free Press.
SJOBERG, G. (1960) The PreIndustrial City. New York: Free Press.
SKINNER, G. W. (1977) "Regional urbanization in nineteenth century China," in G. W. Skinner (ed.) The City in Late Imperial China. Stanford: Stanford University Press.
SMITH, C. A. (1982) "Modern and premodern urban primacy." Comparative Urban Research 11: 79-96.
SPENCER, M. (1977) "History and sociology: an analysis of Weber's *The City.*" Sociology 11, 3:507-525
SPENGLER, O. (1928) The Decline of the West (C. F. Atkinson, trans.). New York: Knopf.
SPREIREGEN, P. D. (1965) Urban Design: The Architecture of Towns and Cities. New York: McGraw-Hill.
STEWART, C. (1958) "The size and spacing of cities." Geographical Review 48: 222-245.
THOMPSON, E. P. (1975) The Poverty of Theory. London: Merlin Press.

TILLY, C. (1984) "History: notes on urban images of historians" in L. Rodwin and R. M. Hollister (eds.) Cities of the Mind. New York: Plenum Press.
TIMMS, D.W.G. (1971) The Urban Mosaic. Cambridge: Cambridge University Press.
TONNIES, F. (1957) Community and Society (C. P. Loomis, trans.). East Lansing: Michigan State University Press.
TOYNBEE, A. [ed.] (1967) Cities of Destiny. London: Thames and Hudson.
VANCE, J. E. (1977) This Scene of Man. New York: Harper's College Press.
VON GRUNEBAUM, G. E. (1976) "The sacred character of Islamic cities," in D. S. Wilson (ed.) Islam and Medieval Hellenism: Social and Cultural Perspectives. London: Variorum Press.
WEBER, M. (1920-1921) Gesammelte Aufsätze zur Religions-soziologie (3 vol.). Tübingen: J.C.B. Mohr (Paul Siebeck).
———(1951) The Religion of China (H. H. Geertz, trans. and ed.). New York: Free Press.
———(1952) Ancient Judaism (H. H. Geertz and D. Martindale, trans. and eds.). New York: Free Press.
———(1958) The City (D. Martindale and G. Neuwirth, trans.). New York: Free Press.
WHEATLEY, P. (1971) The Pivot of the Four Quarters. Edinburgh: Edinburgh University Press.
———(1972) "The concept of urbanism," in P. J. Ucko, R. Tringham, and G. W. Dimbleby (eds.) Man, Settlement and Urbanism. London: Duckworth.
———and T. SEE (1978) From Court to Capital. Chicago: University of Chicago Press.
WHITEHOUSE, R. (1977) The First Cities. New York: Phaidon-Dutton.
WIRTH, L. (1938) "Urbanism as a way of life." American Journal of Sociology XLIV: 3-24.
ZIPF, G. K. (1949) Human Behavior and the Principle of Least Effort. Reading, MA: Addison-Wesley.

CHAPTER 2

ADAMS, R. M. (1960) "Factors influencing the rise of civilization in the alluvium: illustrated by Mesopotamia," in C. H. Karling and R. M. Adams (eds.) City Invincible. Chicago: University of Chicago Press.
———(1966) The Evolution of Urban Society. London: Weidenfeld and Nicolson.
EISENSTADT, S. N. (1963) The Political Systems of Empires. New York: Free Press.
———(1965) "Bureaucracy, bureaucratization, markets and power structure," in S. N. Eisenstadt (ed.) Essays in Comparative Institutions. New York: Wiley.
———(1967) "The city—social center and focus of cultural creativity" (in Hebrew), in City and Community. Proceedings of the 12th Conference of History. Jerusalem: The Israel Historical Society.
———(1971) Social Differentiation and Stratification. Glenview, IL: Scott, Foresman.
———(1978) Revolution and the Transformation of Societies. New York: Free Press.
———(1982) "The Axial Age: the emergence of transcendental visions and the rise of clerics." Archives européennes de Sociologie, pp. 294-314.
MERTON, R. K. (1963) Social Theory and Social Structure (rev. ed.). New York: Free Press.
POLANYI, K., C. M. ARENSBERG, and H. W. PEARSON (1957) Trade and Market in the Early Empire. New York: Free Press.
REDFIELD, R. and M. SINGER (1954) "The cultural role of cities." Economic Development and Cultural Change 3: 53-73.
SHILS, E. (1978) Center and Periphery: Essays in Macrosociology. Chicago: University of Chicago Press.
SIMON, H. (1977) Models of Discovery. Dordrecht: Reidl.
WHEATLEY, P. (1972) "The concept of urbanism," in P. J. Ucko, R. Tringham, and G. W. Dimbleby (eds.) Man, Settlement and Urbanism. Cambridge: Schenkman.

CHAPTER 3 (SOUTHEAST ASIA)

AEUSRIVONGSE, N. (1976) "The *Devaraja* cult and Khmer kingship at Angkor," in K. R. Hall and J. K. Whitmore (eds.) Exploration in Early Southeast Asian History: The Origins of Southeast Asian Statecraft. Michigan Papers on South and Southeast Asia, 11. Ann Arbor: University of Michigan.

BRONSON, B. (1977) "Exchange at the upstream and downstream ends: notes toward a functional model of the coastal state in Southeast Asia," in K. Hutherer (ed.) Economic Exchange and Social Interaction in Southeast Asia: Perspectives from Prehistory, History and Ethnography. Michigan Papers on South and Southeast Asia, 13. Ann Arbor: University of Michigan.

BRONSON, B. and J. WISSEMAN (1978) "Palembang as Srivijaya: the lateness of early cities in Southern Southeast Asia." Asian Perspectives 19, 2: 220-239.

COE, M. D. (1961) "Social typology and tropical forest civilizations." Comparative Studies in Society and History 4, 1: 65-85.

COEDES, G. (1963) Angkor: An Introduction (E. Gardiner, trans.). Oxford: Oxford University Press.

———(1969) The Making of Southeast Asia. Berkeley: University of California Press.

EISENSTADT, S. N. (1973) Traditional Patrimonialism and Modern Patrimonialism. Beverly Hills, CA: Sage.

———(1976) The Form of Sociology—Paradigms and Crises. New York: John Wiley.

FARMER, E. L. et al. (1977) Comparative History of Civilizations in Asia. Reading, MA: Addison-Wesley.

GEERTZ, C. (1963) Peddlers and Princes. Chicago: University of Chicago Press.

GRISWOLD, A. B. (1967) Towards a History of Sukhodaya Art. Bangkok: Fine Art Department.

———(1971) "The inscription of Ramkamhaeng of Sukhothai (A.D. 1292)." Journal of Siam Society 59, 2: 179-228.

GROSLIER, B. and J. ARTHAUD (1966) Angkor: Art and Civilization. New York: Praeger.

HALL, D.G.E. (1968) A History of Southeast Asia. London: Macmillan.

HALL, K. R. (1976) "An introductory essay on Southeast Asian statecraft in the classical period," in K. R. Hall and J. K. Whitmore (eds.) Exploration in Early Southeast Asian History: The Origins of Southeast Asian Statecraft. Michigan Papers on South and Southeast Asia, 11. Ann Arbor: University of Michigan.

KEYES, C. F. (1977) The Golden Peninsula, Culture and Adaptation in Mainland Southeast Asia. New York: Macmillan.

LEUR, J. C. Van (1955) Indonesian Trade and Society. The Hague: W. Van Hoeve.

McGEE, T. G. (1969) The Southeast Asian City. London: Bell.

PYM, C. (1968) The Ancient Civilization of Angkor. New York: Mentor.

SPENCER, G. W. (1983) The Politics of Expansion. The Chola Conquest of Sri Lanka and Sri Vijaya. Madras: New Era Publications.

TAMBIAH, S. J. (1976) World Conqueror and World Renouncer. Cambridge: Cambridge University Press.

WHEATLEY, P. (1971) The Priest of the Four Corners. Chicago: Aldine.

———(1975) "Satyā ta in Suvarnadvipa: from reciprocity to redistribution in Ancient Southeast Asia," in J. A. Sabloff and C. C. Lamberg-Karlovsky (eds.) Ancient Civilization and Trade. Albuquerque: University of New Mexico Press.

———(1983) Nagara and Commandery. Origins of the Southeast Asian Urban Traditions. Department of Geography Research Paper Nos. 207-208. Chicago: University of Chicago.

CHAPTER 4 (LATIN AMERICA)

ALTMAN, I. and J. LOCKHART [eds.] (1976) Provinces of Early Mexico. Los Angeles: Latin American Center Publications, University of California.

BRONNER, F. (1986) "Urban society in Colonial Spanish America: research trends." Latin American Research Review 21, 1: 7-72.

CROUCH, D. and A. MUNDIGO (1977) "The city planning ordinances of the laws of the Indies revisited." Town Planning Review 48: 247-268.

DAVIS, K. (1975) "Colonial expansion and urban diffusion in the Americas," in D. J. Dwyer (ed.) The City in the Third World. London: Macmillan.

EISENSTADT, S. N. (1973) Traditional Patrimonialism and Modern Neopatrimonialism. Beverly Hills, CA: Sage.

———(1978) Revolution and the Transformation of Societies. New York: Free Press.

GAKENHEIMER, R. A. (1967) "The Peruvian city of sixteenth century," pp. 33-70 in G. H. Beyer (ed.) The Urban Explosion in Latin America. Ithaca, NY: Cornell University Press.

HANKE, L. [ed.] (1967) History of Latin American Civilization: Sources and Interpretations. Section VI: Urban Life. Boston: Little, Brown.

HARDOY, J. E. (1973) Pre-Columbian Cities. New York: Walker.

———[ed.] (1975) Urbanization in Latin America—Approaches and Issues. New York: Anchor Books.

HARDOY, J. E. and C. ARANOWICH (1970) "Urban scales and functions in Spanish America toward the year 1600: first conclusions." Latin American Research Review 3: 57-91.

HARING, C. H. (1963) The Spanish Empire. New York: Harcourt Brace Jovanovich.

HARRIS, W. D. (1971). The Growth of Latin American Cities. Athens: Ohio University Press.

HOBERMAN, L. (1977) "Merchants in seventeenth century Mexico City: a preliminary portrait." Hispanic American Historical Review 57: 479-503.

HOSELITZ, B. F. (1960) Sociological Aspects of Economic Growth. New York: Free Press.

JONES, R. and J. D. WIRTH (1978) Manchester and São Paulo. Stanford, CA: Stanford University Press.

KUBLER, G. (1948) Mexican Architecture of the Sixteenth Century. New Haven, CT: Yale University Press.

LOCKHART, J. (1968) Spanish Peru, 1532-1560. Madison: University of Wisconsin Press.

———and E. OTTE [eds.] (1975) The Letters and People of the Spanish Indies: Sixteenth Century. Cambridge, MA: Cambridge University Press.

MARZAHL, P. (1974) "Creoles and government: the Cabildo of Popáyan." Hispanic American Historical Review 54: 636-656.

McCAA, R., S. B. SCHWARTZ, and A. GRUBESSICH (1979) "Race and class in colonial Latin America; a critique." Comparative Studies in Society and History 21: 421-442.

MOORE, J. P. (1954) The Cabildo in Peru under the Hapsburgs: A Study of the Origins and Powers of the Town Council in the Viceroyalty of Peru, 1530-1700. Durham, NC: Duke University Press.

MORENO, A. and C. A. ANAYA, (1975) "Migrations to Mexico City in the nineteenth century: research approaches." Journal of Interamerican Studies and World Affairs 17: 27-42.

MÖRNER, M. (1967) Race Mixture in the History of Latin America. Boston: Little, Brown.

———(1983) "Economic factors and stratification in colonial Spanish America with special regard to the elites." Hispanic American Historical Review 63: 335-369.

MORSE, R. M. (1962a) "Some characteristics of Latin American urban history." American Historical Review 67: 317-338.

———(1962b) "Latin American cities: aspects of function and structure." Comparative Studies in Society and History 4: 473-493.

———(1971) "São Paulo—case study of Latin American Metropolis," pp. 151-185 in F. F. Rabinowitz and F. Trueblood (eds.) Latin American Urban Research, Beverly Hills, CA: Sage.

———(1972a) The Claims of Tradition in Urban Latin America. Rome: Daedalus.

———(1972b) "A prolegomenon to Latin American urban history." Hispanic American Historical Review 52: 359-394.

———(1974) "Trends and patterns of Latin American urbanization, 1750-1920." Comparative Studies in Society and History 16: 416-447.

NUTINI, H. G. (1972) "The Latin American city: a cultural historical approach," pp. 89-97 in T. Weaver and D. White (eds.) The Anthropology of Urban Environments. Washington, DC: Society for Applied Anthropology.

PARRY, J. H. (1973) The Spanish Seaborne Empire. Harmondsworth: Penguin.

PIERSON, W. (1952) "Some reflections on the Cabildo as an institution." Hispanic American Historical Review 5: 585-595.

ROBINSON, D. J. [ed.](1979) Social Fabric and Spatial Structure in Colonial Latin America. Ann Arbor: University Microfilms International.

SARFATI, M. (1966) Spanish Bureaucratic Patrimonialism in America. Berkeley: Institute of International Studies.

SCHAEDEL, R. P., J. E. HARDOY, N. SCOTT, and N. S. KINZER (1978) Urbanization in the Americas from its Beginnings to the Present. The Hague: Mouton.

SEED, P. (1982) "Social dimensions of race: Mexico City, 1753." Hispanic American Historical Review 62: 569-606.

SMITH, R. C. (1955) "Colonial towns of Spanish and Portuguese America." Journal of the Society of Architectural Historians 14, 4: 1-12.

SOCOLOW, S. M. (1978) The Merchants of Buenos Aires, 1778-1810. Cambridge: Cambridge University Press.

———and L. L. JOHNSON (1981) "Urbanization in colonial Latin America." Journal of Urban History 8, 1: 27-60.

STANISLAWSKI, D. (1946) "The origin and spread of the grid-pattern town." Geographical Review 36: 105-120.

———(1947) "Early Spanish town planning in the New World." Geographical Review 37: 94-105.

STEIN, S. J. and B. STEIN (1970) The Colonial Heritage of Latin America. New York: Oxford University Press.

VIOLICH, F. (1944) Cities of Latin America. New York: Reinhold.

YOUNG, E. (1979) "Urban market and hinterland: Guadalajara and its region in the eighteenth century." Hispanic American Historical Review 59: 593-635.

CHAPTER 5 (CHINESE EMPIRE)

BALAZS, E. (1953) Le Traité économique du "Souei-Chou": Etudes sur la Société et l'Economie de la Chine Médiévale. Leiden.

———(1954) "Le Régime de la Propriété en Chine du IVe au XIVe siècles." Journal of World History 1: 669-679.

———(1960) "The birth of capitalism in China." Journal of the Economic and Social History of the Orient 3: 196-217.

———(1964) Chinese Civilization and Bureaucracy: Variations on a Theme. New Haven, CT: Yale University Press.

CHANG, C. L. (1955) The Chinese Gentry. Seattle: University of Washington Press.

CHANG, S. D. (1961) "Some aspects of the urban geography of the Chinese Hsien." Annals of the Association of American Geographers 51: 23-45.

———(1963) "Historical trends of Chinese urbanization." Annals of the Association of American Geographers 55: 109-143.

———(1977) "The morphology of walled capitals," in G. W. Skinner (ed.) The City in Late Imperial China. Stanford, CA: Stanford University Press.

CHÜ, T. T. (1957) "Chinese class structure and its ideology," in J. K. Fairbank (ed.) Chinese Thought and Institutions. Chicago: University of Chicago Press.

EBERHARD, W. (1952) Conquerors and Rulers: Social Forces in Medieval China. Leiden.

———(1956) "Data on the structure of the Chinese city in the pre-industrial period." Economic Development and Cultural Change 4, 3: 253-268.
———(1962) Social Mobility in Traditional China. Leiden: Brill.
———(1967) "Social mobililty and stratification in China," in R. Bendix and S. M. Lipset (eds.) Class, Status and Power. London: Routledge & Kegan Paul.
ELVIN, M. (1973) The Pattern of the Chinese Past. London: Methuen.
FAIRBANK, J. K. [ed.] (1957) Chinese Thought and Institutions. Chicago: University of Chicago Press.
———[ed.] (1968) The Chinese World Order. Cambridge, MA: Harvard University Press.
FEI, H. T. (1953) China's Gentry. Chicago: University of Chicago Press.
FRIEDMAN, M. [ed.] (1970) Family and Kinship in Chinese Society. Stanford, CA: Stanford University Press.
GOLAS, P. J. (1977) "Early Ch'ing guilds," in G. W. Skinner (ed.) The City in Late Imperial China. Stanford, CA: Stanford University Press.
HO, P. T. (1959) "Aspects of social mobility in China, 1368-1911." Comparative Studies in Society and History 1: 330-359.
———(1962) The Ladder of Success in Imperial China: Aspects of Social Mobility. New York: Columbia University Press.
KRACKE, E. A. (1953) Civil Service in Early Sung China, 960-1067. Cambridge, MA: Harvard University Press.
———(1957) "Religion, family and individual in the Chinese examination system," in J. K. Fairbank (ed.) Chinese Thought and Institutions. Chicago: University of Chicago Press.
MASPERO, H. (1950) Etudes Historiques. Paris.
MENZEL, J. M. [ed.] (1963) The Chinese Civil Service: Career Open to Talent? Boston: D. C. Heath.
MOTE, F. W. (1977) "The transformation of Nanking, 1350-1400," in G. W. Skinner (ed.) The City in Late Imperial China. Stanford, CA: Stanford University Press.
NIVISON, D. S. and A. F. WRIGHT [eds.] (1959) Confucianism in Action. Stanford, CA: Stanford University Press.
PULLEYBLANK, E. G. (1960) "Neo-Confucianism and neo-legalism in T'ang intellectual life, 755-805" in A. F. Wright (ed.) The Confucian Persuasion. Stanford, CA: Stanford University Press.
ROZMAN, G. (1973) Urban Networks in Ching China and Tokugawa Japan. Princeton, NJ: Princeton University Press.
SCHURMANN, H. F. (1956) "Traditional property concepts in China." Far Eastern Quarterly 15: 507-516.
SHIBA, Y. (1970) Commerce and Society in Sung China (M. Elvin, trans.). Ann Arbor: University of Michigan, Center for Chinese Studies.
SKINNER, G. W. (1964-1965) "Marketing and social structure in rural China." Journal of Asian Studies 24, 1-3: 3-43, 195-228, 363-399.
———[ed.] (1977). The City in Late Imperial China. Stanford, CA: Stanford University Press.
TAWNEY, R. H. (1962) Land and Labor in China. New York: Harcourt Brace Jovanovich.
TREGEAR, T. R. (1965) A Geography of China. Chicago: Aldine.
TWITCHETT, D. (1966) "The T'ang market system." Asia Major 12, 2: 202-243.
———(1968) "Merchant, trade and government in late T'ang." Asia Major 14, 1:63-95.
Van der SPRENKEL, B. O. (1958) The Chinese civil service: the nineteenth century. Canberra: Australian National University Press.
Van der SPRENKEL, S. (1977) "Urban social structure," in G. W. Skinner (ed.) The City in Late Imperial China. Stanford, CA: Stanford University Press.
WANG, K. (1956) "The system of equal land allotments in medieval times," in E. T. Zen and J. De Francis (eds.) Chinese Social History. Washington, DC: American Council of Learned Societies.
WHEATLEY, P. (1971) The Pivot of the Four Quarters. Edinburgh: Edinburgh University Press.
WHITNEY, J.B.R. (1970) China: Area, Administration and Nation Building. Department of Geography, Research Paper 123. Chicago: University of Chicago.
WILLMOTT, W. E. [ed.] (1972) Economic Organization in Chinese Society. Stanford, CA: Stanford University Press.

WITTFOGEL, K. A. (1957) Oriental Despotism: A Comparative Study of Total Power. New Haven, CT: Yale University Press.

WRIGHT, A. F. (1965) "Symbolism and function: reflections on Changan and other great cities." Journal of Asian Studies 24, 4: 667-679.

———(1967) "Changan," in A. Toynbee (ed.) Cities of Destiny. London: Thames & Hudson.

———(1977) "The cosmology of the Chinese city," in G. W. Skinner (ed.) The City in Late Imperial China. Stanford, CA: Stanford University Press.

YANG, C. I. (1956) "Evolution of the status of dependents," in E. T. Zen and J. De Francis (eds.) Chinese Social History. Washington, DC: American Council of Learned Societies.

YANG, C. K. (1959) "Some characteristics of Chinese bureaucratic behavior," in D. S. Nivison and A. F. Wright (eds.) Confucianism in Action. Stanford, CA: Stanford University Press.

ZEN, E. T. and J. De FRANCIS [eds.] (1956) Chinese Social History. Washington, DC: American Council of Learned Societies.

CHAPTER 6 (RUSSIAN EMPIRE)

BARON, S. H. (1969) "The town in 'feudal' Russia." Slavic Review 28, 1: 116-122.

———(1970) "The origins of seventeenth century Moscow's nemeckaja sloboda." California Slavic Studies 5: 1-18.

BLUM, J. (1971) Lord and Peasant in Russia from the Ninth to the Nineteenth Century. Princeton, NJ: Princeton University Press.

EISENSTADT, S. N. (1971) Social Differentiation and Stratification. Glenview, IL: Scott, Foresman.

FEDOR, T. S. (1975) Patterns of Urban Growth in the Russian Empire during the Nineteenth Century. Department of Geography, Research Paper No. 163. Chicago: University of Chicago.

GERSCHENKORN, A. (1970) Europe in the Russian Mirror: Four Lectures in Economic History. Cambridge, MA: Harvard University Press.

GOHSTAND, R. (1976) "The shaping of Moscow by nineteenth century trade," in M. F. Hamm (ed.) The City in Russian History. Lexington: University Press of Kentucky.

HAMM, M. F. [ed.] (1976) The City in Russian History. Lexington: University Press of Kentucky.

HANCHETT, W. (1976) "Tsarist statutory regulation of municipal government in the nineteenth century," in M. F. Hamm (ed.) The City in Russian History. Lexington: University Press of Kentucky.

HITTLE, M. J. (1976) "The service city in the eighteenth century," in M. F. Hamm (ed.) The City in Russian History. Lexington: University Press of Kentucky.

———(1979) The Service City: State and Townsmen in Russia, 1600-1800. Cambridge, MA: Harvard University Press.

KOENKER, D. (1981) "Peasants, proletarians, and posad people." Journal of Urban History 7, 3: 391-398.

MILLER, D. H. (1976) "State and city in seventeenth century Muscovy," in M. F. Hamm (ed.) The City in Russian History. Lexington: University Press of Kentucky.

PIPES. R. (1974) Russia under the Old Regime. London: Weidenfeld & Nicolson.

RAEFF, M. (1966) Origins of the Russian Intelligentsia: The Eighteenth Century Nobility. New York: Harcourt Brace Jovanovich.

ROZMAN, G. (1976) Urban Networks in Russia, 1750-1800 and Pre-Modern Periodization. Princeton, NJ: Princeton University Press.

SETON-WATSON, H. (1952) The Decline of Imperial Russia. London: Methuen.

TIKHOMIROV, M. (1959) The Towns of Ancient Russia (Y. Sdobnikov, trans.). Moscow: Foreign Languages Publishing House.

CHAPTER 7 (BYZANTINE EMPIRE)

ANDREADES, A. (1935) "Floraison et décadence de la petite propriété dans l'empire byzantin." Mélanges Ernest Mehain, Paris.
——— (1926) "Le recrutement des fonctionnaires et les universités dans l'empire byzantin." Mélanges de Droit Romain, dédiés a Georges Cornil. Paris.
BAER, G. (1970) "The administrative, economic and social functions of Turkish guilds." Israel Journal of Middle East Studies 1: 28-50.
———and G. GILBAR [eds.] (forthcoming) Social and Economic Aspects of the Muslim Waqf.
BARKER, E. [ed.] (1957) Social and Political Thought in Byzantium from Justinian I to the Last Palaeologus. Oxford: Oxford University Press.
BAYNES, N. H. (1955) Byzantine Studies and Other Essays. London: University of London, Athlone.
BECK, H. G. (1956) Vademecum des Byzantinischen Aristokraten. Graz: Verlag Styria.
BOAK, A.E.R. (1929) "The book of the prefect" (translated from Greek). Journal of Economic and Business History 1: 547-619.
BRATIANU, G. (1937) "Empire et démocratie à Byzance." Byzantinische Zeitschrift 37: 87-91.
———(1938) Etudes byzantines d'Histoire économique et sociale. Paris.
BREHIER, L. (1949) Le Monde Byzantin. Les Institutions de l'Empire byzantin. Paris: Albin Michel.
———(1950) La Civilisation byzantine. Paris: Albin Michel.
BURY, J. B. (1910) The Constitution of the Later Roman Empire. Cambridge: Cambridge University Press.
CHARANIS, P. (1941) "Internal strife in Byzantium during the fourteenth century." Byzantion 15: 208-230.
———(1944) "On the social structure of the later Roman Empire." Byzantion 17: 39-57.
———(1948) "The monastic properties and the state in the Byzantine Empire." Dumbarton Oaks Papers 4: 53-118.
———(1951a) "On the social structure and economic organization of the Byzantine Empire in the thirteenth century." Byzantinoslavica 12: 94-153.
———(1951b) "The aristocracy of Byzantine in the thirteenth century," in P. R. Coleman-Norton (ed.) Studies in Roman Economic and Social History in Honour of Allan Chester Johnson. Princeton, NJ: Princeton University Press.
———(1961) "Town and country in the Byzantine possessions of the Balkan Peninsula during the later period of the empire," in H. Birnbaum and S. Vryonis (eds.) Aspects of the Balkans: Continuity and Change. The Hague: Mouton.
———(1966) Observations on the Demography of the Byzantine Empire. Oxford: Thirteenth International Congress of Byzantine Empire 14: 1-19.
CLAUDE, D. (1969) Die Byzantinische Stadt im 6 Jahrhundert. München: Byzantinisches Archiv, 13.
DIEHL, C. (1929) La Société byzantine à l'Epoque des Comnènes. Paris: J. Gamber.
DŌLGER, F. (1953) Byzanz und die Europaische Staatenuelt. Buch-Kunstverlar Ettal.
DOWNEY, G. (1960) Constantinople in the Age of Justinian. Norman: University of Oklahoma Press.
DVORNIK, F. (1946) "The circus parties in Byzantium, their evolution and their suppression." Byzantina-Metabyzantina 1: 119-133.
EISENSTADT, S. N. (1963) The Political Systems of Empires. New York: Free Press.
GIBB, H.A.R. and H. BOWEN (1957) Islamic Society and the West. London: Oxford University Press.
GUILLAND, E. (1947) "La noblesse de race à Byzance." Byzantinoslavica 9: 307-314.
———(1966) "Etudes sur l'hippodrome de Byzance." Byzantinoslavica 27, 1: 289-307; 27, 2: 26-40.
HUSSEY, J. M. (1937) Church and Learning in the Byzantine Empire, 867-1185. New York: Oxford University Press.

———(1957) The Byzantine World. London: Hutchinson's University Library.
JACOBY, D. (1961) "La population de Constantinople à l'époque byzantine: un problème de démographie urbaine." Byzantion 31: 81-109.
JANIN, R. (1964) Constantinople byzantine, Développement urbain et Répertoire topographique. Paris: Institut Français d'Etudes byzantines.
JONES, A.H.M. (1940) The Greek City from Alexander to Justinian. Oxford: Oxford University Press.
———(1964) The Later Roman Empire, 284-602: A Social, Economic and Administrative Survey. Oxford: Oxford University Press.
———(1974) The Roman Empire: Studies in Ancient Economy and Administration. Oxford: Oxford University Press.
KIRSTEN, E. (1958) "Die Byzantinische Stadt." Munich: Berichte zum XI Internationalen Byzantinischen-Kongress: 1-48.
KOLEDAROV, P. S. (1966) "On the initial type differentiation of inhabited localities in the central Balkan Peninsula in ancient times." Etudes Historiques 3.
KURBATOV, G. L. (1971) The Basic Problems of the Internal Development of the Byzantine City (in Russian). Leningrad: Zdanov University.
LEMERLE, P. (1958) "Esquisse pour une histoire agraire de Byzance: les sources et les problèmes." Revue Historique 219: 49-74.
LEWIS, B. (1963) Istanbul and the Civilization of the Ottoman Empire. Norman: Oklahoma University Press.
LIEBESCHUETZ, J.H.W.G. (1972) Antioch, City and Imperial Administration in the Later Roman Empire. Oxford: Oxford University Press.
LOEWENSTEIN, K. (1973) The Governance of Rome. The Hague: Martinus Nijhof.
LOPEZ, R. S. (1945) "Silk industry in the Byzantine Empire." Speculum 20, 1: 1-42.
LUTTWAK, A. (1976) The Grand Strategy of the Roman Empire from the First Century A.D. to the Third. Baltimore: Johns Hopkins University Press.
MARICQ, A. (1949) "La Durée du Régime des Partis populaires à Constantinople." Bulletin de la Classe des Lettres de l'Académie royale de Belgique 25: 63-74.
MENDL, B. (1961) "Les Corporations byzantines." Byzantinoslavica 22: 301-319.
MILLER, D. A. (1969) Imperial Constantinople. New York: John Wiley.
OERTEL, F. (1939) "The economic life of the empire," chap. 7 in Cambridge Ancient History, Vol. 12. Cambridge: Cambridge University Press.
OSTROGORSKY, G. (1954) Pour l'Histoire de la Féodalité byzantine (H. Grégoire, trans.). Brussels: Corpus Bruxellense Historiae Byzantinae.
———(1956) History of the Byzantine State. Oxford: Blackwell.
———(1959) "Byzantine cities in the early middle ages." Dumbarton Oaks Papers 13: 45-66.
RUNCIMAN, S. (1933) Byzantine Civilisation. London: Arnold.
STEIN, E. (1954) "Introduction à l'Histoire et aux Institutions byzantines." Traditio LXXIV: 95-168.
TEALL, J. (1967) "The age of Constantine, change and continuity in administration and economy." Dumbarton Oaks Papers 21.
VASILIEV, A. A. (1952) History of the Byzantine Empire, 324-1453. Madison: University of Wisconsin Press.
VRYONIS, S. (1963) "Byzantine ΔΗΜΟΚΡΑΤΙΑ and the guilds in the eleventh century." Dumbarton Oaks Papers 17: 289-314.

CHAPTER 8 (EARLY PERIODS OF ISLAM)

ASHTOR, E. (1956) "L'Administration Urbaine en Syrie Médiévale." Rivista degli Studi Orientali 31: 73-128.

———(1975) "Républiques Urbaines dans le Proche-Orient à l'Epoque des Croisades." Cahiers de Civilisations Médiévales 18, 2: 117-131.
BAER, G. (1969) "Waqf reform in Egypt," in Studies in the Social History of Modern Egypt. Chicago: University of Chicago Press.
BENET, F. (1964) "The ideology of Islamic urbanization," in N. Anderson (ed.) Urbanism and Urbanization. Leiden: Brill.
BROWN, L. C. [ed.] (1973) From Medina to Metropolis: Heritage and Change in the Near Eastern City. Princeton: Darwin Press.
BRUNSCHWIG, R. (1947) "Urbanisme Médiéval et Droit Musulman." Revue des Etudes Islamiques 16: 127-155.
BULLIETT, R. W. (1972) The Patricians of Nishapur: A Study in Medieval Islamic Social History. Cambridge, MA: Harvard University Press.
CAHEN, C. (1959) Mouvements Populaires et Autonomisme Urbain. Leiden: Brill.
———(1970) "Y a-t-il eu des Corporations Professionnelles dans le Monde Musulman Classique?" in A. H. Hourani and S. M. Stern (eds.) The Islamic City: A Colloquium. Oxford: Cassirer.
De PLANAOL, X. (1959) The World of Islam. Ithaca, NY: Cornell University Press.
———(1970) "The human landscape of Islam," in P. M. Holt et al. (eds.) The Cambridge History of Islam. Cambridge: Cambridge University Press.
ENGLISH, P. (1973) "The traditional city of Herat, Afghanistan," in L. C. Brown (ed.) From Medina to Metropolis. Princeton: Darwin Press.
GIBB, H.A.R. (1962) Studies on the Civilization of Islam. Boston: Beacon.
———and H. BOWEN (1957) Islamic Society and the West. New York: Oxford University Press.
GOITEIN, S. D. (1966) Studies in Islamic History and Institutions. Leiden: Brill.
GRABAR, O. (1976) "Cities and citizens: the growth and culture of urban Islam," in B. Lewis (ed.) The World of Islam. London: Thames & Hudson.
HITTI, P. K. (1973) Capital Cities of Arab Islam. Minneapolis: University of Minnesota Press.
HODGSON, M.G.S. (1974) The Venture of Islam: Conscience and History in a World Civilization. Chicago: University of Chicago Press.
HOLT, P. M., A.K.S. LAMBTON, and B. LEWIS [eds.] (1970) The Cambridge History of Islam. Cambridge: Cambridge University Press.
HOURANI, A. H. and S. M. STERN [eds.] (1970) The Islamic City: A Colloquium. Oxford: Cassirer.
ISMAIL, A. A. (1972) "Origin, ideology and physical patterns of Arab urbanization." Ekistics 33: 113-123.
ITZKOWITZ, N. (1972) Ottoman Empire and Islamic Tradition. New York: Knopf.
JONES, A.H.M. (1940) The Greek City from Alexander to Justinian. Oxford: Clarendon Press.
KARK, R. (1978) "Jerusalem and Jaffa in the nineteenth century as an example of traditional Near Eastern Cities." Studies in the Geography of Israel 10: 75-95 (in Hebrew).
LANDAY, S. (1971) "The ecology of Islamic cities: the case for the ethnocity." Economic Geography 47, 2: 303-313.
LAOUST, H. (1965) Les Schismes dans l'Islam: Introduction à une étude de la religion musulmane. Paris: Payot.
LAPIDUS, I. (1967) Muslim Cities in the Later Middle Ages. Cambridge, MA: Harvard University Press.
———(1969) "Muslim cities and Islamic societies," in I. Lapidus (ed.) Middle Eastern Cities. Berkeley: University of California Press.
———(1973) "Traditional Muslim cities: structure and change," in L. Brown (ed.) From Medina to Metropolis. Princeton: Darwin Press.
LASSNER, J. (1970) The Topography of Baghdad in the Middle Ages. Detroit: Wayne State University Press.
LeSTRANGE, G. (1924) Baghdad During the Abbassid Caliphate. Oxford: Clarendon Press.
LEWIS, B. (1950) The Arabs in History. London: Hutchinson.
———(1973) Islam in History: Ideas, Men and Events in the Middle East. London: Alcove.
———[ed.] (1976) The World of Islam: Faith, People, Culture. London: Thames & Hudson.
MARÇAIS, G. (1955) "Considérations sur les Villes musulmanes et Notamment sur le Rôle du Mohtasib." La Ville 6: 248-262. Bruxelles: Société Jean Bodin.

———(1956) "La Conception des Villes dans l'Islam." Revue d'Alger 2: 517-533.

OSTROGORSKY, G. (1959) "Byzantine cities in the early Middle Ages." Dumbarton Oaks Papers 13: 45-66.

PAUTY, E. (1951) "Villes Spontanées et Villes Créées en Islam." Annales de l'Institut d'Etudes orientales 9: 52-75.

SAARI, E. M. (1971) "Non-economic factors and systems of cities: the impact of Islamic culture on Egypt's urban settlement pattern." Ph.D. thesis, University of Minnesota.

SAUVAGET, J. (1941) Alep. Essai sur le Développement d'une grande ville Syrienne des origines au milieu du XIXe siècle. Paris: Librarie Orientaliste Paul Geuthner.

SCHACHT, J. (1970) "Law and justice," in P. M. Holt et al. (eds.) The Cambridge History of Islam. Cambridge: Cambridge University Press.

STERN, S. M. (1970) "The constitution of the Islamic city," in A. H. Hourani and S. M. Stern (eds.) The Islamic City: A Colloquium. Oxford: Cassirer.

TEKELY, I. (1971) "The evolution of spatial organization in the Ottoman Empire and the Turkish Republic." Ekistics 31: 51-71.

TURNER, B. S. (1974) Weber and Islam: A Critical Study. London: Routledge & Kegan Paul.

Von GRUNEBAUM, G. E. (1946) Medieval Islam: A Study in Cultural Orientation. Chicago: University of Chicago Press.

———(1954) "Studies in Islamic cultural history." American Anthropologist Memoir 76.

———(1955) "The structure of the Muslim town," in G. E. Von Grunebaum (ed.) Islam: Essays in the Nature and Growth of a Cultural Tradition. London: Routledge & Kegan Paul.

———(1962) "The sacred character of Islamic cities," in A. Badawi (ed.) Mélanges Taha Husein. Cairo.

Von SIEVERS, P. (1979) "Military, merchants and nomads: the social evolution of the Syrian cities and countryside during the Classical period, 780-969/164-358." Der Islam 56, 2: 212-244.

WHEATLEY, P. (1976) "Levels of space awareness in the traditional Islamic city." Ekistics 42, 253: 354-366.

———(1981) "I Luoghi dove gli Uomini Pregano Assieme," in P. Wheatley (ed.) La Citta' come Simbolo. Brescia: Morcelliana.

CHAPTER 9 (INDIA)

AKRAM, M. (1965) Muslim Civilization in India. New York: Columbia University Press.

ALI, M. A. (1970) The Mughal Nobility under Aurangzeb. London: Asia Publishing House.

———(1966) The Mughal Nobility under Aurangzeb. London: Asia Publishing House.

APPADORAI, A. (1977) "Kings, sects, and temples in South India, 1350-1700 A.D." Indian Economic and Social History Review 14, 1: 47-73.

BASHAM, A. L. (1959) The Wonder that Was India. New York: Grove Press.

BETEILLE, A. (1965) Caste, Class and Power: Changing Patterns of Stratification in a Tanjore Village. Berkeley: University of California Press.

BHARDWAJ, S. M. (1973) Hindu Places of Pilgrimage in India—A Study of Cultural Geography. Berkeley: University of California Press.

BIARDEAU, M. (1972) Clefs pour la Pensée Hindoue. Paris: Seghers.

BROWN, N. (1961) "The content of cultural continuity in India." Journal of Asian Studies 20: 427-434.

CHAUDHURI, K. N. (1978) "Some reflections on the town and country in Mughal India." Modern Asian Studies 12, 1: 77-96.

COHN, B. S. (1961) "The pasts of an Indian village." Comparative Studies in Society and History 3: 241-249.

———(1971) India: The Social Anthropology of a Civilization. Englewood Cliffs, NJ: Prentice-Hall.
CONLON, I. I. (1970) "Caste and urbanism in historical perspective: the Saraswat Brahmans," in R. G. Fox (ed.) Urban India: Society, Space and Image. Durham, NC: Duke University Press.
DAS GUPTA, A. (1970) "Trade and politics in 18th century India," in P. S. Richards (ed.) Islam and the Trade of Asia. Oxford: Cassirer.
DUMONT, L. (1966) Homo Hierarchicus: Essai sur le Système des Castes. Paris: Gallimard.
———(1970) Religion, Politics, and History in India: Collected Papers in Indian Sociology. Paris: Mouton.
ECK, D. L. (1982) Banaras, City of Light. Princeton, NJ: Princeton University Press.
EL FARUQUI, I.R.A. and D. E. SOPHER (1974) Historical Atlas of the Religions of the World. New York: Macmillan.
FARMER, E. L. et al. (1977) Comparative History of Civilization in Asia. Reading, MA: Addison-Wesley.
FOX, R. G. (1970) "Rurban settlements and Rejput clans in Northern India," in R. G. Fox (ed.) Urban India: Society, Space and Image. Durham, NC: Duke University Press.
———(1977) Urban Anthropology, Cities in Their Cultural Settings. Englewood Cliffs, NJ: Prentice-Hall.
GHOSH, A. (1973) The City in Early Historical India. Simla: Indian Institute of Advanced Studies.
GUMPERZ, E. M. (1974) "City-hinterland relations and the development of a regional elite in nineteenth century Bombay." Journal of Asian Studies 33, 4: 581-601.
HALL, K. R. (1977) "Price making and market hierarchy in early medieval South India." Indian Economic and Social History Review 14, 2: 207-229.
HEESTERMAN, J. C. (1964) "Brahmin, ritual and renouncer." Wiener Zeitschrift fur Die Kunde Sud- und Ostasiens. Special reprint 8.
———(1971) "Kautalya and the ancient Indian state." Wiener Zeitschrift fur Die Kunde Sud- und Ostasiens. Special reprint 15.
———(1980) "Littoral et Intérieur de l'Indie," in L. Blusse, H. L. Wesseling, and G. D. Winius (eds.) History and Underdevelopment. Leiden: Center for History of European Expansion.
———(1985) The Inner Conflict of Tradition. Chicago: University of Chicago Press.
ISHWARAN, K. (1971) Contributions to Asian Studies. Leiden: Brill.
MANDELBAUM, D. G. (1970) Society in India. Berkeley: University of California Press.
MORRISON, B. M. (1970) Political Centers and Cultural Regions in Early Bengal. Tucson: University of Arizona Press.
NAQVI, H. K. (1972) Urbanisation and Urban Centers Under the Great Mughals, 1556-1707. Simla: Indian Institute of Advanced Studies.
NEALE, W. C. (1962) Economic Change in Rural India. New Haven, CT: Yale University Press.
PIGGOT, S. (1950) Prehistoric India. Harmondsworth: Penguin.
POSSEHL, G. [ed.] (1979) Ancient Cities of the Hindus. Durham: Carolina Academic Press.
ROWE, W. L. (1973) "Caste, kingship and association in urban India," in A. Southall (ed.) Urban Anthropology. New York: Oxford University Press.
RUDOLPH, S. M., L. I. RUDOLPH, and SINGH (1975) "A bureaucratic lineage in princely India: elite formation in a patrimonial system." Journal of Asian Studies 34, 3: 717-754.
SINGER, M. [ed.] (1959) Traditional India: Structure and Change. Philadelphia: American Folklore Society.
———and B. S. COHN [eds.] (1968) Structure and Change in Indian Society. Hawthorne, NY: Aldine.
SINGH, K. N. (1968) "The territorial basis of medieval town and village settlement in Eastern Uttar Pradesh, India." Annals of the Association of American Geographers 58, 2: 203-229.
SPEAR, P. (1970) "The Mughal Mansabdari system," in E. Leach and S. N. Mukherjee (eds.) Elites in South Asia. Cambridge: Cambridge University Press.
———(1972) India. Ann Arbor: University of Michigan Press.
SPENCER, G. W. (1969) "Religious networks and royal influence in eleventh century India." Journal of the Economic and Social History of the Orient 12: 42-56.

SPODEK, H. (1973) "Urban politics in the local kingdoms of India." Modern Asian Studies 7, 2: 253-273.
STEIN, B. (1976) Essays on South India. Honolulu: University Press of Hawaii.
THAPAR, R. (1961) Ashoka and the Decline of the Mauryas. Oxford: Oxford University Press.
———(1966) A History of India. Harmondsworth: Penguin Books.
———(1983) From Lineage to State. New Delhi: Oxford University Press.
WEBER, M. (1958) The Religion of India: The Sociology of Hinduism and Buddhism (trans. H. H. Geertz and D. Martindale). New York: Free Press.
WHEELER, M. (1968) The Indus Civilization. Cambridge: Cambridge University Press.

CHAPTER 10 (JAPAN)

BAECHLER, J. (1975) The Origins of Capitalism. Oxford: Blackwell Press.
BEARDSLEY, R. K., J. W. HALL, and R. WARD (1959) Village Japan. Chicago: University of Chicago Press.
BEFU, H. (1968) "Village autonomy and articulation with the state," in J. W. Hall and M. B. Jansen (eds.) Studies in the Institutional History of Early Modern Japan. Princeton, NJ: Princeton University Press.
———(1981) Japan, An Anthropological Introduction. Tokyo: Tuttle.
BLOCH, M. (1964) Feudal Society. Chicago: University of Chicago Press.
DORE, R. P. (1965) Education in Tokugawa Japan. Berkeley: University of California Press.
DUUS, P. (1969) Feudalism in Japan. New York: Knopf.
EDWARDS, W. (1983) "Event and process in the founding of Japan: the horse-rider theory in archeological perspective." Journal of Japanese Studies 9 (2): 265-297.
GUTSCHOW, N. (1976) Die Japanische Burgstadt. Paderborn: Schöningh.
HALL, J. W. (1955) "The Castle Town and Japan's modern urbanization." Far Eastern Quarterly 15, 1: 37-56.
———(1962) "Feudalism in Japan—a reassessment." Comparative Studies in Society and History 5: 15-51.
———(1966) Government and Local Power in Japan, 500-1700. Princeton, NJ: Princeton University Press.
———(1970) Japan from History to Modern Times. London: Weidenfeld & Nicolson.
———and M. B. JANSEN (1968) Studies in the Institutional History of Early Modern Japan. Princeton, NJ: Princeton University Press.
———and P. J. MASS [eds.] (1974) Medieval Japan: Essays in Institutional History. New Haven, CT: Yale University Press.
HAUSER, W. B. (1974) Economic Institutional Change in Tokugawa Japan. Cambridge: Cambridge University Press.
———(1977-78) "Osaka, a commercial city in Tokugawa Japan." Urbanism Past and Present 5: 23-32.
JOHNSTON, N. (1969) "Nara—the imperial capital of Japan." Town Planning Review 40, 1.
KIDDER, E. (1977) Ancient Japan. Oxford: Elsevier-Phaidon.
LEDYORD, G. (1975) "Galloping along with the horseriders: looking for the founders of Japan." Journal of Japanese Studies 1, 2: 217-255.
McCLAIN, J. (1980). "Castle towns and daimyo authority in the years 1583-1630." Journal of Japanese Studies, pp. 267-299.
———(1982) Kanazawa, A Seventeenth Century Japanese Castle Town. New Haven, CT: Yale University Press.

MORRIS, D. V. (1981) "The city of Sakai and urban autonomy," in G. Elison and L. Smith (eds.) Warlords, Artists and Commoners, Japan in the Sixteenth Century. Honolulu: University Press of Hawaii.
MOURER, R. and Y. SUGIMOTO [eds.] (1980) "Japanese Society, reappraisals and new directions." Social Analysis 5-6. (Adelaide, Australia).
MURAKAMI, Y. (1984) "The Ie society as a pattern of civilization." Journal of Japanese Studies 10, 2: 279-364.
NAJITA, T. and A. KOSHMAN [eds.] (1982) Conflict in Modern Japanese History. Princeton, NJ: Princeton University Press.
NAJITA, T. and I. SCHEINER [eds.] (1978) Japanese Thought in the Tokugawa Period. Chicago: University of Chicago Press.
NAKAMURA, H. (1964) Ways of Thinking of Eastern People: India, China, Tibet and Japan. Honolulu: East-West Center Press.
NAKANE, C. (1970) Japanese Society. London: Weidenfeld & Nicolson.
PASSIN, H. (1968) "Japanese society." pp. 236-250 in D. L. Sills (ed.) International Encyclopedia of the Social Sciences. New York: Macmillan/Free Press.
REISCHAUER, E. O. (1981) The Story of a Nation. Tokyo: Tuttle.
———and A. CRAIG (1978). Japan, Tradition and Transformation. Boston: Houghton Mifflin.
ROZMAN, G. (1973) Urban Networks in Ch'ing China and Tokugawa Japan. Princeton, NJ: Princeton University Press.
RYOSUKU, I. (1978) "Japanese feudalism." Acta Asiatica 35: 1-29.
SCHEINER, I. (1978) "Benevolent lords and honorable peasants: rebellion and peasant consciousness in Tokugawa Japan," in T. Najita and I. Scheiner (eds.) Japanese Thought in the Tokugawa Period. Chicago: University of Chicago Press.
SMITH, R. J. (1960) "Pre-industrial urbanism in Japan: a consideration of multiple traditions in feudal society." Economic Development and Cultural Change 9, 1: 241-257.
SMITH, T. C. (1959) The Agrarian Origins of Modern Japan. Stanford, CA: Stanford University Press.
WAKITA, H. (1980) "Cities in Medieval Japan." Acta Asiatica 40: 28-52.
———and S. B. HANLEY (1981) "Dimensions of development: cities in fifteenth- and sixteenth-century Japan," pp. 295-326 in J. W. Hall, N. Keiji, and K. Yamamura (eds.) Japan before Tokugawa. Princeton, NJ: Princeton University Press.
WEBB, H . (1968) The Japanese Imperial Institution in the Tokugawa Period. New York: Columbia University Press.
WHEATLEY, P. and T. SEE (1978) From Court to Capital: A Tentative Interpretation of the Origins of the Japanese Urban Tradition. Chicago: University of Chicago Press.
WILKINSON, T. O. (1965) The Urbanization of Japanese Labor, 1868-1955. Amherst: University of Massachusetts Press.
YAMAMURA, K. (1974) "The decline of Ritsuru system—hypothesis on economic and institutional change." Journal of Japanese Studies 1(1): 3-39.
YAZAKI, T. (1963) The Japanese City. Tokyo: Japan Publishing Trading Co.
———(1968) Social Change and the City in Japan. Tokyo: Japan Publications.

CHAPTER 11 (MEDIEVAL EUROPE)

AMMANN, H. (1956) "Wie Gross Was Die Mittelalterliche Stadt?" Studium Generale 9: 503-506.
———(1963) "Vom Lebensraum der Mittelalterliche Stadt. Eine Studie an Schwaebischen Beispielen." Berichte Zur Deutschen Landeskunde 31: 284-316.
BAREL, Y. (1977) La Ville médiévale: Système social, Système urbain. Grenoble: Presses Universitaires de Grenoble.

BELOCH, K. J. (1937-1941) Bevölkerungsgeschichte Italiens. 3 volumes. Berlin.
BLOCH, M. (1961) Feudal Society (L.A. Manyon, trans.). Chicago: Chicago University Press.
BRUNNER, O. (1963) "Souveraenitaetsproblem und Sozialstruktur in den Deutschen Reichsstaedten der freuhen Neuzeit." Vierteljahrschrift fuer Sozialund Wirtschaftegeschichte 50: 328-360.
———(1968) Neue Vege der Werfassungs und Sozialgeschichte. Revised edition. Göttingen: Vandenhoeck und Ruprecht.
BURKE, G. L. (1956) The Making of Dutch Towns: A Study in Urban Development from the 10th to the 17th centuries. London: Cleaver-Hume Press.
CAM, H. M. (1954) "Mediaeval representation in theory and practice." Speculum 29, 2: 347-355.
DeVRIES, J. (1984) European Urbanization 1500-1800. London: Methuen.
DICKINSON, R. E. (1964) City and Region: A Geographical Interpretation. London: Routledge & Kegan Paul.
ENNEN, E. (1953) Fruehgeschichte der Europaeischen Stadt. Bonn: Foehscheid.
———(1967) "The different types of formation of European towns," pp. 174-182 in S. L. Thrupp (ed.) Early Medieval Society. New York: Appleton-Century-Crafts.
FRÖLICH, K. (1933) "Kirche und Staedtisches Verfassungsleben im Mittelalter." Seitschrift der Savigny-Stiftung fuer Rechtsgeschichte, Kanonistische Abteilung, 22: 188-287.
GUTKIND, E. A. (1964) International History of City Development. Vol. I: Urban Development in Central Europe. New York: Free Press.
HAASE, C. (1963) "Die Mittelalterliche Stadt als Festung. Wehrpolitische-Militaerischen Stadt." Studium Generale 16: 379-390.
HEER, R. (1968) The Intellectural History of Europe. Vol. I: From the Beginnings of Western Thought to Luther. Garden City, NY: Doubleday.
HEITZ, C. (1963) Recherches sur les Rapports entre Architecture et Liturgie à l'Epoque carolingienne. Paris: Service d'Edition et de Vente des Productions de l'Education nationale.
HERLIHY, D. (1958) Pisa in the Early Renaissance, A Study of Urban Growth. New Haven: Yale University Press.
———(1978) "The distribution of wealth in a Renaissance community: Florence 1427," in P. Abrams and E. A. Wrigley (eds.) Towns in Societies. Cambridge: Cambridge University Press.
HIBBERT, B. A. (1978) "The origins of the medieval town patriciate," in P. Abrams and E. A. Wrigley (eds.) Towns in Societies. Cambridge: Cambridge. University Press.
JONES, P. S. (1965) "Communes and despots—the city-state in Late Medieval Italy." Transactions of the Royal Historical Society, London, 5th Series 15: 71-96.
KEYSER, E. (1941) Bevölkerungsgeschichte Deutschlands (2nd ed.). Leipzig.
"La Ville. Deuxième Partie: Institutions Economiques et Sociales." (1955) Bruxelles: Recueils de la Société Jan Bodin, VII.
LE GOFF, J. (1957) Les Intellectuels au Moyen-Age. Paris: Editions du Seuil.
———[ed.] (1968) Hérésies et Sociétés dans l'Europe pré-industrielle, 11è-18è siècles. Paris. Mouton.
———(1972) "The town as an agent of civilization, c.1200-c.1500," pp. 71-106 in C. M. Cipolla (ed.) The Fontana Economic History of Europe, Vol. I: The Middle Ages. London: Collins/Fontana Books.
LESTOCQUOY, J. (1952) Les Villes de Flandre et d'Italie sous le Gouvernement des Patriciens, XIè-XVè siècles. Paris: Presses Universitaires de France.
LOT, F. (1945-1954) Recherches sur la Population et la Superficie des Cités remontant à la période Gallo-Romaine (3 vols.). Paris.
MARTINES, L. [ed.] (1972) Violence and Civil Disorder in Italian Cities, 1200-1500. Berkeley: Berkeley University Press.
MASCHKE, E. (1973) "Die Unterschichten der Mittelalterlichen Staedte Deutschlands," pp. 345-354 in C. Haase (ed.), Die Stadt des Mittelalters. Vol. III: Wirtschaft und Gesellschaft. Darmstadt: Wege der Forschung, Band CCXLV, Wissenschaftliche Buchgesellschaft.
MOLS, R. (1954-1956) Introduction à la Démographie historique des Villes d'Europe du XIVè au XVIIIè siècles (3 vols.). Gembloux, Belgium: J. Duculot.
MÜLLER, W. (1961) Die Heilige Stadt, Roma Quadrate, Himmlisches Jerusalem, und die Mythe vom Weltnabel. Stuttgart.

MUMFORD, L. (1961) The City in History. London: Secker and Warburg.
MUNDY, J. H. and P. REISENBERG (1958) The Medieval Town. Princeton: Princeton University Press.
NICHOLAS, D. (1978) "Structures du Peuplement, Fonctions Urbaines et Formation du capital dans la Flandre médiévale." Annales Economies-Sociétés-Civilisations. 33, 3: 501-527.
O'DEA, J., T. O'DEA, and C. J. ADAMS (1972) Religion and Man: Judaism, Christianity and Islam. New York: Harper & Row.
PIRENNE, H. (1946) Medieval Cities (F. D. Halsey, trans.). Princeton: Princeton University Press.
PIZZORNO, A. (1973) "Three types of urban social structure and the development of industrial society," pp. 121-138 in G. Germani (ed.) Modernization, Urbanization and the Urban Crisis. Boston: Little, Brown.
PRAWER, J. and S. N. EISENSTADT (1968) "Feudalism," in D. L. Sills (ed.) The International Encyclopedia of the Social Science. New York: Macmillan/Free Press.
RÖRIG, F. (1967) The Medieval Town. London: Batsford.
RUSSELL, J. C. (1948) British Medieval Population. Albuquerque.
———(1958) "Late ancient and medieval population." Philadelphia: Transactions of the American Philosophical Society 43, No. 3.
———(1972) Medieval Regions and Their Cities. Newton Abbot: David and Charles.
SAALMAN, H. (1968) Medieval Cities. New York: Braziller.
SMITH, C. T. (1967) An Historical Geography of Western Europe Before 1800. London: Longmans.
VAN DER WEE, H. (1975-1976) "Reflections on the development of the urban economy in Western Europe during the Late Middle Ages and Early Modern Times." Urbanism Past and Present 1: 9-15.
WALEY, D. (1969) The Italian City Republics. New York: McGraw-Hill.
WERHULTS, A. (1977) "An aspect of the question of continuity between antiquity and Middle Ages: The origin of the Flemish cities between the North Sea and the Scheldt." Journal of Medieval History 3: 175-206.
ZUCKER, P. (1959) Town and Square From the Agora to the Village Green. Cambridge: MIT Press.

CHAPTER 12

COE, M. D. (1961) "Social typology and tropical forest civilization." Comparative Studies in Society and History 4: 65-85.
FINLEY, M. I. (1973) The Ancient Economy. Berkeley: University of California Press.
HEILBRONER, R. L. (1968) The Making of Economic Society (2nd ed.). Englewood Cliffs, NJ: Prentice-Hall.
JONES, E. L. (1981) The European Miracle. Cambridge: Cambridge University Press.
MUNDY, J. H. and P. REISENBERG (1958) The Medieval Town. Princeton: Princeton University Press.
PIRENNE, H. (1946) Medieval Cities (F. D. Halsey, trans.). Princeton: Princeton University Press.
REDFIELD, R. and M. SINGER (1954) "The cultural role of cities." Economic Development and Cultural Change 3: 53-73.
STEIN, S. J. and B. STEIN (1970) The Colonial Heritage of Latin America. New York: Oxford University Press.
STRAYER, J. R. (1974) "Notes on the origin of English and French export taxes." Studia Gratiana 15: 399-422.
WATSON, A. M. (1974) "The Arab agricultural revolution and its diffusion, 700-1100." Journal of Economic History 34: 8-35.
WHITE, L., Jr. (1962) Medieval Technology and Social Change. Oxford: Oxford University Press.

Index

About the Authors

S. N. EISENSTADT is Professor of Sociology at the University of Jerusalem, where he has been a faculty member since 1946. He has served as a Visiting Professor at numerous universities, including Harvard, M.I.T., Chicago, Michigan, Oslo. Zurich and Vienna. He was a Fellow of the Center of Advanced Studies in Behavioral Sciences and of the Netherlands Institute of Advanced Studies. He is a member of the Israeli Academy of Sciences and Humanities, Foreign Honorary Fellow of the American Academy of Arts and Sciences, Foreign Member of the American Philosophical Society, Foreign Associate of the National Academy of Sciences, and Honorary Fellow of the London School of Economics. His publications include *From Generation to Generation* (Free Press, 1956); *The Political System of Empires* (Free Press, 1963); *Israeli Society* (Basic Books, 1968); *Tradition, Change and Modernity* (Basic Books, 1973); *The Form of Sociology,* with M. Curelaru (John Wiley, 1976); *Revolutions and the Transformation of Societies* (Free Press, 1978); and *Patrons, Clients and Friends*, with L. Roniger (Cambridge University Press, 1984); and *The Transformation of Israeli Society* (Weidenfeld & Nicholson, 1985).

A. SHACHAR is a professor of geography at the Hebrew University of Jerusalem, where he has been a faculty member since 1961. His areas of teaching and research are urban geography and regional development. He has frequently served as a visiting professor at the Graduate School of Architecture and Urban Planning at the University of California, Los Angeles, and at the Federal University in Rio de Janeiro. Professor Shachar has published extensively on urban systems, regional inequalities, urban ecology, and in the urban geography of Israeli cities. He is coauthor of the *Atlas of Jerusalem* and *Urban Geography of Jerusalem*. He also serves as adviser on urban and regional development to various international organizations.